Laurier L. Schramm

Technological Innovation – An Introduction

AF695013

Also of interest

Technology Strategies.
Turning Technological Change into Competitive Advantage
Thomas Durand, 2025
ISBN 978-3-11-139277-6, e-ISBN (PDF) 978-3-11-139797-9

Innovation Management.
In Research and Industry
Edited by: Carolina Machado and J. Paulo Davim, 2025
ISBN 978-3-11-914707-1, e-ISBN 978-3-11-220669-0

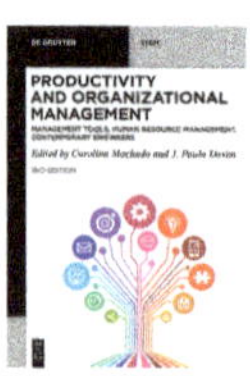

Productivity and Organizational Management.
Management Tools, Human Resource Management,
Contemporary Engineers
Edited by: Carolina Machado and J. Paulo Davim, 2025
ISBN: 978-3-11-914732-3, e-ISBN 978-3-11-220677-5

Global Markets, Diverse Economie.
Integrating Economics, Culture and ESG Strategies
Marcus Goncalves, 2025
ISBN: 978-3-11-151004-0, e-ISBN 978-3-11-151028-6

Managing Innovation in the Digital World
Esther Baldwin, Martin Curley, 2026
ISBN: 978-1-5015-1578-1, e-ISBN 978-1-5015-0733-5

Laurier L. Schramm

Technological Innovation – An Introduction

Innovation Models, Innovation Ecosystems, Management and Measurement of Innovation

2nd, Revised and Extended Edition

DE GRUYTER

Author
Dr. Laurier L. Schramm
Laurann Ventures Inc.
511 Braeside View
Saskatoon, SK S7V 1A6
Canada
lschramm@sasktel.net

ISBN 978-3-11-914725-5
ISBN 978-3-11-220672-0 (PDF)
ISBN 978-3-11-220697-3 (EPUB)

Library of Congress Control Number: 2026932366

Bibliographic information published by the Deutsche Nationalbibliothek
The Deutsche Nationalbibliothek lists this publication in the Deutsche Nationalbibliografie; detailed bibliographic data are available on the Internet at http://dnb.dnb.de.

© 2026 Walter de Gruyter GmbH, Genthiner Straße 13, 10785 Berlin, Deutschland

All rights reserved. No part of this publication may be reproduced, translated, stored in a retrieval system, or transmitted in any form or by any means, electronic, mechanical, photocopying, recording or otherwise, without prior written permission from the publisher. Text and datamining for commercial purposes requires the publisher's permission.

De Gruyter and Walter de Gruyter GmbH are part of De Gruyter Brill.
www.degruyterbrill.com

Questions about General Product Safety Regulation:
productsafety@degruyterbrill.com

Cover illustration: KamiPhotos/iStock/Getty Images Plus

Preface

"Innovation" has been around for a long time. The term "*innovation*" dates back to at least the 5th century BCE (Before the Common Era), although originally with a different meaning and a negative connotation [1]. The more specific term "technological innovation," and the purposeful management of technological innovation by organizations, both date from the early 20th century BCE. As the conduct of technological innovation has grown and spread, much has been written on the subject, particularly from the points of view of economics and business management. For the practitioner of applied research and development, there are mainly very specific books and articles on aspects such as science, technology, and engineering, the new product development process (NPD), and intellectual property, but with little focus on the broader world within which these components are important.

This book represents a consolidation, update, and expansion on two previous books:

- ***Technological Innovation – An Introduction*** (2018, ISBN: 978-3-11-043827-7), and
- ***Innovation Technology – A Dictionary*** (2017, ISBN: 978-3-11-043824-6).

I have tried to provide an introduction and overview of the world (and the process) of technological innovation without overly dwelling on any one specific aspect. Even at a high level, this field contains a massive body of specialized terminology, only some of which can be explained in a book of modest size and scope. This edition contains an updated version of the companion dictionary mentioned above, which provides brief explanations for nearly 1,400 terms and acronyms that may be encountered in a study of the fundamental principles, application approaches and strategies, and commercial aspects of technological innovation [2].

The origin of the first edition of this book lies in materials created for modules of the Canada School of Innovation developed by Innoventures Canada Inc. (I-CAN), which was originally targeted at Canada's research and technology community, but eventually expanded to a much broader audience. The material has since been expanded and updated in concert with innovation workshops presented by the author in subsequent years, up to and including 2024. Its evolution benefited from the many questions, contributions, and discussions from and with the participants in these courses and workshops, to whom I am grateful for their enthusiasm, questioning, and interest.

https://doi.org/10.1515/9783112206720-202

Contents

Acronyms, contractions, and abbreviations

With the section(s) where first used and/or explained shown in parentheses.

3M	Minnesota Mining and Manufacturing (1.1)
ABG	Academia, business, and government (as in ABG Model; 5.1)
AI	Artificial intelligence (6.2)
AIST	Advanced Industrial Science and Technology (Japan; 10.1)
AMA	American Management Association (7.3)
ARC	Alberta Research Council (Canada; now InnoTech Alberta; 10.1)
ARIZ	Algorithm of inventive problems solving (Russian acronym; 11.1)
ATAR	Awareness, trial, availability, repeat (as in ATAR Model; 4.6)
BCE	Before the Common Era (Preface)
BERD	Business-enterprise funded R&D (8.1)
BI	Behavioural intention to use (in TAM; 4.7)
BI	Business intelligence (7.5)
BMI	Battelle Memorial Institute (USA, 10.1)
BOP Innovation	Innovation at the Bottom of the Pyramid (11)
BRIC(S/K)	Brazil, Russia, India, China, and South Korea (11)
CAMARTEC	Centre for Agricultural Mechanization and Rural Technology, Tanzania (10.1)
CapEx	Capital expense (11.4)
CBM	Circular business model (6.2)
CDA	Confidential disclosure agreement (7.5)
CE	Circular economy (6.2), or Concurrent engineering (11)
CI	Competitive Intelligence (7.5)
CITER	*Centro Infornaazione Tessile di Emilia Romagna* (Italy, 10.1)
Clean-Tech	Clean technology (or technologies; 6.1)
CMI	Cambridge-MIT Institute (U.K./U.S.; 5.2)
CNRS	*Centre national de la recherche scientifique* (France; 10.1)
COSTECH	Tanzania Commission for Science & Technology (10.1)
CRI	Commercial Readiness Index (4.5)
CRIQ	Centre de recherche industrielle du Québec (Canada; 10.1)
CSIRO	Commonwealth Scientific and Industrial Research Organization (Australia; 10.1)
CRISPR	Clustered Regularly Interspaced Short Palindromic Repeats (9.2)
CSIR	Council of Scientific and Industrial Research (India; 10.1)
CTEs	Critical technology events (2.3, 4.5)
CVCA	Customer value-chain analysis (4.4)
DFA	Design for assembly (A DFX tool; 4.4)
DFM	Design for manufacture (A DFX tool; 4.4)
DFMA	Design for manufacture and assembly (A DFX tool; 4.4)
DFX	Design for excellence (4.4)
DHL GCI	DHL Global Connectedness Index (11.5)
DIKW Hierarchy	Data-information-knowledge-wisdom hierarchy (11.5)
EARTO	European Association of Research and Technology Organizations (5.2)
DPI	Domestic product of industry (11.6)
EBIT	Earnings before interest and taxes (11.6)
EBITA	Earnings before interest, taxes, and amortization (11.6)
EBITD	Earnings before interest, taxes, and depreciation (11.6)

https://doi.org/10.1515/9783112206720-204

EBITDA	Earnings before interest, taxes, depreciation, and amortization (7.7)
EBITDAR	Earnings before interest, taxes, depreciation, amortization, and restructuring or rent costs (11.6)
ESG	Environmental, social, and governance principles, management, and/or metrics (7.3)
ETRI	Electronics and Telecommunications Research Institute (South Korea, 10.1)
EVA	Economic value-added (4.8)
FBE	Fuzzy back-end (11.7)
FEF	Front-end fuzziness (11.7)
FER	Financial Effectiveness Ratio (8.3)
FFE	Fuzzy front-end (in TechSG Process; Sect. 4.8)
FFF Capital	Friends, family, and fools capital (11.7)
FTE	Full-time equivalent (11.7)
G7	The Group of Seven (major advanced economies; 9.1)
GAAP	Generally accepted accounting principles (11.8)
GBAORD	Government budget appropriations or outlays for R&D (8.1)
GCI	Global Competitiveness Index (8.1)
GDP	Gross domestic product (1.3)
GEO	Geostationary Orbit (9.2)
GERD	Gross domestic expenditure on R&D (8.1)
GIL	Government-innovation lab (5.2)
GNERD	Gross national expenditure on R&D (11.8)
GNP	Gross national product (11.8)
GOVERD	Government intramural expenditure on R&D (11.8)
GSO	Geosynchronous Orbit (11.8)
GTS	*Godkendt Teknologisk Serviceinstitut* (Denmark, 10.1)
GUI	Graphical user interface (7.7)
HEO	High Earth Orbit (9.2)
HERD	Higher education expenditure on R&D (11.9)
HKPC	Hong Kong Productivity Centre (10.1)
I18N	"I-eighteen letters-N" (11.10)
I-CAN	Innoventures Canada Inc.[1] (5.2)
ICI	Index of corporate innovation (8.2)
ICT	Information and communications technology (1.3; 6.2)
(ICV)	Internal corporate venturing (4.8)
ILO	Industry liaison office (5.2)
IMEC	Inter-University Micro Electronics Centre (Belgium; 5.2)
INO	National Optics Institute (Canada; 10.1)
IoT	Internet of things (6.2)
IP	Intellectual property (7.5)
I-Pipe	Innovation pipeline (7.7)
IPA	*Institut für Produktionstechnik und Automartisierung* (Germany, 10.1)
IPO	Initial public offering (as in IPO of shares; 4.6)
IP Mining	Intellectual property portfolio mining (11.10)
IP Portfolio Mining	Intellectual property portfolio mining (11.10)

1 At the time of writing, I-CAN was in the process of rebranding as Canadian Association of Industrial Research and Technology Organizations (CAIRTO).

IPR	Intellectual property rights (11.10)
IRAP	Industrial Research Assistance Program (as in NRC-IRAP; 5.2)
ISF	Innovation system foresight (11.10)
ISO	International Organization for Standardization (1.1)
ITC	Industrial Technology Centre (Canada; 10.1)
ITRI	Industrial Technology Research Institute (Taiwan; 5.2)
IVF	*Institutet für Verkstadsteknisk Forskning* (Sweden, 10.1)
KBE	Knowledge-based economy (5.1)
KETs	Key enabling technologies (9.2)
KIBS	Knowledge-intensive business services (5.2)
KICs	Knowledge-integration communities (5.2)
KIS	Knowledge-intensive services (5.2)
KIST	Korea Institute of Science and Technology (10.1)
KM	Knowledge management (11.12)
K-Waves	Kondratieff waves (6.1)
L10N	"L-ten letters-N" (11.13)
LEO	Low Earth Orbit (9.2)
LEPEST Analysis	Legal, environmental, political, economic, social and technological analysis. (11.13)
MBTI	Myers Briggs type indicator (2.5)
ME	Medium-sized enterprise (11.14)
MEO	Medium Earth Orbit (9.2)
MER	Mandate effectiveness ratio (8.3)
MFP	Multifactor productivity (8.1)
MGI CI	McKinsey Global Institute Connectedness Index (11.14)
MIT	Massachusetts Institute of Technology (5.2),
MNC	Multinational company (11.14)
MNEs	Multinational enterprises (5.1)
MSTI	Main science and technology indicators (8.1)
MTA	Material testing agreement (7.5)
NDA	Non-disclosure agreement (7.5)
NGO	Non-government organization (5.1)
NIH	Not invented here (as in NIH Syndrome; 7.6)
NIS	National innovation system (5.3)
NPD	New product development (as in NPD Process; 4.4, 4.8)
NPV	Net present value (4.8)
NPVI	New Product Vitality Index (8.2)
NRC	National Research Council of Canada (5.2)
NSI	National system of innovation (11.15)
NSTDA	National Science and Technology Development Agency (10.1)
NWO	Dutch Research Council (Netherlands, 13)
OECD	Organisation for Economic Co-operation and Development (1.1)
OEM	Original equipment manufacturer (11.16)
OpEx	Operating expense (11.16)
P&G	Procter & Gamble (7.7)
P&ID	Piping and instrumentation diagram (11.17)
PACE Process	Product and cycle-time excellence process (11.17)
PEST Analysis	Political, Economic, Social and Technological Analysis. (11.17)
PESTEL Analysis	Political, Economic, Social, Technological, Environmental, and Legal Analysis (11.17)
PESTLE Analysis	Political, Economic, Social, Technological, Legal, and Environmental Analysis (11.17)

PEU	Perceived ease of use (in TAM; 4.7)
PFD	Process flow diagram (11.17)
P-KIBS	Professional knowledge intensive business services (5.2)
PLC	Product life-cycle (11.17)
pMTA	Plant material testing agreement (11.17)
Project SAPPHO	*See* SAPPHO
PU	Perceived usefulness (in TAM; 4.7)
R&D	Research and development (1.1)
RCI	Resource-constrained innovation (11.19)
RDC	Research & Development Corp. (Canada; 10.1)
RIS	Regional innovation systems (11.19)
RISE	Research Institutes of Sweden (10.1)
ROI	Return on investment (11.19)
ROI2	See ROII
ROII	Return on innovation investment (8.2)
ROInn	See ROII
RoPDE	Return on product development expense (8.2)
RPC	Research and Productivity Council (Canada; 10.1)
RTI	RTI International (United States; formerly Research Triangle Institute, 10.1)
RTO	Research and technology organization (4.1),
S&T	Science and technology (5.1)
SAPPHO	Scientific Activity Predictor from Patterns with Heuristic Origins (as in "Project SAPPHO"; 7.4)
SENAI-ISI	*Serviço Nacional de Aprendizagem Industrial – Institutos de Inovação* (Brazil; 10.1)
SG&A Expenditures	Selling, general, and administrative expenditures (11.20)
SIDO	Small Industries Development Organization, Tanzania (10.1)
SIN Model	System integration and networking model (of innovation; 3.1)
SINTEF	*Stiftelsen for industriell og teknisk forskning* (Norway; 10.1)
SISIR	Singapore Institute of Standards and Industrial Research (10.1)
SLEPT Analysis	Social, legal, economic, political, and technological analysis (11.20)
SMBs	Small- and medium-size businesses (11.20)
SmE	Small-sized enterprise (11.20)
SMEs	Small- and medium-sized enterprises (4.5, 5.1)
SMED	Single minute change of a die (1.4)
SOI	Sustainability oriented innovation (7.3).
SRC	Saskatchewan Research Council (Canada; 7.5)
SRI	SRI International (United States; formerly Stanford Research Institute, 10.1)
SROI on R&D	Social return on investment in R&D (8.1)
STEEP	Social, technological, economic, environmental, and political) (as in STEEP analysis; 7.5)
STEEPLE Analysis	Social, technological, economic, environmental, political, legal, and ethical analysis (11.20)
STEEPLED Analysis	Social, technological, economic, environmental, political, legal, ethical, and demographic analysis (11.20)
STEEPV Analysis	Social, technological, economic, ecological, political, and values analysis (11.20)
STEER Analysis	Social, technological, economic, ecological, and regulatory analysis (11.20)
STEP Analysis	Social, technological, economic, and political analysis (11.20)
SWOT	Strengths, weaknesses, opportunities, and threats (as in SWOT analysis; 4.6)
TAM	Technology acceptance model (4.7)

TAMA	Technology Advanced Metropolitan Area Association (Japan; 5.2)
TechSG	Technology Stage-Gate™ (as in TechSG Process; 4.8)
TEEPSE	Technological, economic, environmental, political, social, and/or ethical (as in TEEPSE futures in technology foresight; 7.5)
TEMDO	Tanzania Engineering and Manufacturing Design Organization (10.1)
TFP	Total factor productivity (8.1)
TH	Triple Helix (as in Triple-Helix Model or Theory; 5.1)
TIC	Technology and Innovation Centre (5.2) or Technological Innovation Committee (7.3)
TIPS	Theory of inventive problem solving (or TRIZ, the Russian acronym; 2.1)
TIRDO	Tanzania Industrial Research and Development Organization (10.1)
T-KIBS	Technological knowledge intensive business services (5.2)
TMI	Technology market intermediaries (5.2)
TNO	Nederlandse Organisatie voor Toegepast Natuurwetenschappelijk Onderzoek (Netherlands, 5.2)
TPP	Technological product and process (*see* 11, Innovation (20th–21st century))
TRAM	Technology readiness and acceptance model (integrated TRI and TAM; 4.7)
TRI	Technology Readiness Index (4.7)
TRIZ	See TIPS
TRL	Technology readiness level (4.4, 4.5)
TRM	Technology road-mapping (7.5)
TTM	Time to market (11.21)
UAV	Unoccupied aerial vehicle (9.2)
UILO	University-industry liaison office (5.2)
U.S. NRC	United States National Research Council (2.0)
Vencap	Venture capital firm (4.6)
VoC	Voice of the customer (7.2, 8.3)
VTT	*Valtion teknillinen tutkimuskeskus* (Finland, 10.1)
VUCA	Volatility, uncertainty, complexity, and ambiguity (7.6)
WAITRO	World Association of Industrial and Technological Research Organizations (5.2)
WEF	World Economic Forum (8.1)

About the author

Laurier L. Schramm, Ph.D., P.Chem., C. Dir., FCIC, FCAE, has over 45 years of R&D experience spanning all four sectors: industry, not-for-profit, government, and academia. He has served as President and CEO of the Saskatchewan Research Council, Vice-President with the Alberta Research Council, President and CEO of the Petroleum Recovery Institute, and Senior Research Scientist with Syncrude Canada Ltd. For much of this time he served in parallel as Adjunct Professor of Chemical and Petroleum Engineering, and before that Adjunct Professor of Chemistry, both with the University of Calgary. His research interests include applied colloid-, interface-, and nanoscience. His leadership and management interests have included applied research, technology development and deployment, and technological innovation.

Dr. Schramm holds 17 patents and has published 20 other books, and over 400 other scientific/technological publications or proprietary reports. Many of his inventions have been adopted into commercial practice. He was awarded one of the first NSERC-Conference Board Synergy Awards for Best Practices in University-Industry R&D Partnership, and his work on the development of oil-tolerant foams for enhanced oil recovery was judged to be a Milestone of Canadian Chemistry in the 20th Century by the Canadian Society for Chemistry. He has received other national and international awards for his work, including the Queen Elizabeth II Platinum Jubilee Medal, Queen Elizabeth II Diamond Jubilee Medal, Saskatchewan Centennial Medal, and Queen Elizabeth II Golden Jubilee Medal. He is a Fellow of the Chemical Institute of Canada, Fellow of the Canadian Academy of Engineering, and an honourary Member of the Engineering Institute of Canada, among others.

Among other community service contributions, he served for over two decades on numerous committees and panels of the Natural Sciences and Engineering Research Council of Canada, the Canada Foundation for Innovation, and Sustainable Development Technology Canada, among others. He has been a member of several national or international expert advisory panels, and has served on the Boards of Directors or executive/management committees of numerous applied R&D organizations. He was also a co-founder of Innoventures Canada Inc. and a co-founder of Canada's Innovation School.

https://doi.org/10.1515/9783112206720-205

Acknowledgements

Thanks to John McDougall, past-President of the National Research Council of Canada, and co-founder with me of the Canada School of Innovation. John's many presentations on Canada's innovation system inspired me to study more broadly and think more deeply about the world of technological innovation. Thanks also to Trevor Cornell, Chief Operating Officer of the former Industrial Technology Centre (ITC), and Dr. Denis Beaulieu, past Vice-président of Centre de recherche industrielle du Québec (CRIQ). Trevor and Denis were co-founders with John and I of Innoventures Canada Inc., a network of Canadian research and technology organizations (RTOs), and both contributed to numerous foundational discussions about the process of technological innovation and its place in the broader world of regional and national innovation ecosystems.

With regard to this second edition specifically, I thank Eric Cook and Dr. Craig Aumann for additional information and many more discussions related to the evolving worlds of technological innovation and RTOs in particular.

Many thanks also to Ann Marie Schramm, Wanda Nyirfa, and Eric Cook for reading and commenting on early drafts of this book and/or its forerunners. The preparation of this book was made possible through the support of my family, Ann Marie, Katherine, and Victoria, all of whom have provided consistent encouragement and support.

Even in the modern electronic and Internet age, there remains a need for major research libraries with substantive collections of scientific, engineering, and technical books and periodicals. In the preparation of this book, my work was greatly assisted by the collections of the libraries of the University of Calgary, Carleton University, Massachusetts Institute of Technology (MIT), University of Alberta, University of Saskatchewan, University of Regina, McGill University, and University of Toronto.

Thanks also to the editorial staff of de Gruyter, particularly Karin Sora, Jessika Kischke, and Purva Santhya Pottin (via Integra).

Laurier L. Schramm
December, 2025

https://doi.org/10.1515/9783112206720-206

1 What is this "innovation" everyone is talking about?

1.1 Technological innovation defined

Everyone seems to be talking about innovation these days. It's on the news, it's on board-meeting agendas, and it's driving over 2 billion Google search results (Figure 1.1). The term "*innovation*" has become ubiquitous in business, government, academic, media, and public discussions and appears in an ever-increasing plethora of company names, government policies, and programmes (Table 1.1).

Figure 1.1: The term "*innovation*" seems to "pop up" almost everywhere.

Unfortunately, as the term has become more commonplace, it has increasingly been used with a variety of meanings, some inconsistent, some dated, some vague, and some incorrect [3, 4]. This has been lamented by many authors. In 2009, Jerry Courvisanos commented that "*Innovation is the buzzword of the new century . . . The problem is the word has become so ubiquitous that it now simply means anything 'new,' and mostly is just a form of 'spin-doctoring*'" [5]. When this book was being written in 2025, the problem had only become worse.

Based on a review of the business and technological literature up to 2010, de Castro et al. [6] found at least 37 distinctly different perspectives on the meaning of the

https://doi.org/10.1515/9783112206720-001

Table 1.1: Illustrative listing of named innovation terms.

20th- to 21st-century innovation	Inbound open innovation	Innovation system theory
Adjacent innovation	Inclusive innovation	Innovation trap
Administrative innovation	Incremental innovation	Innovation voucher programme
Adopt and adapt innovation	Indigenous innovation	Innovation union
Architectural innovation	Induced innovation	Innovative firm
Blowback innovation	Industrial innovation	Innovative industrial cluster
BOP innovation	Innovation 2.0	Innovative regional cluster
Bottom-up innovation	Innovation activities	Innovator
Catalytic innovation	Innovation and sophistication factors	Institutional innovation
Chain of innovation	Innovation at the bottom of the pyramid	Integrated innovation
Closed innovation	Innovation barrier	*Jugaad* innovation
Cognitive innovation	Innovation black box	Knowledge-induced innovation
Collaborative innovation	Innovation bridge	Lean innovation
Commercial innovation	Innovation cluster	Local innovation
Competence-destroying innovation	Innovation continuum	Management innovation
Competence-enhancing innovation	Innovation deficit	Marketing innovation
Component innovation	Innovation diffusion model	Market-pull innovation
Concept-push innovation	Innovation-driven economy	Messy fireworks innovation
Consumer innovation	Innovation ecosystem	Modular innovation
Continuous innovation	Innovation ecosystem entities	Multiparty innovation
Core innovation	Innovation ecosystem models	National innovation system
Cost innovation	Innovation expenditures	Needs-pull innovation
Customer-oriented innovation	Innovation foresight	Negative innovation
Demand-induced innovation	Innovation funnel	Non-commercial innovation
Demand-pull innovation	Innovation impacts	Non-technological innovation
Design-driven innovation	Innovation incubator	Open innovation
Discontinuous innovation	Innovation inputs	Organizational innovation
Discovery-push innovation	Innovation intermediary	Other innovation activities
Disruptive innovation	Innovation killer	Outbound open innovation
Downstream innovation	Innovation management	Outcome-driven innovation
Early modern innovation	Innovation metrics	Participative innovation
Eco-innovation	Innovation models	Platform innovation
Ecological innovation	Innovation outcomes	Political innovation
Economic innovation	Innovation outputs	Pontin's 1st and 2nd rules of innovation
Ecosystem innovation	Innovation paradoxes	Principles of innovation
Educational innovation	Innovation park	Process innovation
Evolutionary innovation	Innovation pipeline	Product innovation
Forward innovation	Innovation project	Radical innovation
Fundamental innovation	Innovation reach	Recombinant innovation
Gandhian innovation	Innovation sandbox	Regional innovation cluster (hub)
Generations of innovation	Innovation S-curve	Regional innovation system
Government innovation lab	Innovation snail	Resource-constrained innovation
Hyper-innovation	Innovation strategy mapping	Return on innovation
Hypo-innovation	Innovation system	Reverse innovation
Imitative innovation	Innovation system entities	

Table 1.1 (continued)

Revolutionary innovation	Socio-institutional innovation	TPP innovation
Routine innovation	Soft innovation	Transformational innovation
Schumpeterian innovation	Strategic innovation	Trickle-up innovation
Science-push innovation	Sustaining innovation	Upstream innovation
Self-determined innovation	Synthetic innovation	User innovation
Service innovation	Technology-push innovation	Waves of innovation
Shanzhai innovation	Technological innovation	*Zizhu Chuangxin*
Social innovation	Tenets of innovation	

term "*innovation.*" A 2011 blog post on innovation by Steven Bell begins by commenting that "*We are subject to a non-stop barrage of information about innovation*" [7]. Also, in 2011, Eunika Mercier-Laurent noted that the term "*innovation*" has become so broadly and differently interpreted (polysemous), and so fashionable, that it has become commonly used in marketing, sales, and branding promotions [8].

So, what is innovation anyway?

Innovation in earlier times. The term "*innovation*" comes from the Greek word *kainotomia,* which is derived from *kainos*, or "*new,*" and seems to date back to the 5th century BCE [1]. The term seems to have originally referred to new thoughts, sometimes with a neutral or positive connotation, but more often with a negative connotation [1]. Godin [1, 9] and Kotsemir and Abroskin [10] distinguish between two eras or *episteme,*[1] within which the term "*innovation*" has been understood and used quite differently. In the early-modern era, from the 1500s through to the 1800s, innovation meant "*introducing novel change,*" particularly with regard to religious, societal, and/or political change.

In those early times, the term "*innovation*" seems to have quickly taken on a very negative, pejorative connotation. Some specific early 18th-century examples of books railing against "*modern innovations*" and "*modern innovators*" are references [11–13].

In this era, whether in religion, society, politics, or other fields, the use of labels like innovation and innovator was usually meant to imply that the changes were unwanted, unnatural (being apart from the "*natural order of things*"), revolutionary, and/or dangerous, as in "*introducing change into the established order*" [1]. Machiavelli warns of this in his 1513 book *The Prince* (Chapter 6 in reference [14]):

> "*. . . it ought to be remembered that there is nothing more difficult to take in hand, more perilous to conduct, or more uncertain in its success, than to take the lead in the introduction of a new*

1 The body of knowledge and understanding that is generally taken to be intellectually certain at any particular time.

> *order of things, because the innovator has for enemies all those who have done well under the old conditions, and lukewarm defenders in those who may do well under the new. This coolness arises partly from fear of the opponents, who have the laws on their side, and partly from the incredulity of men, who do not readily believe in new things until they have had a long experience of them. Thus it happens that whenever those who are hostile have the opportunity to attack they do it like partisans . . ."*

In this era, whether in the context of politics or religion, introducing changes (innovation) was the purview of the political and religious leaders, and no one else. In contrast, the terms "*reformation*" or "*restoration*" were frequently used to describe positive, moderate, natural-order-restoring changes within religions or governments (with the consent of the leaders). It was only near the end of the 19th century that innovation began to be more often associated with positive, constructive changes [10].

Innovation in the 20th century. In the second era, a new meaning was introduced and gained traction. The term "*innovation*" was used by the economist Joseph A. Schumpeter in the 1930s to describe the conversion of ideas and knowledge into new and commercially successful products and services [15, 16]. Here, "commercially successful" means meeting the needs of customers, the purchasers, and ultimate end-users, in a way that encourages them to take up the new approach and have it diffuse through the marketplace.

Schumpeter's use of the term "*innovation*" to describe the introduction of novelty to the commercial marketplace is a technological, commercial, and economic meaning of the term, and it was meant to be a positive, but still disruptive, connotation. Schumpeter also coined another important term, "*creative destruction*," to describe the process by which innovative products and services displace old ones.

Schumpeter's theory was that continuing processes of innovation and creative destruction are needed for economic growth. The terms "*technological innovation*"[2] or "*commercial innovation*" are often used to distinguish the 20th-century Schumpeterian definition of innovation from that of earlier centuries, and it is this meaning of technological innovation that represents the subject of this book. Technological innovation can be viewed as a process and/or as a result. de Castro et al. [6] provide a review of these uses and the authors advocating for one or the other usage. The best practice is probably to simply be clear when technological innovation is being considered as a process and when it is being considered as a result.

> Technological innovation is the conversion of ideas and knowledge into new and commercially successful products, processes, and services.

2 The term "*technological innovation*" came into use in the 1950s, representing a merger of the work of Maclaurin and Schumpeter, and may have been coined by Maclaurin, who frequently referred to "*technological change*." The term "*technovation*" is derived from the technological and innovation, but has never really caught on in the technological innovation world.

In their paper "A Critical Look at Technological Innovation Typology and Innovativeness Terminology," Garcia and Calantone point out some key features of this definition [17]:

> *"It is important to elucidate that an invention does not become an innovation until it has processed through production and marketing tasks and is diffused into the marketplace . . . The solution to a basic scientific puzzle or the invention of a new "product" only in a laboratory setting makes no direct economic contribution . . . A discovery that goes no further than the laboratory remains an invention. A discovery that moves from the lab into production, and adds economic value to the firm (even if only cost savings) would be considered an innovation. Thus, an innovation differs from an invention in that it provides economic value and is diffused to other parties beyond the discoverers."*

Technology and technological capacity. "*Technology*" is also a very broad term, representing any of a knowledge of how to effectively use a product, process, or service ("*Know-how*"); a way of conducting or controlling a manufacturing activity (*a practice or process*); or a thing to be manufactured, used, or consumed (*product*). In this context, a *tool* can be considered to be a product. Each of these kinds of technology can be protected as IP, and each can be commercialized. "*Technological capacity*" (also termed "*intellectual capital*") is the amount of technological knowledge in an organization, region, or country, divided by the size of the labour force or the population [18]. The rate of growth of an organization's, region's, or nation's technological capacity depends on the rate at which new technology is produced (the "*rate of technological progress*") and the rate at which technology is disseminated (the so-called "*rate of replication*").[3]

Technological sophistication. Terms like "*high technology*" and "*low technology*" usually refer to the technological sophistication of a new product, process, or service, although such terms have also been applied to entire industries, sectors, and even regions [19, 20]. Technological sophistication, in this context, is often assessed based on direct (and sometimes also indirect) research and development[4] (R&D) intensity. Four categories of technological sophistication are often distinguished, based on some measure of R&D intensity, such as R&D investments divided by value-added and/or gross production values:

---

3 Since some new technologies are incremental additions, whereas others replace earlier technologies that are thus made obsolete, it is sometimes appropriate to distinguish between the "*net rate of replication*" and the "*gross rate of replication.*"

4 Research and development (R&D) in a broad sense covers discovery research, applied research, and experimental development. Discovery research is experimental and/or theoretical investigation undertaken to acquire new knowledge and/or understanding of facts and/or phenomena – without any particular use or application in mind. Applied research is experimental and/or theoretical investigation undertaken to acquire new knowledge and/or understanding of facts and/or phenomena – but directed with a specific use or application in mind. Experimental development is systematic work, drawing on existing knowledge gained from research and/or practical experience, aimed at producing new or improved materials, products, processes, systems, or services.

- High technology (also termed *"high-tech"*),[5] such as aerospace, pharmaceuticals, and instruments;
- Medium-high technology (also termed *"medium-high-tech"*), such as chemicals, electrical equipment, and motor vehicles.
- Medium-low technology (also termed *"medium-low-tech"*), such as refined petroleum products, metals and metal products, and shipbuilding; and
- Low technology (also termed *"low-tech"*), such as food, pulp and paper, and textiles.

Such categorizations are both generalized and relative (*see* reference [21]). A "*high-tech*" industry can produce "*low-tech*" products, and vice versa.

Technological innovation. Technological innovation is not new; the conversion of ideas and knowledge into new and commercially successful products, processes, and services has been going on since the beginning of commerce. What are somewhat new are the focus on how to get more successful products, processes, and services into the marketplace when there are so many there already, as well as the linkage between the introduction of new products, especially "*game-changing*" products, into the marketplace and the health, sustainability, and growth potential of entire economies. Systematic studies of innovation practices in the context of economic health, sustainability, or growth do not seem to have occurred until about the time the term "*innovation*" itself was redefined by Schumpeter in the 1930s [15, 16]. Examples of such systematic studies of technological innovation are the works of C. F. Carter and B. R. Williams in the 1950s [22, 23] and Jacob Schmookler in the 1960s [18].

Several types of technological innovation have been defined based on the nature of the technology; hence, "*product innovation,*" "*process innovation*"[6] [24], *and* "*service innovation*" [25]. "*Marketing innovation*" and "*organizational innovation*" refer to improvements in the practice of marketing or the operations of an organization, respectively, in ways that ultimately translate into improved sales and/or margins for the products, processes, or services that the organization is selling in the marketplace [26]. Several types of technological innovation have been defined based on outcomes, such as "*disruptive innovation,*" "*evolutionary innovation,*" and "*incremental innovation.*" The various types of technological innovation will be discussed further in Chapter 2.

Innovation in modern times. In the 21st century, the term "*innovation*" has been extended to include introducing novelty into realms beyond those of religion, politics, and technology, such as organizational processes, marketing, and social structures. An example is the ideation, development, and deployment of new and improved internal processes within an organization, even if these have little or no connection to

5 Also termed *"Advanced Technology."*

6 Some examples of process innovations are Henry Ford's automobile assembly line manufacturing process and FedEx's overnight long-distance parcel delivery process. For some case studies, *see* reference [24].

commercialization or the marketplace. This new meaning, again with a positive connotation, has been termed *"non-commercial innovation"* (or, *"non-technological innovation"*) to distinguish it from technological innovation and from the pre-20th-century meaning of the term *"innovation"*. The Organization for Economic Co-operation and Development (OECD) *Oslo Manual*[7] definition of innovation includes both technological and non-commercial innovation:

> *"An innovation is the implementation of a new or significantly improved product (good or service), or process, a new marketing method, or a new organisational method in business practices, workplace organisation or external relations"* [26].

The new ISO[8] standard, ISO 56000:2025, puts it more simply:

> Innovation is a *"new or changed entity realizing or redistributing value"* [27].

Nevertheless, not everything can be categorized as innovation, even in the broadest sense of the term (Table 1.2).

Table 1.2: Some things that are not innovation.

Is anything not innovation?
Some things that are NOT, by themselves, technological innovation: – Ideas – Accidents – Discoveries – New knowledge – Research – Creativity – Inventions – Prototypes – Continuous improvements

It will be clear from the foregoing introduction that research and innovation are two different kinds of activities, and, as will be discussed in later sections, there can be a connection between the two, but this is an extremely common area of confusion. A rather pithy distinction between research and (technological) innovation comes from Dr. Geoffrey Nicholson, of 3M fame, who said [28]:

> *"Research is the transformation of money into knowledge.*
> *Innovation is the transformation of knowledge into money."*

7 An OECD document providing recommended methods for collecting and interpreting data on innovation.

8 International Organization for Standardization: Geneva.

Innovation studies. Individual studies related to technological innovation, such as those by Schumpeter, Carter and Williams, and Schmookler, were the norm between the 1930s and 1950s. By the 1960s, however, innovation studies, and particularly technological innovation studies, had emerged as a distinct multi-disciplinary field of research, and the amount of literature on the topic began to expand dramatically [29]. The field is multi-disciplinary because it involves not only the natural sciences, engineering, business, and economics but also cognitive science and sociology, for example. Commenting on both the breadth and quantity of literature available on innovation (by 2009), Jan Fagerberg wrote:

> *"Two decades ago, it was still possible for a hard-working student to get a fairly good overview of the scholarly work on innovation by devoting a few years of intensive study to the subject. Not anymore. Today, the literature on innovation is so large and diverse that even keeping up-to-date with one specific field of research is very challenging"* [29].

Section 11 of this book (the dictionary) includes a selection of brief biographical introductions to some of the pioneers in the field (Table 1.3).

Table 1.3: Some famous names in innovation.

3M Inc.
Aghion, Philippe
Altshuller, Genrich
Bell, Alexander Graham
Bush, Vannevar
Chesbrough, Henry W.
Christensen, Clayton
Cooper, Robert
Drucker, Peter
Edison, Thomas A.
Etzkowitz, Henry
Foster, Richard
Holland, Maurice
Howitt, Peter
Kondratieff, Nikolai
Maclaurin, (William) Rupert
Mokyr, Joel
Porter, Michael
Rogers, Everett
Schmookler, Jacob
Schumpeter, Joseph
Solow, Robert
Tesla, Nikola
Verhulst, Pierre

1.2 Non-commercial innovation

Although not the focus of this book, two of the principal types of non-commercial innovation, *"organizational innovation"* and *"social innovation,"* will be discussed briefly in this section.

"Organizational innovation" refers to the ideation, development, and implementation of new and improved internal processes within an organization. Such new organizational processes should have some kind of efficiency or productivity benefit, even if they are internal and have little or no connection to commercialization or the marketplace. Two key sub-categories are administrative and strategic innovation. Some case studies for organizational innovation are provided in reference [24] and Section 10.2.

In *"administrative innovation,"* new knowledge is applied to the development and implementation of new and improved organizational structures and administrative processes.
- An example of this is in business restructuring, which could include mergers or acquisitions to improve overall performance (*"business structure innovation"*),
- Another example is the *total quality management* processes (or *"business process innovation"*), pioneered by W. Edwards Deming in the United States in the 1980s [30],
- Another form of administrative innovation is *"business culture innovation,"* in which an organization sets about to change its internal culture in order to improve its organizational performance.

In *"strategic innovation"* (also termed *"business model innovation"*), new knowledge is applied to the development and implementation of new and improved organizational (business) strategies. Ideally, such a new strategy would create new value for the organization and for its customers or other stakeholders, but it could be either or both. Strategic innovations are probably the least common because they usually involve dramatic changes in how an organization operates and therefore have to be led, or at least supported, by chief executive officers (CEOs). An example of this is the strategy change at General Electric (GE) under CEO Jack Welch in the United States in the 1980s [31]. Another example is provided by Amazon.com, whose business strategy changed the breadth of product availability, distribution, and promotion compared with traditional bookstores, while also reducing overhead costs and prices [32].

"Social innovation" refers to the ideation, development, and deployment of new and improved solutions, processes, or practices within society in ways that may have no direct connection to commercialization or the marketplace. These could take place in such areas as education, healthcare, social services, or the performing arts [8]. Some examples are as follows:
- *"Socio-institutional innovation,"* relating to improved internal processes within a community organization that produce some kind of efficiency or productivity benefit, such as in producer co-operatives.

- *"Cognitive innovation,"* relating to improved ways of thinking and conceptual models, such as lateral thinking, for example.
- *"Educational innovation,"* relating to improved education models and processes, such as e-learning.
- *"Ecological innovation,"* (or *eco-innovation)* relating to improvements in environmental protection, such as a new recycling programme.

Doner [33] provides some additional examples. Similarly to technological innovation, social innovation can involve new or improved institutional products, processes, or services and can be incremental or transformative. Some case studies on social innovation are referenced in Section 10.2.

1.3 Schumpeter, Solow-Swan, and innovation theories of economic growth

Mention has already been made of the work of Austrian economist and political scientist Joseph A. Schumpeter, who is probably best known for his theory of "*creative destruction*" as a necessary component of sustainable and/or growing economies and for having championed the concepts and roles of entrepreneurship and technological innovation [15, 16].

Schumpeter's original description of economic growth involved this creative destruction in the marketplace, driven by the introduction into the marketplace of "*game-changing*" new products, processes, or services by entrepreneurs or inherently entrepreneurial companies. The creative destruction involved the innovative companies succeeding over and/or replacing previous dominant companies that did not innovate. This is termed "*Schumpeter Mark I innovation.*"

In later work, Schumpeter recognized another mode of technological innovation, in which existing, dominating (usually mature and large) companies could maintain their competitive position by themselves introducing game-changing new products, processes, or services into the marketplace. This is termed "*Schumpeter Mark II innovation,*" and these large-company technological innovation activities have been referred to as "*creative accumulation*" (or "*creative agglomeration*") with reference to the maintenance of competitive position by building, evolving, and (usually) broadening their portfolio of innovative products, processes, or services.

From a broad, market perspective, the Mark I and II innovation pathways are usually viewed as being complementary, and they both refer to technological innovation.

From the point of view of an entire economy, then, the function of technological innovation is to introduce novelty (new products and services) into the economy. Otherwise, an economy stagnates because technological innovation is needed for long-term economic growth [34]:

- Industries that innovate tend to grow more rapidly, as do clusters within an industry.
- The capacity to undertake change is essential to creating and benefitting from technological innovation.
- Those that innovate tend to prosper at the expense of their less able competitors: this applies to companies, regions, and entire countries.

Most of Schumpeter's work was published in the 1930s–1950s. The importance of technological innovation in economic health and growth has also been more recently affirmed. The American economist Robert Solow developed a macroeconomic theory of economic growth that is now termed the "*Solow-Swan Growth Model*,"[9] for which Solow received the 1987 Nobel Prize in Economic Sciences [35]. The Solow-Swan Growth Model separates the contributions to economic growth into increases in inputs (labour and capital) and technological progress and predicts that sustainable economic progress requires labour-enhancing technological progress in order to increase output without needing more labour or capital. In his 1957 paper, Solow calculated that about four-fifths of the growth in US output per worker was attributable to such technological progress (*see* references [35–37]).

More recently, another model of economic growth has emerged. The "*endogenous growth theory*" (or "*new growth theory*") assumes that economic growth is the result of internal (endogenous) rather than external (exogenous) factors. There are various versions of this theory, but the main internal growth factors are investments in human capital, technological innovation, and knowledge as they relate to productivity [38–40]. As a result, the theory also gives importance to such things as industrial business strategies and government policies insofar as they affect the three principal drivers. The Endogenous Growth Theory is different from external (exogenous) economic growth theories, such as the Solow-Swan Growth Model, which are based on the accumulation of physical capital and growth of the labour force, although both kinds of theories assume a strong connection between continuing technological innovation and overall economic growth. As is the case for many models and theories, there is a continuing debate about the applicability of the endogenous growth theory, which has been criticized for being nearly impossible to check and considering "*too much of it involved making assumptions about how unmeasurable things affected other unmeasurable things*" [41].

Meanwhile, the 2025 Nobel Prize in Economic Sciences was awarded to Dutch economist Joel Mokyr "for having identified the prerequisites for sustained growth through technological progress," and the other half jointly to French economist Phil-

---

9 It is termed the "*Solow-Swan Growth Model*" because essentially the same model was independently developed and published in the same year (1956) by Trevor W. Swan. It is also referred to as the "*Neoclassical Growth Model.*"

ippe Aghion and Canadian economist Peter Howitt "for the theory of sustained growth through creative destruction" [42].

To repeat a reflection already presented above: technological innovation isn't new, it's been going on since the beginning of commerce, and it's been fairly well described since the work of Schumpeter and Solow. What is somewhat new is the present-day focus on how to get more successful products/services into the marketplace when there are so many there already, and also on how to increase the pace. These are being driven by the market changes brought about by the high rate of scientific, engineering, and technological changes, globalization, and the impact of information and communications technology (ICT) [43].

Some of the elements required in technological innovation are to develop new product/service concepts, acquire the knowledge to be able to put those concepts into practice, get a new product/service developed and introduced into the marketplace, have it be something the market wants to buy, and be able to sell it at a price that exceeds all of the development, production, marketing, distribution, and sales costs (Figure 1.2). This is not an easy task. But when these things have been accomplished, technological innovation has been produced.

Although the product or service has to be new and commercially successful, none of the other elements have to be new. You could take an old idea, apply existing knowledge, and use known techniques to accomplish all the other steps, and it would still be technological innovation as long as it is new and commercially successful in the marketplace.

Furthermore, and this may seem like heresy, R&D is not always needed to produce technological innovation [44], and even if R&D is conducted, it may not result in technological innovation. When R&D is used to develop new ideas, that's not enough to be called technological innovation yet. If they are taken a step further and developed into a new product or service, that's not technological innovation yet either. It's only when a new product/service is commercially successful that technological innovation has been produced. Of course, R&D really is quite often needed in order to get to technological innovation; it's just that R&D by itself isn't enough.

We can restate Schumpeter's key concepts as follows:

- technological innovation brings new life into the economy,
- without such new life, economies stagnate, but
- with such new life, economies can maintain long-term economic health and grow.

The current market wisdom is that businesses and industries that can create and deploy technological innovation (with all those new and commercially successful products, processes, and services) will tend to grow and prosper at the expense of their competitors. When industries grow and prosper, they frequently spend more money, build or expand their operations, and hire more people (Figure 1.3). That benefits

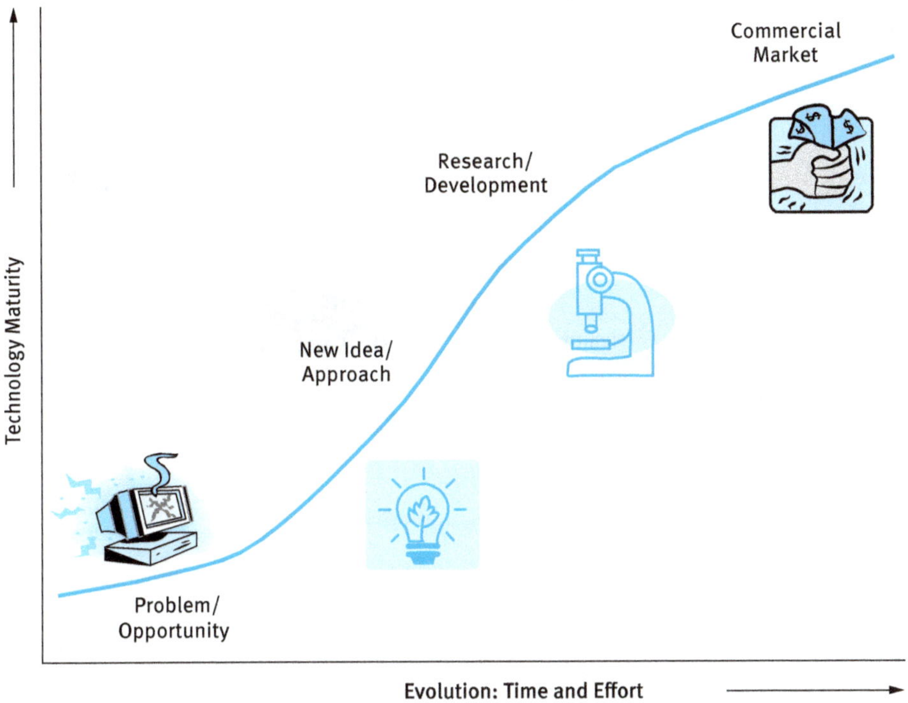

Figure 1.2: Some milestones along the process of technological innovation.

other businesses and creates private sector jobs. *That's one means by which everyone can benefit from technological innovation: jobs, but there is another.*

The innovation theory of economic growth applies not only to companies but also to regions and entire countries. When industries in a region grow and prosper, the broader economy in that region tends to grow and prosper. Prosperous companies and employed people both contribute revenue to their governments through various forms of fees and taxes. This is why governments at all levels have become so interested in technological innovation: more technological innovation should lead to more and healthier industries and other businesses, which should lead to more jobs, and all of these should lead to increased government revenue. All that new government revenue can then be deployed into the provision of more and/or better services to the public in such areas as health, safety, education, environmental protection, and even support for the technological innovation system itself. *This provides a second means by which everyone can benefit from technological innovation: increased support for vibrant and safe communities in which people can live, work, raise families, and play.* As the regions of a country become stronger, so does the country as a whole.

The most common reasons for government investment in research, development, industrial productivity improvements, and technological innovation are to encourage

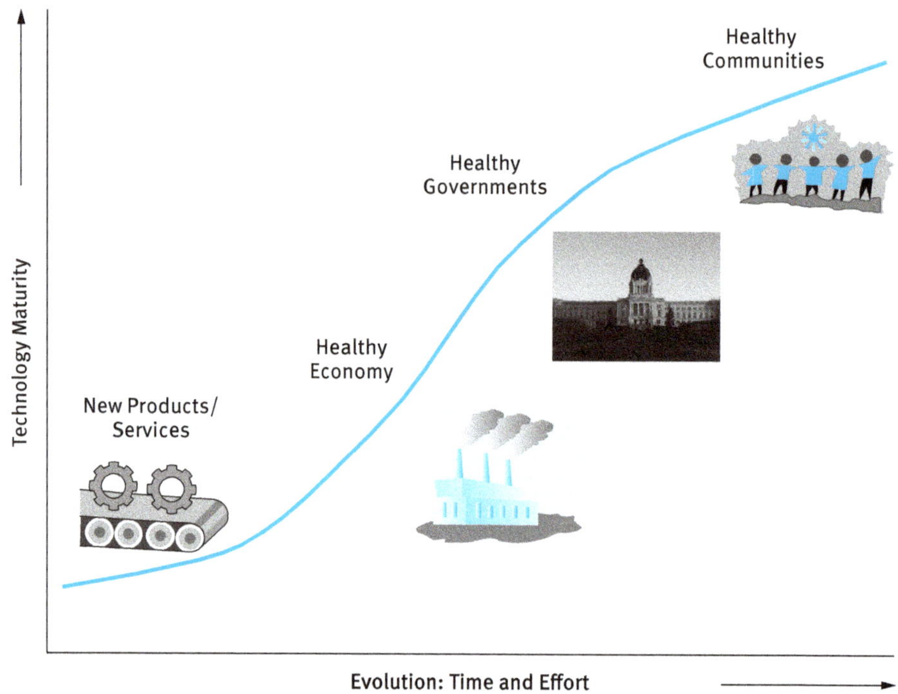

Figure 1.3: Some beneficiaries of technological innovation.

and support advanced education and skills development in the workforce and overall economic health and growth, in support of a high and rising standard of living [43, 45]. Much of the justification for such government investment is based on Schumpeter’s and Solow’s works and the endogenous growth theory. More recent analyses, such as those of Michael Porter [45], continue to conclude that the continuous pursuit by industry of higher productivity through improved products and processes is crucial to a nation’s competitive advantage and standard of living. A government policy approach might be to invest in sufficient research, development, and technological innovation to ensure that the total of government and private sector investments provides a reasonable return on investment (ROI) in terms of the overall economic benefits achieved in the region [43]. The social rate of return on these kinds of investments is often expressed as the long-term increase in gross domestic product (GDP) divided by the amount invested. Studies of OECD countries have produced estimates of the national social rate of return on R&D investments in the range 50% to 65% [43].

A cautionary note that is often overlooked is that capital markets are inherently competitive. While the economic theories and models that show how technological innovation can enable and support (or drive) an economy tend to be broadly appealing to industries, governments, and societies, the winners win at the expense of losers. While this tends to be generally accepted within industries and in international relations, it is not as broadly accepted or appreciated within governments and regions. This issue periodically appears in the media, especially when economic updates show that some regions in a national economy have advanced while others have languished or even retreated, despite government economic growth policies and programmes that may have been intended, or have been perceived to have been intended, to promote economic growth across an entire nation. In a C.D. Howe report on technological innovation and competitiveness, Peter Howitt wrote [46]:

> ". . . portraying technological change as a uniformly beneficial process that raises everyone's standard of living, . . . ignores a critical social aspect of the growth process: technological change is a game with losers as well as winners."

Some innovation milestones are arranged along a timeline in Figure 1.4. Most of these will be discussed in the next three chapters. Many additional such milestones are presented by Kotsemir and Abroskin [10].

1.4 Some examples of technological innovation

Some significant historical technological innovations include the following:

Interchangeable parts in clock manufacturing. In the United States, in about 1800, Eli Terry applied the idea of interchangeable parts to the manufacture of clocks using wooden parts [47]. By using machines to mass produce accurate parts, clocks could be built at reduced cost because the clock assemblers no longer had to be skilled artisans. Such clocks could also be repaired at reduced cost for the same reason. By 1814, Terry was using the same principles to mass produce clocks having brass and steel parts. The technological innovation of interchangeable parts manufacturing diffused across many industries and had become a standard textbook principle by the early 1900s.

Mass production. Also in the United States, around the turn of the 20th century, Henry Ford developed the Model T automobile, which was inexpensive and contained numerous improvements over other early automobiles (Figure 1.5). It was launched in 1908. His biggest innovation, however, was the 1913 combination of standardization, mass production, and moving assembly lines in the manufacturing plants. This enabled reductions in production time and cost, which, in turn, enabled Ford to sell the Model Ts at greatly reduced prices. This was a disruptive innovation that revolutionized the personal transportation market. It almost immediately caused the demise of the horse-and-buggy industry, and by just prior to 1920, Ford had gained half of the US automobile market as well.

Mechanical refrigeration. Between the mid-1850s and early 1900s, a number of mechanical refrigeration systems for the production of ice were invented, involving a wide range of refrigerants, such as ammonia, carbon dioxide, ether, methyl chloride, ethyl chloride, and sulphur dioxide, among others [48]. Some of these inventions, when commercialized, were able to begin competing with the use of natural ice for refrigeration. The machinery involved was originally quite large, and used industrially in ice-making, dairy, and meat-packing plants, but by the 1930s and 1940s, compressor innovations had enabled smaller, lighter, and less expensive refrigeration ma-

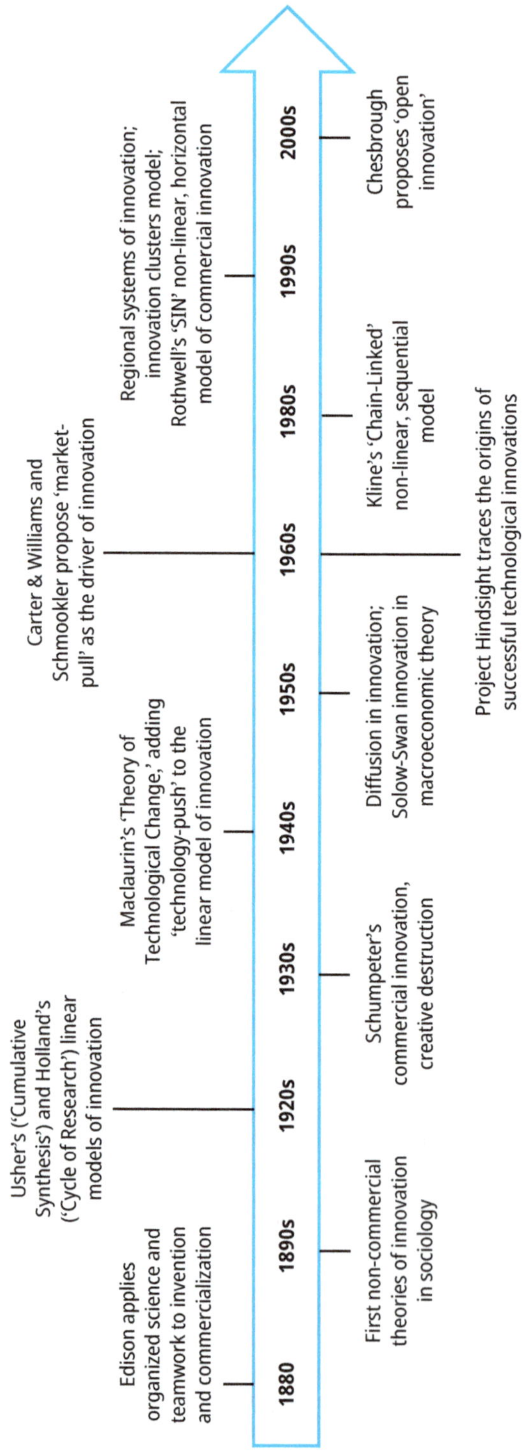

Figure 1.4: A timeline of some innovation milestones.

Figure 1.5: 1910 photo of a Model T Ford. (Public domain image courtesy Wikimedia Commons.).

chines [48]. Also required, however, were commercialized innovations in electric motors, control systems, less hazardous refrigerants,[10] and even the introduction of alternating current utility power systems – all of which came into the market in the first few decades of the 1900s [49]. The first mass-produced refrigerator in the United States was the Domelre (DOMestic ELectric REfrigerator, later renamed ISKO) between about 1916 and 1925. Meanwhile, Frigidaire had been created in 1918 and soon purchased by General Motors, which devoted its entire Research and Engineering Division to developing a more practical household refrigerator [49]. By 1925, they had improved the product and cut its price in half (from $1,000 to $500), making it a "bestseller" by 1926 and having sold a million units by 1928 [49]. GE and others soon entered the field, and an industry was born. By 1933, Frigidaire was able to reduce the price to less than $100 with its Meter Miser refrigerator.

Multiple product lines. In Japan, in the 1950s, Shigeo Shingo developed the single minute change of a die (SMED) system to reduce the time and cost required to changeover a manufacturing machine from producing one product to another [47]. Reducing the fixed cost of changeovers made smaller production runs economic and reduced the costs of both inventory and the space required to store the inventory. This helped reduce the amount of land needed for the manufacturing of multiple product lines.

The above represent only a few illustrative examples. A somewhat longer list of famous, revolutionary technological innovations is given in Section 10.3.

10 A huge breakthrough in this area was the development of chlorofluorocarbons (trade-named Freon) by the General Motors Research Laboratory in 1928.

1.5 Understanding innovation

Despite, or perhaps because of its increasing use in everyday life, there is a lack of understanding of the meaning of the word innovation and also of how it happens. In a recent editorial in the science journal *Nature*, Andrew Kusiak asserts [50] that despite all the time, money, and effort expended on creating innovation:

- *"There is no deep understanding of the innovation process"*
- *"There is no unified theory or reliable model for innovation"*
- *"We simply do not know how the innovation happens."*

He is partly right. Part of the problem is that creating innovation comes from creative thinking and is more of an art than it is science or engineering. The literature has many books and articles describing key philosophies, attitudes, environments, or practices that are advocated as being conducive to encouraging, accelerating, or improving the success rates in innovation, but there does not seem to be a single generally accepted set of such principles.

However, much is known about where technological innovation comes from and how it is created, although there is not a single pathway to innovation but a range of pathways. In his 2009 book, *Change by Design*, Tim Brown observed [51] that

> *"The myth of innovation is that brilliant ideas leap fully formed from the minds of geniuses. The reality is that most innovations come from a process of rigorous examination through which great ideas are identified and developed before being realized as new offerings and capabilities."*

In fact, there are a number of innovation development processes that have been used, and there are multiple models that attempt to describe the technological innovation "process." Some of these are introduced in Chapters 2 and 3.

Something that often comes with a lack of understanding is fear, and fears related to technological innovation have been expressed publicly from time to time. Jerry Courvisanos' summary [52] of the main negative societal perceptions of innovation includes fears that innovation may create and set loose on society:

- Monsters (the *"Frankenstein hypothesis"*). Such concerns have been frequently expressed in the media, in recent years, with the proliferation of new products based on, or including, nanoparticles[11] (see Chapter 10 in reference [53]);
- Job losses and unemployment due to any or all of mechanization, process efficiency improvements, robotics, and artificial intelligence (*"technological unemployment"*);

11 Hence, the term *"Frankenparticles."*

- Technologies that are advanced and marketable but environmentally and/or socially unsustainable; and
- The driving or determining by technology of a society's social structure and cultural values ("*technological determinism*").

These fears are not unfounded, as each of the risks noted above is real. The challenge, as always, is to ensure that scientific, engineering, and technological knowledge is used wisely and well.

2 Linear models of technological innovation

"Invention, it must be humbly admitted, does not consist in creating out of void, but out of chaos . . ."
Mary W. Shelley, British Author
In her Introduction to *Frankenstein*, 1831 Edition.

The process of technological innovation is neither simple nor linear. Some non-linear models will be discussed in later sections, but many aspects of the process can be introduced using simple linear models as a conceptual guide.

In 1880, Thomas Edison became one of the first North American inventors to apply the principles of organized science and teamwork to the processes of invention and innovation, working with many researchers and employees [54]. This concept had actually been put into practice in Germany at about the same time, as industrial research laboratories were created by such firms as Bayer, BASF, and Hoechst between 1877 and 1883 [55, 56]. Nevertheless, at the beginning of the 20th century, organized research, development, and technological innovation were still quite rare, but this was about to change. One example of this is provided by the U.S. National Research Council (U.S. NRC), which was formed in 1916 and seems to have quite quickly become interested in the role of research in helping industry advance (hence the term "*industrial research*") [56].

If R&D, invention, and technological innovation were to become organized and managed, then it was important to seek to understand where inventions and innovations come from. A number of people then set out to figure this out by such means as looking for shared factors and devising rules, looking backwards from successful inventions and innovations, and analyzing failed invention and innovation processes, why many inventions don't get patented, why many patented inventions don't get successfully commercialized, and why many research programmes and product trials aren't successful.

In the 1920s, Abbott Payson Usher proposed his "*Cumulative Synthesis*" model, which may be the first simple, linear model for the innovation process [57]. In it, four steps are followed:

1. "*Perception of the problem,*" involving the recognition of a need and the problems associated with its fulfilment;
2. "*Setting the stage,*" in which the technical and financial elements necessary for the solution are acquired;
3. "*Primary act of insight,*" in which the essential solution to the problem is found; and
4. "*Critical revision and development,*" in which the solution is fully realized and made practical.

https://doi.org/10.1515/9783112206720-002

Usher's model sets out the early steps of problem recognition, method of attack, inventive and/or creative steps, plus the notions of development and revision, but stops short at the completion of a practical, working model.

In 1928, Maurice Holland, the first head of Engineering and Industrial Research at the U.S. NRC, published what was probably the first complete description of the steps involved in the *"cycle of research"* leading to technological innovation and industrial growth [58–60]. The steps, according to Holland, were the following[12] [60]:

1. Discovery in pure science research
2. Applied science
3. Invention
4. Industrial research
5. Industrial application
6. Standardization
7. Mass production

This is another linear model of innovation, and Holland also recognized the importance of the "time lags" inherent in each stage [60].

Holland was one of the early champions of systematic research, to which contributions are made by numerous researchers, as opposed to the independent, lone-wolf-type of inventor. Systematic R&D and the linear model of technological innovation were fairly new concepts at the time, as in 1928, industrial research was still a relatively new tool of industry in North America: some of the earliest and most famous U.S. industrial research laboratories had been established by GE (in 1900), Du Pont (in 1902), AT&T (in 1907), and Eastman Kodak (in 1912) [56, 61].

The linear model is useful for describing many of the key steps in the innovation process (and in documenting them), but it doesn't lend itself to understanding the process in real time. That is, *"linear models describe what happened but not how it happened, and tend to reinforce the belief in a kind of orderliness which does not exist"* [10]. Nevertheless, although Holland's model later turned out to be an oversimplification, the concepts of a development pathway and a systematic approach were critical to the evolution of the modern approach to technological innovation.

2.1 The "technology-push" model of innovation

Building on the industrial research cycle work of Maurice Holland, the first model for the process of innovation was developed by Rupert Maclaurin in the 1940s [62]. His

12 A version of this list was repeated later in a report by the U.S. National Research Council, "Research – A National Resource II. Industrial Research," U.S. Government Printing Office: Washington, 1941.

"*theory of technological change*" laid out a sequential series of steps beginning with research and ending with commercialization.[13] This is now known as the "*technology-push*"[14] model of technological innovation[15] [62]. Maclaurin also suggested a number of measures that could be used as indicators of progress in each element of the sequence.

A very simple linear technology-push model could look like the following [63]:

Research → Development → Production → Marketing

A slightly more developed model is shown in Figure 2.1. In this model, discovery research produces an idea, applied research develops the idea into something that potentially addresses a market need or want, proof of concept shows that it could actually work, development engineering leads to some kind of model that does work, further lab and/or field testing shows that it can work under real-world conditions, further development engineering and field prototype testing show how it could be commercialized, commercial demonstration shows a fully working implementation, and commercial deployment represents market acceptance and use. These remain the essential steps involved in technological innovation.

Even in modern times, different linear technology-push models generally vary mostly in the number of steps involved. For example, Badulin's "*innovation snail*" model involves a linear series[16] of 12 stages, beginning with a product idea and continuing through development, business planning, capitalization, production, market saturation, and then either declining sales or transitioning to another, newer product [64].

An example of a technological innovation that was understood to have evolved in a linear technology-push manner is the dynamo (a kind of electrical generator) [58]:
- Discovery of chemically driven, steady-flow electricity by Alessandro Volta in 1779;
- Application of knowledge of electricity to develop the electromagnet by William Sturgeon in 1825;
- Invention of the dynamo by Michael Faraday in 1831; and
- Commercialization by various companies in the 1870s and 1880s.

Technological innovation can begin with discovery. A discovery is a new addition to our body of knowledge. Such an addition could be developed through the formula-

13 His actual steps were pure science, invention, innovation, finance, and acceptance (or diffusion) [62].

14 Also termed the *"first-"* or *"first-generation model of innovation."*

15 It may have been Maclaurin who coined the term *"technological innovation."*

16 Badulin's sequence was drawn in an almost open-circular shape, having the appearance of a snail's shell; hence, the name Innovation Snail, but it is a linear model of technological innovation.

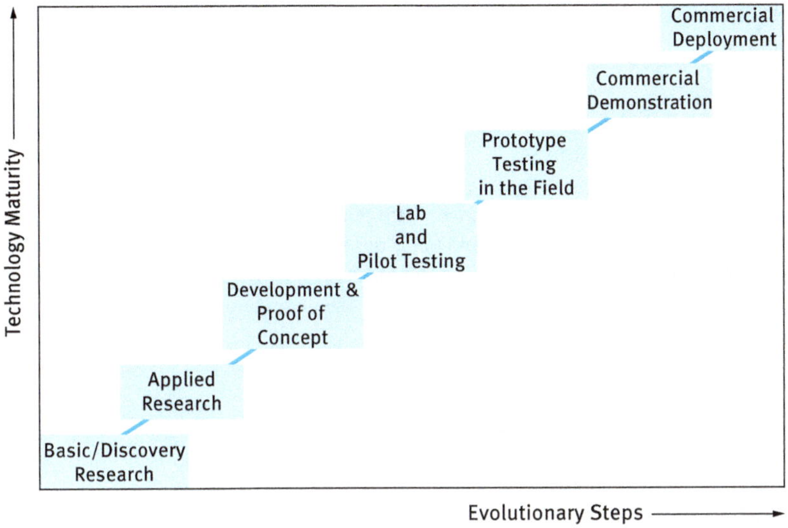

Figure 2.1: A simple technology-push, linear model of technological innovation.

Figure 2.2: New ideas and discoveries: classic components of technological innovation.

tion and testing of new theories, or through the observation and study of new phenomena (Figure 2.2).

One can discover:

- new physical things such as previously undiscovered elements, species, or stars;
- new ideas, such as new concepts that lead to new theories; or

- new understanding, such as new or improved understanding of the behaviour of natural phenomena.

Someone might see or make a new discovery and then think of a way to develop a technological innovation. For example:

> Velcro® was invented in 1941 by Swiss engineer George de Mestralhe who examined natural burrs (burdock seeds) sticking to his clothes and to his dog's fur (Figure 2.3). When he discovered the hooks and how they attached to anything resembling a loop, he thought of a way to create a new product [65].

Figure 2.3: Burdock plant. (Courtesy pxhere, image #1155059, CC0 public domain.).

> In the petroleum industry, there is a story that "thermal cracking" was invented in 1861 by the operator of a petroleum still, who accidentally allowed the still to overheat. It was then discovered that heavier vapours had re-condensed, returned to the still, and been redistilled, resulting in what seemed to be an increased yield of kerosene but later turned out to be thermally cracked gasoline [18, 66].

Discoveries are frequently the result of scientific or engineering research. Scientific research is the systematic investigation of hypotheses and theories (proposed explanations of things based on available facts and understandings) with the goal of generating new knowledge and understanding of things and phenomena. Engineering research is conducted somewhat similarly, but with the goal of generating new knowledge and understanding of how to make and/or control things. Some technological innovations that have arisen from engineering studies include developing and applying results from material testing to create flush riveting for airplanes and from wind tunnel testing to create advanced airplane propellers [67–69]. Some descriptions of technological results from scientific and engineering research are given in Table 2.1.

Table 2.1: Technological results produced by scientific and engineering research.

Technology-producing activities	Technological results
Scientific discovery research	Knowledge and understanding of things and phenomena, without regard to their utility.
Applied scientific research	Knowledge and understanding of useful and potentially useful things, phenomena, and processes.
Engineering research	Knowledge and understanding of how to control things, phenomena, and processes.
Inventive activities	Novel (invented) processes and products.
Development engineering	Models and prototypes of novel, practical, and controlled processes and products.

Technological innovation can begin with an invention. An invention is a newly created thing, such as a new synthetic molecule, software program, device, or process. Inventions are often derived from one or more discoveries. The most common invention strategies include the following:

- Empirical trial and error, usually in an industrial setting, such as in American inventor Thomas A. Edison's inventions (discussed later in this section). Ridley [70] argues that trial-and-error is a critical component of most developments that lead to technological innovations.
- Scientific theoretical and/or experimental work, usually in R&D laboratories, such as in Canadian medical scientists Frederick Banting and Charles Best's discovery of insulin.
- Beginning with the scientific approach to narrow the field, and then following with empirical trial and error to get to something that works. For example, the modern rocket evolved from theoretical work of Russian rocket scientist Konstantin Tsiolkovsky, followed by the empirical liquid fuel work conducted independently by German-American engineer Wernher von Braun and American engineer Robert Goddard.
- Adopting an existing technology and then using one of the above strategies to adapt it to the innovator's needs. For example, American inventor Samuel Morse's invention of the telegraph included adopting the practice of stagecoaches to frequently change horses at relay stations and adapting this idea to the design

of signal-boosting relay stations that enabled Morse code signals to be sent over long distances.

A scientist or engineer might think of a way to develop a technological innovation from the result(s) of an R&D project.

For example, 3M Post-it® Notes were invented in 1974 by US scientist Art Fry, who realized that a low-tack, reusable adhesive could be made based on the "failed" results of another scientist, Spencer Silver, who had been trying to develop a new super-adhesive [71].

Whereas an invention is the first occurrence of an idea or concept for a new product or process, technological innovation involves taking it into commercial practice. In this sense, some technological innovations can be described as successfully commercialized discoveries or inventions.

Morse's invention, mentioned above, is an example of the systematic approaches to invention that can be more successful and more efficient than brainstorming and/or trial-and-error methods. These generally involve looking at a problem or opportunity in unconventional ways. In the 1960s, Soviet engineer and inventor Genrich Altshuller developed a systematic approach to invention based on a comprehensive analysis of the patented solutions to hundreds of thousands of previously solved inventive problems. He identified a series of approaches that, taken together, were key to the solution of the majority of these problems. Altshuller called his approach the "*Theory of Inventive Problem Solving*" (TIPS, or TRIZ the Russian acronym) [72]. The TRIZ approach involves several groups of methods that can be used to look at a problem in ways that, either individually or in combination, frequently lead to finding an inventive solution (see also Section 4.3).

It has often been observed that many, if not most, discoveries and inventions tend to be made more or less simultaneously by multiple (independent) discoverers and inventors, respectively. This has been referred to as the "*Multiple Discovery Theory*."[17] Some commonly cited examples of this include the theory of the origin (evolution) of species and the formulation of calculus. Merton postulated that this phenomenon could be the result of a combination of the prerequisite knowledge and tools becoming available at a similar point in history and of having a substantial number of people pursuing similar investigations at about the same point in history, possibly in response to similar social and intellectual forces [73]. This has also been called the "determinist theory," stressing the role of social, military, and economic forces in driving technological change [57]. Epstein points out that, in addition, most inventions are improvements of previous inventions [74].

An opposing theory is the "*Heroic Theory of Discovery or Invention*," which is the hypothesis that the principal discoverers and inventors of most discoveries and inven-

17 Or the "*Multiple Invention Theory*," or "*Simultaneous Discovery (or Invention) Theory*."

tions, respectively, are rare *"greats"* or *"geniuses."* It is certainly true that many, if not most, inventions make use of prior advances in science and engineering, and such advances can provide the stimulus for further advances. Schmookler observes that *"Those who know the problems and opportunities may not know the science, and those who know the science may not know the problems or opportunities. By the time either group learns what it lacks, the problems or opportunities may have vanished"* [18].

Once an invention has been made, the inventor generally has some kind of use in mind for it, but its usefulness may not be readily apparent to others. It may even seem silly to others. An illustration of this point is given by Pontin's First Rule of Innovation: *"Any sufficiently radical invention seems ridiculous to most people when they first encounter it"* [75]. Therefore, additional development, or at least marketing, will be needed. A very simple invention-driven, linear technology-push model could look like the following [52]:

Inventing → Incubating → Promoting → Sustaining

In *"technology-push"* innovation, invention should drive the development and commercialization of products, processes, or services into the marketplace, thus realizing technological innovations. Since it is technology-push, most or all of this process is conducted without regard for or knowledge of the existence of a market niche or need, so the ultimate product, process, or service may or may not be commercially successful. In fact, although some are wildly successful, the vast majority of technology-push innovation attempts fail.

An example of a successful technology-push technological innovation is the mass-market personal computer (1977: Commodore PET™, Apple™, and Tandy TRS-80™). It was only after the personal computer was commercialized and became successful in the marketplace that consumer demand was created, which led to this technological innovation becoming disruptive and highly successful. Other successful examples include the Sony Walkman™ portable audio cassette player (1979), and the Apple iPad™ tablet computer (2010).

Innovation requires more than just discovery and/or invention. From the point of view of technological innovation, once something has been invented, there are still a number of questions to be answered (adapted from [76]):

1. Does it work, or can it be made to work?
2. Does it work better than existing technology?
3. Will anybody buy it?
4. Will enough sell at a price that will yield an adequate profit?

Only if the answers to all four questions are likely to be "yes" will it make sense to do the R&D that will be required to commercialize the invention.

The next step involves whether and how to protect the invention. Some inventions are simply released to the public, some are described in materials that are copyrighted, some are held as trade secrets, and others are patented. A patentable invention is an invention that meets a patent system's requirements for novelty, usefulness, and significance. That is, to be patentable, an invention must be original enough not to be obvious to someone skilled in the field of the particular invention, it must be more than a minor improvement on the prior inventions in the field, and it must be useful. This is usually demonstrated by a practical application ("*Reduction to Practice*") of the concept. Reduction to Practice is one of the key differences between an invention and a patentable invention.

Also, in practice, most inventions are only minor improvements on the prior art and, therefore, not patentable. Even most patented inventions do not have significant commercial value. Kline and Rosenberg note that the "*overwhelming majority of inventions recorded at the US Patent Office were never introduced on a commercial basis*" [63].

Furthermore, most patentable inventions are not completely invented by a single person either. Holland found that "*less than five percent of patents that reach the commercial stage are the result of individual, independent inventors*" and concluded that "*historians and philosophers who trace the records of basic inventions back to the 'only firsts' do not take into account the fact that a great invention is not the completed result of a single [person] – it is the resultant of many inventions, the composite of a number of realized ideas merged into a workable whole*" [58, 59]. It usually requires considerable R&D, multiple people, and a multitude of inventions to realize a patentable invention that has significant commercial value.

Even for a commercially valuable invention, further development and pilot or field testing (beyond the patenting stage) are generally needed in order to translate it into a new product or service launch (i.e., technological innovation) in the marketplace. An illustration of this point is given by Pontin's Second Rule of Innovation: "*The first attempt to commercialize an invention almost never succeeds*" [75].

The pace of technological innovation can be slow. In general, the various elements of the linear model of innovation do not proceed at the same pace. Most often, the discovery rate exceeds the invention rate, and the invention rate greatly exceeds the technological innovation rate.

The main reasons that the invention rate exceeds the technological innovation rate are as follows:

- The rapidly expanding body of scientific and engineering knowledge enables many invention alternatives to be evaluated theoretically, whereas technological innovation alternatives usually need to be tested by experimentation and pilot testing, which are much slower processes.
- The amount of private- and public-sector funding available for discovery research (leading to new discoveries and understanding) massively exceeds the amount available for applied R&D (leading to new technologies and inventions). Although both are frequently needed to get to technological innovation, the slower rate of technology development creates a bottleneck in the process.

- The knowledge base derived from discovery research is usually promptly and readily accessible through the published literature, whereas the knowledge derived from applied R&D is often held proprietary to the originators in order to prevent competitors from copying or adopting and adapting it. The reduced degree of access to new technology creates another bottleneck in the process.

The discovery rate exceeds the invention rate, and the invention rate greatly exceeds the innovation rate:

$v_{\text{discovery}} > v_{\text{invention}} \gg v_{\text{innovation}}$

It often takes several decades to go from an invention to a technological innovation [28, 77]. There can be many reasons for this:

- It may still be necessary to identify a market need or want (in technology-push innovation).
- It may be necessary to experiment to find ways to make the invention practical and implement it.
- It may be necessary to develop or acquire knowledge and/or capabilities in development, engineering, production, markets, distribution, financing, and so on.

A common occurrence is to hit a so-called "*Innovation Barrier*" (also termed "*Bottleneck*" or "*Reverse Salient*"), which can be any challenge or obstacle to the successful process of innovation, up to and including successful technology commercialization [28]. This usually means that a critical component is lacking or insufficiently developed, which could be technological, financial, manufacturing, cultural, or regulatory, for example. Such barriers are more specific, however, than the general requirements to conduct R&D, experimental development, prototyping, pilot testing, demonstration, preproduction, financing, marketing, sales, and deployment. In *The Innovation Illusion* [78], Erixon and Weigel describe numerous factors acting to block, or at least hinder, successful innovation in modern societies. By definition, all innovation barriers have to be successfully identified and managed before innovation (commercialization) is possible. When this is accomplished, one sees references to having achieved a breakthrough, or a so-called "*New-to-the-World Solution.*"

> An example of an innovation barrier in the developing of the first airplane was having to wait for the invention and commercialization of the internal combustion engine, even though the basic idea for the airplane can be traced back to da Vinci's famous notebooks of the 1400s.

Despite the slow pace of individual innovation processes, technology is changing fairly rapidly at a societal level. This only heightens the pressure on companies. Clayton Christensen coined the phrase "Technology Mudslide Hypothesis" in describing the difficulties companies face in trying to cope with the "*relentless onslaught of tech-*

nology change" in their respective markets[18] as being like "*trying to climb a mudslide raging down a hill. You have to scramble with everything you've got to stay on top of it, and if you ever once stop to catch your breath, you get buried*" [79].

Limitations of the technology-push model. Notwithstanding the examples given above, the innovation literature shows repeatedly that the technology-push model of innovation doesn't actually work very often, and only in a few cases have technology-push processes led directly to successful technological innovation [18, 63, 67, 80–83]:

> A classic example is Thomas Edison's first patented invention, an electric vote-tabulating machine that could tally votes in a legislative assembly virtually instantaneously (U.S. Patent 90,646, June 1, 1869). Unfortunately, after inventing and developing it, he found that governments didn't actually want it, and it was never commercialized. After this experience, Edison is said to have resolved "*never again to spend time on an invention until he was sure a sound market was in prospect*" [63]. His second strategy was much more effective, and much later, he is quoted [84] as saying "*I find out what the world needs. Then, I go ahead and invent it.*" See Figure 2.4.

> A more recent example of the problem is given by the Concorde jet, recognized by many as a great engineering achievement, but which was also a very expensive commercial failure. Although the Concorde can cross the Atlantic Ocean in about half the time required by a Boeing 747, its fuel costs per passenger-kilometre are at least 15 times as much [63].

Another issue is that some technological innovations are developed using only existing knowledge and/or ideas, but using them in new ways, and without doing any discovery or applied research at all ("*Synthetic Innovation*"). An example of this is when people develop something new for themselves out of their own imagination and interests, and without necessarily having any thought or even interests in technological innovation. These "*consumer-innovators*" make inventions or modify existing products to create the first concepts or even the first prototypes of new technological innovations, especially in the area of products. For example, the skateboard was reportedly first developed and built by children for their own use by hammering roller skate wheel assemblies onto wooden boards [85].

In this mode of technological innovation, the consumer-innovators develop some kind of product that they want or need. Next, other people either take an interest, or do not, possibly leading to copying and adoption by others. If enough copying and adoption take place, then eventually, a commercial enterprise will take note and develop a commercializable version that can be produced and marketed. Although this mode of technological innovation does happen, the success rate is not very high. Von Hippel et al. found that out of over 5,000 potential consumer-innovations in the

18 This assumes that all of the competing companies in a market niche constantly have to move "upwards" in terms of technological innovations, or else they will slide "downhill" and lose market share.

Figure 2.4: Thomas Edison in his West Orange, New Jersey laboratory, 1901. (Courtesy U.S. Library of Congress and Wikimedia. Public domain photo originally by Underwood & Underwood, 1901.).

United Kingdom, the United States, and Japan, only 8% were actually adopted by other individuals and/or companies [85]. (Success rates are discussed further in Section 2.5.)

Linear and sequential models, like the technology-push ("supply-side") model, have some value in terms of distinguishing broad driving forces in innovation, but the process of innovation itself should not be viewed as being sequential, or linear, or as simply being concerned with individual inventions or other creative solutions. With hindsight, these have become increasingly referred to as first-generation models of technological innovation. Model evolutions beyond the first generation (see also Table 4.2) will be discussed further in the next few sections, and also in Chapters 3 and 4.

2.2 Degrees of technological innovation

Technological innovations can range from those that are quickly developed and deployed but are modest in scope and impact to those that take significantly more time and effort to realize but change entire marketplaces. There have been several categorizations proposed for degrees of technological innovation. The following is intended to be illustrative of these:

> *Incremental innovation.* This is a modest product/process/service improvement that is (frequently) not overly expensive or risky and is relatively easily developed and deployed, usually within a few years. Incremental innovations often fit within a company's financial means and risk tolerances and, therefore, can be self-funded. An example is incremental improvements to an existing commercial process for which the market is already well-established. The continuous improvement movement within businesses of all kinds can produce such incremental innovations. Also termed "*Cosmetic* or *Survival Innovation.*"
>
> *Adjacent innovation.* This represents the commercialization of a product, process, or service that is already being sold in one market into a new market, particularly where the new market is, in some sense, "near" to the established market. Like incremental innovations, adjacent innovations are relatively easily developed and deployed in a short period of time because the product/process/service already exists; however, the marketing and sales into a new market can increase both costs and risks. Adjacent innovations are also frequently self-funded. An example is Procter & Gamble's Swiffer mop/duster product line.
>
> *Evolutionary innovation.* This is a more impactful technological innovation than the two above that competes with earlier commercial product, process, or service(s) but does not significantly change (disrupt) markets or even market niches. Evolutionary innovations can be quite expensive and risky and can take more than five years to develop and deploy, partly because they can be quite R&D intensive. A market potential "rule of thumb" (the "20/30 Rule") is that such a technology should perform its function 20% better and 30% less expensively than the pre-existing, competing technologies [76]. Evolutionary innovations usually significantly help companies enhance their competitive positions in a marketplace. An example is the introduction of hybrid electric vehicles in the automotive industry.
>
> *Disruptive innovation (or "radical innovation").* This is a technological innovation that displaces earlier commercial product, process, or service(s) and significantly changes (disrupts) a market niche or even an entire market. Disruptive innovations are very often expensive, risky, and time-consuming (more than five years) to develop and deploy, partly because they are usually very R&D intensive. Due to the high costs and risks involved, a company may need government support, venture capital investment, and/or partnerships with universities and research and technology organizations (RTOs). An example is the electronic calculator, which made the slide rule obsolete. Another example is the electric typewriter, which made the manual typewriter obsolete.

There are a number of static and dynamic models of innovation. The static models describe the major kinds of technological innovation and apply them to organizations as they exist at a certain point in time. Examples of static technological innovation models are:

- Schumpeter's Mark I Innovation (Section 1.3);
- Schumpeter's Mark II Innovation (Section 1.3);
- Abernathy and Clark's incremental-radical model of innovation; and
- Henderson and Clark's model of evolutionary innovation.

These are by no means the only such models. Kotsemir and Abroskin provide additional examples [10].

The *"Abernathy and Clark Incremental-Radical Model"* [86] distinguishes between organizations that develop incremental innovations, such as those that are already incumbent in a marketplace, and organizations that develop disruptive innovations, such as new entrants to a marketplace. They defined forms of evolutionary innovation and distinguished between a technological innovation's effect on an organization's technological knowledge and resources, and its effect on the scale of the technological advance and whether the competing products, processes, or services remain somewhat competitive or are made obsolete.

In the Abernathy-Clark Model, organizations that simply enhance their technological knowledge and resources are most likely to achieve modest incremental innovation[19] at best, whereas those that develop completely new technological knowledge and/or resources and use them to make huge, game-changing (i.e., market changing) technological advances are most likely to achieve disruptive innovation[20] (*see* Figure 2.5). In between these extremes fall two categories of organizations. One category represents organizations that develop completely new technological knowledge and/or resources but only use them to make more competitive products that do not displace or render obsolete their competing products. Such companies achieve evolutionary innovations.[21] The other category represents organizations that develop advances to their technological knowledge and/or resources and are able to introduce products that do displace or render obsolete their competing products in a small market niche. Such companies achieve *"niche innovation."*

The Abernathy-Clark Model has been used to explain how incumbent companies tend to be in a good position to implement incremental innovations, as they can leverage their existing technological knowledge and resources, whereas new companies entering a marketplace tend to be in a good position to implement disruptive innovations, as they do not need to change their existing technological knowledge and/or resources in order to do so.

---

19 Abernathy and Clark originally referred to this as "regular innovation." I have substituted the term incremental innovation.

20 Abernathy and Clark originally referred to this as "architectural innovation," but in other literature, this term has very different meanings. I have substituted the term disruptive innovation.

21 Abernathy and Clark originally referred to this as "revolutionary innovation," but in much of the literature, this term is a synonym for disruptive innovation. In order to avoid confusion, I have substituted the term evolutionary innovation.

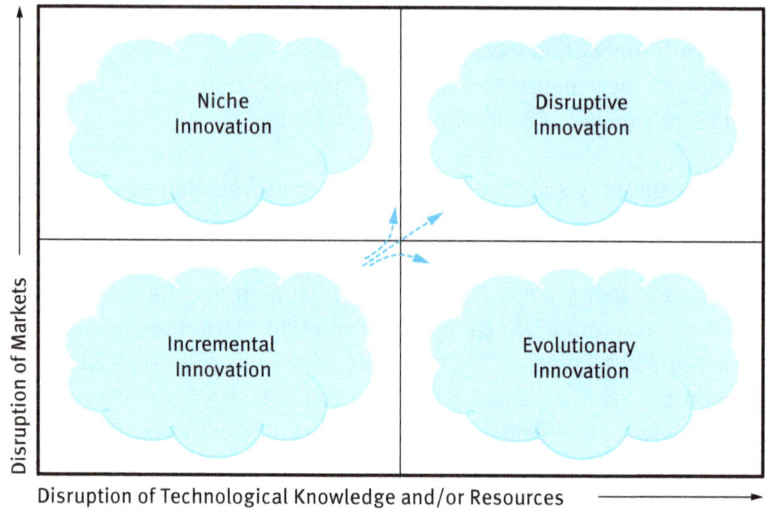

Figure 2.5: A *"Transilience Map"* illustrating Abernathy and Clark's forms of technological innovation and their influence on an organization's prior technological knowledge and resources (horizontal axis) and on the competitive marketplace (vertical axis).

The Abernathy-Clark Model illustrates two of the most important attributes of a technological innovation: namely, its influence on technology and its influence on the marketplace. This model also brought to light what is sometimes referred to as the *incremental-radical dichotomy,*[22] which is that the large, successful, incumbent companies in an established marketplace tend to find it extremely difficult to implement disruptive innovations because, by their very nature, they require the creation of new technological knowledge and/or resources that render the old ones obsolete, and they involve the introduction of such radically new products, processes, or services into the marketplace that they render the previous ones obsolete – including those of the innovating company! To risk disrupting a well-established and successful business, or line of business, by introducing a new product that may or may not be successful at all, goes against the mindset, risk tolerance, and strategy of most such companies.

Henderson and Clark [87] expanded prior innovation models by defining forms of evolutionary innovation and distinguishing between an organization's product component knowledge and its architectural knowledge (i.e., knowledge of the linkages among product components). In the *"Henderson-Clark evolutionary innovation model,"* organizations having low degrees of both component and architectural knowledge are most likely to achieve modest incremental innovations at best, whereas those with high degrees of both component and architectural knowledge are most likely to achieve disruptive innovations (see Figure 2.6). In between these extremes fall two categories of organ-

22 Also termed *"The Innovator's Dilemma,"* see Section 4.9.

izations. One category represents organizations having a low degree of component knowledge but a high degree of architectural knowledge and, therefore, are most likely to achieve *"architectural innovations,"* in which improvements are made to the linkage(s) between the components in a product, but not to the components themselves [6, 87]. The other category represents organizations having a low degree of architectural knowledge and a high degree of component knowledge and, therefore, are most likely to achieve *"modular innovations,"* in which improvements are made to one or more of the components in a product, but not among the linkages between those components [6, 87]. The Henderson-Clark Model has been used to explain how incumbent organizations in a marketplace can achieve substantial innovations that fall short of being disruptive.

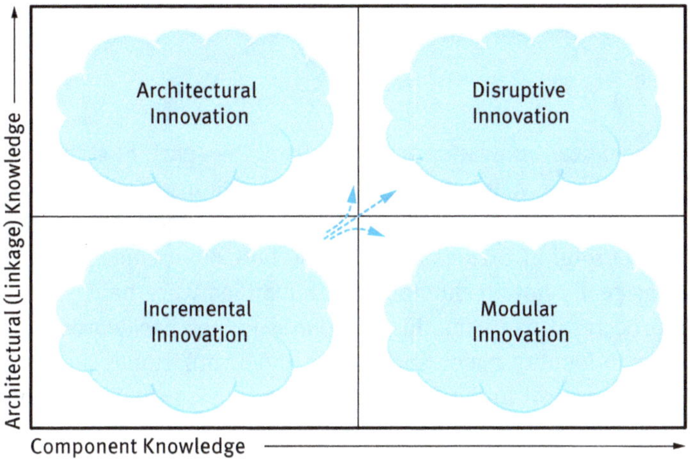

Figure 2.6: Illustration of Henderson and Clark's forms of evolutionary innovation.

Platform innovations provide an illustration of Henderson-Clark innovation. Platform innovation occurs when a new core technology, based on a technological breakthrough, leads to multiple new products, processes, or services. Some examples include:

- the technological breakthrough of electronic imaging enabled an entire generation of digital camera products to be introduced, whose core technology was very different from the previous celluloid film technology;
- signal compression enabled more cable television channels to be transmitted over the existing cable infrastructure;
- computers (mainframe and personal);
- software platforms (such as operating systems and databases);
- drug delivery devices;
- satellites; and
- smart phones.

It is possible to have a platform technology and create modular and/or architectural innovations within it, as long as the core technological breakthrough remains. For example, in digital data recording, magnetic tapes and a succession of smaller and smaller-sized floppy disks involved different materials and components but were all based on the same breakthrough core technology (magnetic-head recording).

The above models, Schumpeter Mark I, Schumpeter Mark II, Abernathy and Clark, and Henderson and Clark, are illustrative of static technological innovation models. There are others (see, e.g. Popadiuk and Choo [88] and de Castro et al. [6]). These static models serve to illustrate the major kinds of technological innovation in the context of organizations as they exist at a certain point in time, but organizations are not frozen in time. Dynamic models of innovation incorporate multiple evolutionary processes, and the concept of lifecycles, and are discussed further in Chapter 3.

2.3 Hindsight: where do most innovations come from?

The "push" model of technological innovation proceeds from the spark of a creative idea, a flash of understanding, or a moment of discovery through a series of applied R&D stages and into a commercializable product, process, or service. As just discussed, this model has been useful in various ways, but in the 1960s, people began to question whether this was really how technological innovations were being developed in practice. One approach to answering this question is to look backwards from major innovations in order to identify major approaches and/or milestones that may have been important or even critical to the ultimate development of the innovation. Unfortunately, not many such studies have been reported.

In "*Project Hindsight*," however, this tracing-back to the original key advances in knowledge was actually carried out [89, 90]. The US government traced the technology-advance origins behind successful military systems innovations developed between 1945 and 1962. They found that only 0.3% came from basic scientific research, whereas 99.7% came from applied R&D and engineering. Of the applied work, only about 10% came from universities, whereas about 90% came from government labs and/or industry labs (Figure 2.7). Not surprisingly, then, only about 5% of the innovations came from solutions developed without any particular need or opportunity in mind ("technology-push"), and about 95% came from targeted work with a specific market opportunity or need in mind ("market-pull").

"Project Hindsight" found that for technological innovations created by or for the U.S. military between 1945 and 1962:

- 99.7% of the original key advances in knowledge came from applied research and development and engineering (0.3% from basic scientific research).
- 90% of the applied work came from government and/or industry labs (10% from universities).
- 95% of the innovations came from "market-pull" (5% from "technology-push").

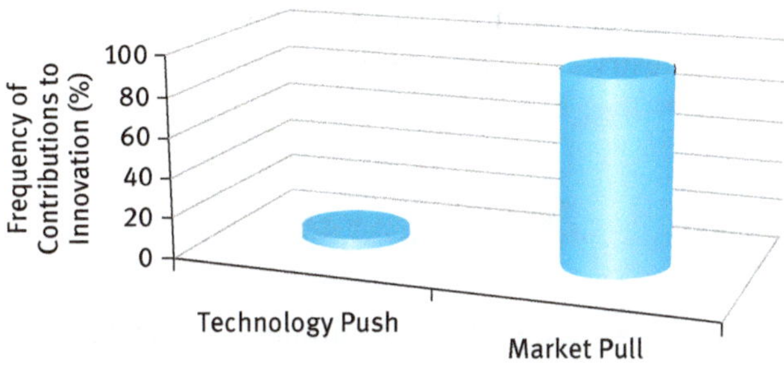

Figure 2.7: Illustration of the origins of successful technological innovations as identified in "*Project Hindsight.*" From data reported in Sherwin and Isenson [89, 90].

Project Hindsight was only one of several such studies on the origins of successful technological innovations. In the 1960s, similar studies were conducted in other U.S. military services and in the non-military economies of the United States and the United Kingdom, all of which concluded that specific demands (or "needs") drive the vast majority of successful technological innovations [83]. For other descriptions of military technological innovations and their history, *see* references [91–93].

In 2004, a modern follow-up project called "Project Hindsight Revisited," conducted another tracing-back to original key advances in knowledge for four key examples of modern military innovations [92, 93]. In this work, the authors examined *critical technology events* (CTEs), defined as "ideas, concepts, models, and analyses that had a major impact on the development" of a particular innovation. Although the 2004 study found that there had evolved new ways of developing and deploying technologies, it again found that more than 90% of the critical technology advances (CTEs) were made in government labs and/or industry labs [92].

In another "hindsight-type" study, in the 1980s, Mansfield [94] surveyed 76 major US industrial companies and found that only about 10% of their technological innovations, as viewed by the innovating companies, could not have been developed without recent academic research findings (recent meaning within the previous 15 years).

Other studies of the components of successful innovations have come to the same conclusion: that only in a few cases have scientific discoveries directly led to successful technological innovation [18, 63, 67, 80–83]. As a result, the technology-push model has been described as "*not only wrong but backward*" [95].

Although discovery research is not needed for the vast majority of technological innovations, the conclusion to be drawn from these hindsight studies is *not* that there is no need for "*discovery research*" (basic scientific research). Discovery research is absolutely needed in order to advance our broad base of scientific knowledge and understanding. The conclusion to be drawn from the hindsight studies is that there is also a critical need for mission-oriented, applied research, development, and commercialization in order for the entire process to work. Taken all together, one can think of a system of approaches, all of which are ultimately needed.

The components of the processes, and principal institutions involved, including universities, RTOs, governments, and industry are discussed in more detail in later sections of this book.

2.4 The "market-pull" and "concept-push" models of innovation

Technological innovation can begin with market-pull. Although glamorous, the leaps from discoveries or inventions directly to technological innovations are actually quite rare. As the studies cited in the previous section have shown, it turns out that most technological innovations actually come from market-pull: someone develops a practical solution to a specific industry problem or opportunity, and they, or someone else, commercialize it. For example:

- About two thousand years ago, the ancient Romans developed concrete by combining aggregate and water with a cement made from lime and volcanic ash. This was developed in response to their need for a strong and durable building material with which to construct buildings, roads, and bridges.
- In the paper industry, "*wet-strength paper*" was apparently invented in 1942 by a J.B. Butler, whose experiences in military exercises led him to conceive a need for waterproofed maps for military operations [18].

An argument can be made that technological innovation begins when someone identifies a problem or opportunity and has an idea for how to address it in a way that is practical and commercializable. In the "*transcendentalist model*" of the innovation process, the innovation arises from an instantaneous act or inspiration of genius. The technological origins, however, may well trace further back to some kind of advance(s) in knowledge, understanding, or wisdom that provided the key enabler(s) for the specific technological innovation. In the "*mechanistic model*" of the innovation process, innovation arises from an accumulation of numerous smaller steps or ideas that occur over a reasonably long period of time.

Abbott Payson Usher proposed a model called "*Cumulative Synthesis*," by which a sequence of Edison-like steps describes the process of creating something novel, practical, and useful (i.e., a technological innovation) [96, 97]. Some of these smaller steps are novel, but some may not be. Usher's model is a hybrid of the transcendentalist and mechanistic models in that, somewhere in the process, there is still a critical act of inspiration and/or insight. One formulation of Usher's Cumulative Synthesis steps is "*perception of the problem*," followed by "*setting the stage*" (in which the elements necessary for the solution are acquired), followed by the "*primary act of insight*" (in which the solution to the problem is found), followed by "*critical revision and development*" (in which the solution is fully understood and made practical) [96, 97].

The previous paragraphs have essentially introduced the second "linear" model of technological innovation:[23] the "*Market-Pull*" model. The term "*market-pull*" seems to have originated from the 1950s work of C. F. Carter and B. R. Williams [22, 23], the work of economist Jacob Schmookler[24] [18] in the early 1960s, and the works of others who studied the process of innovation in the 1950s [66, 96], 1960s [98, 99], and 1970s [81, 100, 101].

Schmookler describes searching through a thousand recorded and significant inventions in the United States, in four different industries (petroleum refining, paper making, railroading, and farming), for any suggestion in the literature that a particular scientific discovery led to the creation of the invention. In the vast majority of cases for which the initiating stimuli were identified, the stimulus was the conception of a technical problem or opportunity in economic (market) terms. In the remaining cases, the stimulus was an accidental discovery from which the solution of a technical problem or opportunity was conceived, also in economic terms. He reports [18] that "*in no single instance is a scientific discovery specified as the factor initiating an important invention in any of these four industries.*" Thus, Schmookler found the main driving force to be what is now usually termed market-pull. Similarly, out of over 1,800 successful technological innovations reviewed by Marquis [102], the vast majority (over 75%) was "a result of perceived market needs" rather than "perceived technical opportunity."

In market-pull innovation, marketplace opportunities drive the search for new product, process, or service concepts, for which innovators search the existing body of knowledge, inventions, and technologies. Since it is market-pull, the market niche or need is clear, but the search for appropriate knowledge, invention, and/or product conceptualization and development may not be successful. Even the commercialization may or may not be successful in the marketplace. Examples of successful market-pull technological innovations include mass-market digital photo editing software (1980s) and mass-market digital cameras (1990s), both of which were originally developed for speciality niche-markets.

Considering the terms "*technology-push*" and "*market-pull,*" Godin and Lane have drawn an analogy with the terms "supply" and "demand" in economic theory [83]. This underscores the idea that both are relevant, either can lead to limitations on the degree and rate of success, and either may dominate in a given situation:

> Another example of market-pull technological innovation is the steam-powered ship. There was already a market demand for faster ships than could be provided by sails alone, and the number and types of sails on sailing ships had reached their technology limit (see technology S-curves below, in Sections 3.2 and 3.4).

23 Also termed the *"second-generation model of innovation"* or *"Demand Pull"* innovation.

24 Schmookler's work actually emphasized the role of economic conditions in driving invention (not technological innovation *per se*), but his focus was on patented inventions that are successfully commercialized, that is, patented technological innovations.

There are other forms of technological innovation that are neither market-pull nor technology-push. For example, the interactive model is a linear innovation model in which each step of the process is influenced, to some degree, by "push" and "pull" influences [98, 103].

Another example is termed "*Concept-Push*" innovation, meaning that the innovation relates to an entirely new product or service concept that was not previously imagined, and which prospective customers didn't previously know that they might want and/or need.[25] Roberto Verganti refers to this latter form as "*design-driven innovation*" [104]. Where they are successful, concept-push innovations are frequently breakthrough innovations.

Examples of concept-push innovations include
- The Nintendo *Wii* entertainment system, which combined video gaming with physical activity and social interaction
- The *Swatch*, which transformed time-pieces into fashion accessories
- Apple's combination of the *iPod* and *iTunes*, which went past creating a modern way to carry around music (replacing the MP3 players, which had, in turn, displaced the *Walkman*) and provided a new kind of system by which customers could not only store, transport, and play music but also more easily sample and buy music, virtually anywhere

As these examples illustrate, concept-push innovations need a special kind of creative visioning.

The death of linear innovation models. Maclaurin had warned in 1955 that the stages of a linear innovation model do not proceed in a sequential fashion, that they may not even be linked, and that even if they are linked, the process is highly discontinuous [105]. However, it wasn't until the 1970s that it became fairly broadly recognized that the technology-push and the market-pull models are both oversimplified. As a result of this, more complex models began to evolve. With hindsight, the market-pull models have become increasingly referred to as second-generation models of technological innovation (see also Table 4.2). In more recent years, the processes of innovation and also the broader innovation systems (both regional and global) have become recognized as complex, adaptive systems that tend to be dynamically evolving over time [103, 106]. Such systems both operate and evolve in non-linear ways. Illustrations of this will be discussed in the next several chapters.

---

25 This is an example of "*Future Pull,*" in which the innovation development is driven mostly by a sense or vision of future benefits, needs, or demands. This term is sometimes used to provide contrast with the terms technology-push and market-pull.

2.5 Success rates in technological innovation

Both the technology-push and market-pull innovation pathways work, but the market-pull pathway is much more likely to be successful than the technology-push pathway. This is partly because the success rate for technology-push innovation is very low, and partly because of the risk of producing a product that the marketplace doesn't really need or want (for more on the latter reason, see Section 4.7).

Regardless of the approach to technological innovation, the failure rates along the way are extremely high. A relatively recent Deloitte study [107] found that *"only 4.5% of all innovation efforts met ROI goals established by the companies that funded them."* In other words, over 95% of the innovation efforts failed.

In the early 1900s, Thomas Edison said, *"Genius is one percent inspiration, ninety-nine percent perspiration"* [108]. But it's worse than that. One bright, new idea in isolation has a low probability of being successfully commercialized. When questioned about his apparent lack of results in attempting to develop a practical nickel-iron battery in 1910, Edison responded, *"Results! Why, man, I have gotten a lot of results! I know several thousand things that won't work"* [109]. Such reflections illustrate not just the experience of one inventor, but the norm. It turns out that, on average, a huge amount of work and several thousand failures almost always mark the landscape between a bright new idea and a commercial success.

A graphic illustration of the difficulty of making technology-push work is given by the research of Stevens and Burley [110], who found that, in most industries, it takes about 3,000 initial, undeveloped ideas to produce a single successful commercial product. They also identified the improving success rates as an idea survives various intermediate staged-gates of development, as illustrated in Figure 2.8. Stevens and Burley found that, regardless of the industrial sector, only about 60% of new product launches succeed. At earlier stages of the development process, the success rates are even lower.

- Only 25% of products from the major development stage succeed
- Only 11% of products from the early stages of development succeed, even after detailed analysis
- Only 0.8% (1 in 125) of products from the small project stage succeed, even after a patent is granted
- Only 0.3% (1 in 300) of products from the developed idea stage succeed, even after a patent disclosure is made
- Only 0.03% (1 in 3,000) of products from the original/raw idea stage succeed

This means that to get two new products, processes, or services added to an organization's lines of business, on average, 6,000 new ideas would be needed, and they would have to be generated reasonably quickly because the development and filtering processes take a lot of time, as will be discussed later. Others have found similar trends

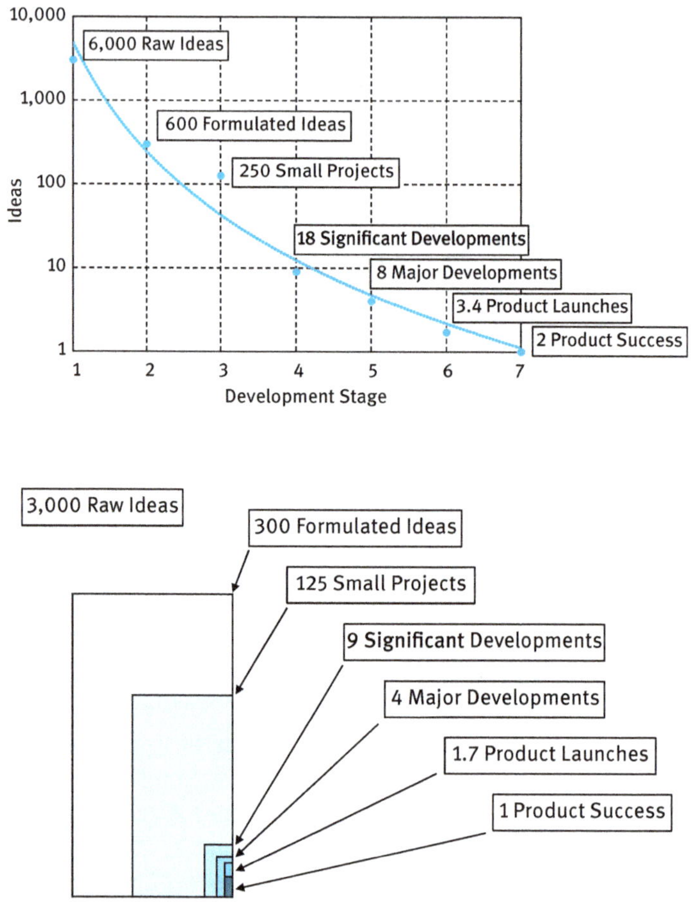

Figure 2.8: Illustration of a "*universal success curve*" (upper) and the concept that it can take 3,000 raw ideas to develop a single commercial success. From data reported in Stevens and Burley, 1997 [110]. The dotted line in the upper figure represents a power law fit of the reported data.

to that just described, but with somewhat different numbers for different industries. For example,

- Von Hippel et al. found that, out of over 5,000 potential consumer-innovations in the United Kingdom, United States, and Japan, only 8% were actually adopted by other individuals and/or companies [85].
- A study for the pharmaceutical industry estimated that it could take at least 6,000 ideas to produce a single commercially successful product [111].
- An estimate for the Information Technology (IT) sector is that "*96% of all innovations do not return their capital cost, and 66% of new products fail within two years*" [112].

- In an evaluation of more than 5,000 innovations over a 15-year period, Tuff and Wunker [113] found that 95.5% were not successful (with "success" being defined as returning their cost of capital).

There are many reasons why the failure rates are so high. Some of these are shown in Table 2.2 (see also Sections 3.4 and 3.5).

Table 2.2: Some reasons for failure in technological innovation [78, 81, 114].

Reason	**Explanation**
Lack of capital	The development costs exceed predictions, and without sufficient demonstrated progress, refinancing is unsuccessful.
Momentum stalls	As development proceeds from one stage to another, continued efforts and rework cause the development team(s) to lose energy and/or confidence in the innovation.
Failed demonstration	The demonstration of the product, process, or service is unsuccessful.
Poor cash flow	Early sales revenues are low, causing cash flow problems.
Premature scaling	Attempting to scale-up to production scale before the working-prototype has been sufficiently tested.
Misunderstanding the market	Incorrect assumptions about the market opportunity.
Poor customer experiences	Lack of customer satisfaction leads to complaints, returns, and/or poor reviews.
Supply chain limitations	Supply chain limitations cause failure to meet production and/or quality requirements.
Market timing mismatch	The right product, process, or service is introduced too early or too late to realize the market opportunity.

These statistics underscore the practical reality that it can take a lot of time, effort, money, and risk to get from the new idea stage all the way to a commercially successful product, process, or service. As a result, any steps that could shorten the cycle time, reduce the failure rate, and/or reduce the development costs could pay back substantially. Amazingly, despite over 50 years of industrial evolution since the introduction of staged-gate processes (see Section 4.8), these success rates remain substantially unchanged. Part of the reason for this is that, despite decades of empirical research "*there still exists no precise prescription or recipe for successful [technological] innovation*" [98].

Speaking of time, even when successful, technological innovations can take a long time to realize. Éric Favre invented what became the Nespresso machine in the 1970s (the first patent on the special capsules was in 1976; see Figure 2.9). It took the development team until 1986 to develop and launch the first marketable machine with the special coffee capsules. Even then, it took time to discover a successful marketing approach (their first attempts were to sell the units to the hospitality industry, particularly hotels, but this did not work out for them). When they changed their business model to what has been termed the *"razor-and-blade"* approach (in which the durable base unit is sold at a low-margin price, and then the consumable components at a high-margin price)[26] they found a combination that worked with individual consumers – that was in 1989. In subsequent years, the company developed improved machines, improved capsules, and a barcode system for the capsules that enables the machine to automatically select the correct brewing parameters. By 1991, there were 126 Nespresso system patents worldwide, expanding to 366 by 2000, 513 by 2003, 948 by 2007, and 2,125 by 2011. The Nespresso story is also one of market-pull innovation, as the inventor's work was in direct response to his perception of unmet consumer needs [115–117].

Figure 2.9: The special capsules at the core of the Nespresso machine. (Courtesy pxhere, image #1188203, CC0 public domain.).

Technological innovations do not arise out of pure luck, but neither do they come from a formula or template, and they are almost never developed in isolation. There has to be a process, and that process requires interaction with the external environment: other people, other knowledge, suppliers, customers, markets, and so on (see reference [24]). This is discussed further below, particularly in Section 3.1. Similarly, managing the hand-offs from R&D to organizational units that will carry a new development into commercialization and launch needs to be more than just another routine process. Some of the approaches that have been used will be described in later

26 This approach has been used, for example, by Kodak and others when cameras with roll-film were popular, and by Hewlett-Packard and others for laser- and ink-jet printers.

sections of this book, but there are more things involved than just tools and processes.

Researchers pondering the success rate question have turned from looking for process factors to human factors. In most cases, many different people and/or teams of people are involved in the process of successful technological innovation, regardless of the specific pathway(s) followed. Not surprisingly, it turns out that, to paraphrase an old saying, "*It's all about the people . . .* " As the notions of push and pull pathways were being developed and explored in the 1960s, so too did "*coupling,*" as in "*coupling research and production,*" an expression of the idea that different kinds of people would have to be able to effectively interact in order for knowledge to be transferred along some or all of the stages along a developmental pathway [83]. The "*coupling model of innovation,*" described in the next section, is sequential but with feedback loops and technology-push and market-pull combinations. Over time, the term coupling was gradually superseded by terms such as communication and interaction.

The research shows that engaging people with highly developed creative abilities (sometimes called "*rainmakers*"[27] or "*wizards*") throughout the new product, process, and service development process (not just at the new idea and invention stages) can increase the success rates by as much as a factor of nine [118, 119]. Such people tend to be good at problem identification and problem solving and frequently serve as catalysts within teams. Of course, highly creative spirits, by their very nature, can also be somewhat disruptive in team environments and aren't necessarily best-suited to other important tasks, like management and programme execution, but by adding some people with highly developed creative abilities to a team, one can maintain disciplined staged-gate processes while achieving a large positive impact when it comes to developing technological innovations.

Some other examples of stereotypes and their roles in the technological innovation process are as follows [98]:

- The "*technological gatekeeper,*" a person that, on the one hand, acts as a conduit to external knowledge by maintaining a watching brief on relevant parts of the science and technology (S&T) world, staying on top of the literature and technical conferences, and being knowledgeable of and/or actually linked to key external S&T individuals (experts). On the other hand, this person acts as the communicator of such information to others involved in the technological innovation process.
- The "*product champion,*" a person that, as the label suggests, champions the innovation process throughout its stages of development with a clear goal of produc-

27 The Rainmaker Index refers to a "creative potential" scale that is based on the Myers Briggs Type Indicator (MBTI) personality-type indicator system [119], and which highly ranks people who score highly in the personality preferences intuitive (N) and thinking (T) of the MBTI system [118].

ing a successfully launched new product, process, or service in the marketplace. Regardless of the kind of organization, this person has to be able to guide the project through the "*Valley of Death*" (*see* Section 4.4) and all other obstacles in its path.

- The "*leading-edge customers,*" engaging prospective customers throughout the innovation process, is often identified as a means of improving the success rates of innovation processes or projects [98]. The idea is that continuous or semi-continuous feedback from prospective customers will help an innovation project integrate necessary technological advances while maintaining a connection to market-pull. In this context, Rothwell [98] describes the benefits of interacting with what he terms "*leading-edge customers,*" who are
  - Early adopters of technology (see Section 4.7)
  - Experienced users of previous innovative products
  - Able to provide forward-looking advice on desired new products or product features or benefits
  - Potential testers of prototypes
  - A source of potential post-launch improvement ideas

3 Some non-linear models of innovation

Concept-push innovation, mentioned earlier, is interesting in another sense. As a reminder, concept-push innovation relates to an entirely new product or service concept that was not previously imagined and which prospective customers didn't previously know that they might want and/or need. Naturally, this kind of approach can have an extremely high risk of failure akin to that of the *technology-push* approach, but Roberto Verganti has shown that successful concept-push organizations follow "*a strategy and a process that leverage the rich and multifaceted network of a firm's outsiders, looking beyond customers to those 'interpreters' – such as scientists, customers, suppliers, intermediaries, designers, artists – who deeply understand and shape the markets they work in*" [104]. This approach, of engaging customers and other stakeholders, substantially improves the probability of developing successful technological innovations.

The concept-push approach illustrates a pathway to innovation that relies heavily on interactions with a number of kinds of external stakeholders, including customers, suppliers, intermediaries, domain experts, and others, interactions that may occur discretely, semi-continuously, or even continuously along the pathway to innovation. In the next section, a few other models for the interactive development of technological innovation are introduced.

3.1 The coupling, integrated, and systems integration and networking models

It was noted in Chapter 2 that the simple one-way, push or pull, linear models of innovation have some value but are unrealistic. Christopher Freeman and others have pioneered a view that technological innovation should be thought of as an interactive process, rather than a linear process that always begins with R&D [120, 121]. A very simple model for this is "*agile development*," a product development approach that involves collaboration among self-organizing, cross-functional teams (see, e.g. Kerzner [122]). This approach facilitates adaptive planning, evolutionary development, responsiveness, and continuous improvement. The agile development concept seems to have originated in the computer software industry.

There are many reasons to take a much broader view of interactive product development, however, including the following:

- Firstly, technological maturity does not usually advance in a linear, or even a sequential fashion with respect to development activities, time, and/or effort.
- Secondly, technology-push and market-pull factors and a range of other parties, from suppliers, to customers, to partners of various kinds, can all be involved (and sometimes must be involved) throughout the process.

https://doi.org/10.1515/9783112206720-003

- Thirdly, the flow is not always one-way, and various kinds of reverses, feedback loops, and waves of development may need to be engaged in the process.
- Fourthly, innovation processes are complex, unpredictable, and fraught with failures, re-calibrations, and re-starts along the journey. It has been said that: "*Even the best innovation concept is partly wrong*" [81].
- Finally, the ultimate applications of the innovation may have been impossible to foresee from the beginning, or even throughout most of the process: technological innovations frequently create benefits in industries far removed from the one for which they were first developed and/or commercialized.

The non-linear innovation models attempt to deal with the fact that technological innovation processes can be highly irregular, involving jagged lines of activity, and highly uncertain, and they are sometimes referred to using metaphors such as "*fireworks*" or "*messy fireworks,*" hence the term "*Messy Fireworks Innovation.*"

Evolution to the awareness and use of better models (than the linear models) in the innovation world is not helped by the low evolution of the language of innovation. The ubiquitous terms "*research and development*" and "*R&D*" imply a sequence of steps: research and then development. Nevertheless, there are better models for the innovation process. The primary non-linear models are the "*coupling model,*" "*integrated model,*" and "*systems integration and networking model.*"

The "*coupling model*" is sequential, drawing on technology-push and/or market-pull at each stage, but it is non-linear in that the various kinds of communication and interaction among contributors create feedback loops. Over time, the term coupling was gradually superseded by terms such as "*communication,*" "*interaction*" [123], and "*chain-link.*" The term "*coupling*" has also been used in innovation with reference to the coupling of ideas in the minds of, for example, discoverers and inventors.

An example of a nonlinear coupling model is the "*chain-linked model*" proposed by Stephen Kline in 1985 [63, 67]. This begins with the identification of a potential market and then proceeds through the conception of a design (possibly, but not necessarily, involving research), development of the design, prototyping and testing, then redesigning and retesting, then production, marketing, and distribution. This principal pathway, illustrated in Figure 3.1, forms a central "*chain-of-innovation.*"

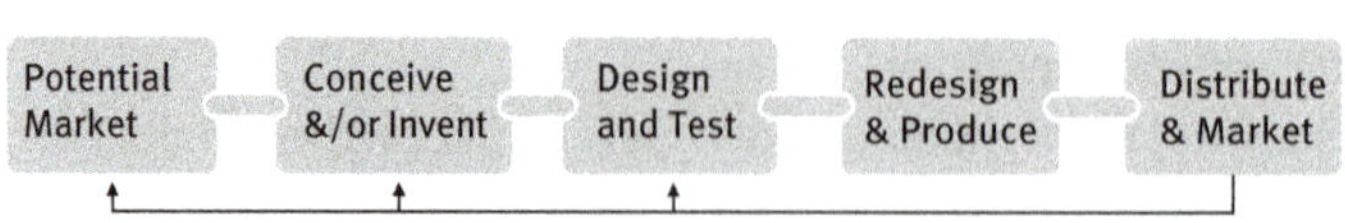

Figure 3.1: Illustration of the central chain-of-innovation underlying the "*chain-linked model*" of innovation.

Supplementing and interconnecting with this principal pathway are a number of feedback loops, including loops that feed-in knowledge and research results of various kinds, such that there is not a single pathway but at least five major possible pathways from beginning to end (illustrated in Figure 3.2). The importance of iterations in the development of technological innovations has been emphasized by Fitzgerald et al. [121].

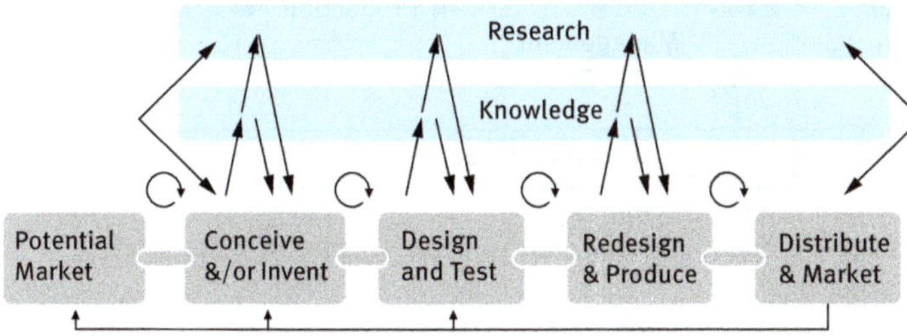

Figure 3.2: Illustration of the "*chain-linked model*" of innovation proposed by Kline, in which feedback loops and alternative developmental pathways are superimposed on the central chain-of-innovation.

In the chain-linked model, research processes, for example, can feed most of the other elements in the process. It has been found for many industries [63] that initial inventions are frequently rather crude, and subsequent improvements are often needed in order to achieve substantial technological progress, so performance (user) feedback loops can be critical to achieving significant innovation. Discovering limitations and failures can be important parts of the innovation process, so feedback on these learnings is important. Sometimes, the key to a successful technological innovation is the ability to obtain and quickly respond to customer feedback by feeding this information back into the innovation process and using it to develop and launch improvements to the product, process, or service: this is one of several kinds of feedback loops in the chain-linked process. Finally, with so many pathways and feedback loops, this model includes and transcends the technology-push and market-pull models.

Some examples of non-linear innovation processes can be found in case studies of the locomotive engine, *ca.* 1804–*ca.* 1829, and of Bessemer's iron to steel conversion process, *ca.* 1854–*ca.* 1878, conducted by Usher in the early 1920s and 1950s [96, 124]. These examples show that, although the original innovations were important, subsequent improvements (including additional inventions) had to be developed and incorporated before the overall technological innovations were fully successful in industry. As such, they also provide examples of the evolution of technology during its diffusion.

The "*integrated innovation model*" involves parallel, rather than sequential, development, with simultaneous R&D, prototype development, and manufacturing activities, for example. Integrated innovation processes encompass integrated development teams, engagement with suppliers and customers, and possibly engagement with partner organizations (horizontal collaboration). An example of a model constructed along these lines is shown in Figure 3.3.

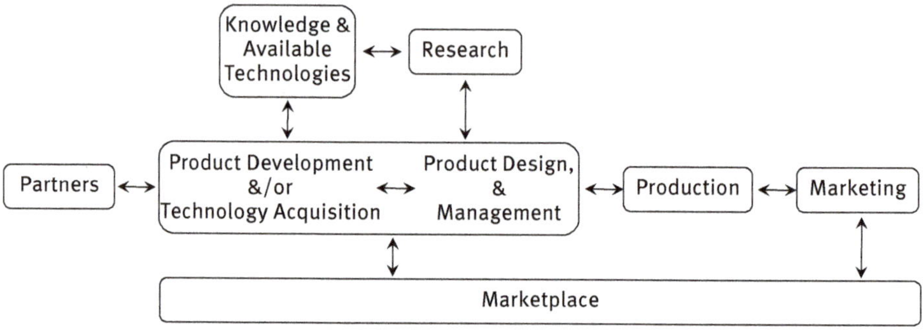

Figure 3.3: Illustration of an integrated, open innovation process model. Adapted from Conseil de la science et de la technologie, Québec [125].

The *"system integration and networking model"* (SIN innovation model) involves parallel development integrating horizontal linkages with customers, suppliers, and other partners (e.g. collaborators, marketers, and others) throughout the process. An interactive helix-like illustration of the SIN Innovation Model is shown in Figure 3.4, in which multiple key organizations are continuously or semi-continuously working together on development aspects along the entire innovation process, from beginning to end. The vertical bars in this figure are intended to illustrate specific interaction "linkages" in the case of semi-continuous collaboration. Some other helix-like illustrations are used in describing innovation ecosystem models in Section 5.2.

Figure 3.4: An interactive helix-like illustration of the SIN innovation model.

An increasingly common feature of such helix-like ecosystems is the ever-increasing number of people that are almost continuously connected to other people and the world of open data, making it easier than ever before to exchange and build-on ideas [112].

A variation on this theme has been called *"Innovation 2.0."* This is a SIN innovation model that involves parallel development, integrating horizontal linkages among communities, governments, industry, research organizations, and universities, with the aim of creating and deploying disruptive innovations that drive economic growth, improved quality of life, and reduced resource use and environmental impacts [112].

An example would be a community's adoption of a smart traffic flow-adjusting system that responds to real-time traffic and air quality sensor readings [112].

Rothwell [98, 126] identified five generations of models of technological innovation, also termed the *"Rothwell models of innovation."* The first two are the two broad categories of linear models of innovation discussed in Chapter 2, while the other three are the non-linear models just described just above:

- 1st Generation: technology-push model
- 2nd Generation: market-pull model
- 3rd Generation: coupling model
- 4th Generation: integrated model
- 5th Generation: systems integration and networking model

Of these, the 5th-generation model was presented (at the time) as a model for the future, with the incorporation of expert systems, simulation modelling, and even artificial intelligence (AI).[28] A 6th-generation model is now frequently discussed, which adds "open innovation" (see Section 4.2 and Table 4.2) and/or the influence of interactions in a broader innovation ecosystem [127] (see Chapter 5). For a somewhat different view and categorization of the generations of innovation models, see Marinova and Phillimore [128] and the case study by Barbieri and Álvares [127].

3.2 Technology "S-curves"

Another category of non-linear technological innovation models is that of technology *"S-curves."* In mathematics, the simplest function producing an S-shaped curve is the logistic function, with the resulting curves being termed *"Logistic curves."* The first technological application of the S-curve concept appears to have been in 1845 by Pierre François Verhulst, who used it to illustrate stages in the development of yeast fungi colonies: a phase of initial growth and achievement of critical mass, a phase of rapid (approximately exponential) growth, a plateau-like phase of maturity, and a phase of decline due to depletion of the system's resources. Kucharavy and De Guio describe the most general application of S-curves to physical phenomena as follows: *"The essential meaning of this function is 'the rate of growth is proportional to both the amount of growth already accomplished and the amount of growth remaining to be accomplished"* [129].

Since Verhulst's time, S-curves have been used to describe many other kinds of phenomena, sometimes involving four phases, as in the example just given, and other

28 This has been referred to as the electronification of innovation [98].

times with only the first three phases (i.e. without the decline phase).[29] In fact, S-curves have been used in so many fields that they are commonly referred to under a wide variety of names [2, 129, 130] (see also S-curves in Section 11). An example is Lehfeldt's 1916 "Normal Law of Progress" [130].

With a slight re-statement of the phases, S-curves can be used to describe the development of new technologies and/or the lifecycles of products, processes, or services.[30] Specifically, the "*Technology S-Curve*" concept was introduced by Richard Foster in 1986 as a visualization and planning tool to help R&D managers [131, 132]. The typical stages of a basic technology S-curve are the following:

1. A phase of hypotheses, experimentation, addressing fundamental issues, and trial-and-error.
2. A phase of rapid maturing of the new technology as the fundamental issues become resolved, new approach(es) begin to take shape, individual advances begin to cluster, and/or a breakthrough occurs.
3. A plateau as the technology matures and physical limitations are reached. According to Foster, the evolutionary approach ultimately "*is doomed to fail*" in the face of discontinuities brought about by competing technologies or products/services [131, 132].

As each phase leads to the next in sequence, the technology S-curve has also been termed the "*innovation continuum*" or the "*technology growth curve.*" The vertical axis is usually technology maturity and/or performance, while the horizontal axis is usually some kind of measure of cumulative R&D effort over time, summarized as R&D effort, investment, and/or time (see Figure 3.5) [131–133]. The only difference between Figures 2.1 and 3.5 is in the non-linearity of the rate of progress from each stage to the next.

A number of mathematical models have been developed for, or applied to, technology S-curves. One of the simplest S-curve models is the Pearl logistic curve model, which relates the technology maturity (y) to time (t) as

$$y = y_\infty / \left(1 + \alpha \cdot \exp^{\beta t}\right)$$

where the maximum value for y is given by y_∞ and the shape of the S-curve is determined by the adjustable parameters α and β.

Another, slightly more complex, S-curve model is the Gompertz logistic curve model, which relates the technology maturity (y) to time (t) as

29 It is the three-phase kind, having an approximately sigmoid, or "S" shape, that gave rise to the name "S-curve."

30 Another kind of technology S-curve can be used to represent technology diffusion. Technology diffusion S-curves are discussed further in Section 4.7.

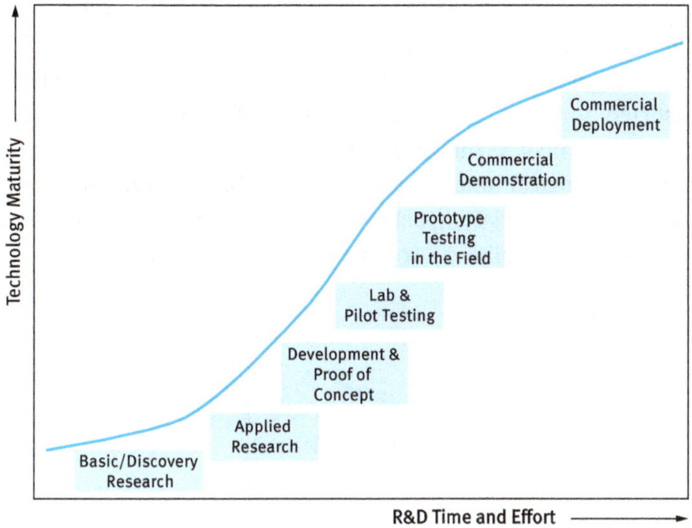

Figure 3.5: Illustration of a technology S-curve.

$$y = y_\infty \cdot e^{\alpha \cdot \exp(-\beta t)}$$

where, again, the maximum value for y is given by y_∞ and the shape of the S-curve is determined by the adjustable parameters α and β. In this model, the growth curve is not symmetrical about the inflection point. Instead, it sharply increases up to the inflection point, after which the rate of growth slows.

Some other mathematical models for S-curves are discussed by Ryu and Byeon [134] and by Phillips [135].

S-curve models can be used to make predictions about the amount of time needed to advance a technology, or even the technology maturity level of a region or country, from one level of maturity to another, although such predictions involve making huge assumptions about the nature of the S-curve and the values of the adjustable parameters in whichever model is chosen. Some examples are provided by Modis [136] and Ryu and Byeon [134].

A key concept is that all technologies eventually reach some kind of physical limitation. One theory for the plateauing phenomenon holds that a new technology in any particular field is perfectible, that continuing developments cause it to approach perfection relatively quickly, leaving diminishing returns for future potential advances, and that, as a result, the technological maturity plateaus (i.e. the inventive potential becomes exhausted). This has been termed *"the perfectibility hypothesis"* and has been used to attempt to explain why it is so common for empirical technology S-curves to plateau. A counter-theory is provided by Schmookler [18], who concludes that *"the s-shaped long-run growth curve for individual industries, in which output tends to grow at a declining rate, usually reflects demand, not supply conditions . . . a given percentage cut in costs probably does not become progressively more difficult or costly to achieve over time. Rather, the return from achieving it declines.*"

The process depicted in Figure 3.5 can be extended to illustrate the diffusion (see Section 4.7) throughout society that can follow a highly successful commercialization process, as shown in Figures 3.6 and 4.8. The greater the diffusion, the greater the social impact, showing that technological innovation success can be measured in terms of social impact in addition to commercial impact [81]. This, of course, is not new: for examples of technology diffusion in Neolithic Europe, see reference [137].

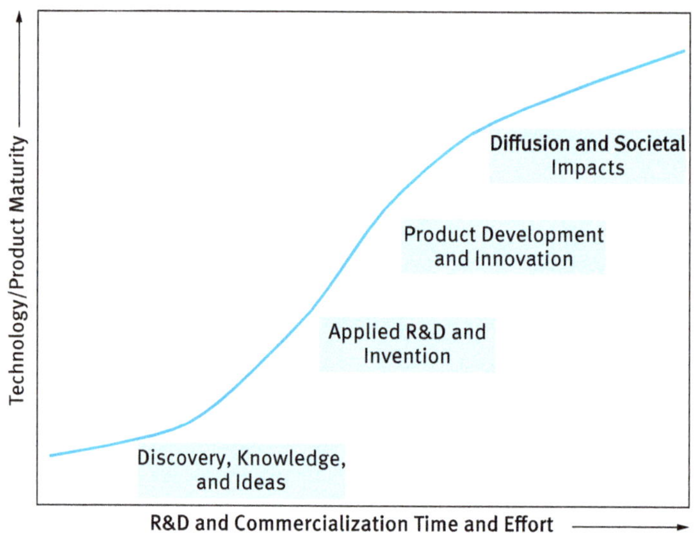

Figure 3.6: Illustration of an extended technology S-curve.

A key concept in technology S-curves is that all technologies eventually reach some kind of physical limitation to their advancement.

Some other kinds of S-curves are introduced in the next three sections, and yet another kind will be discussed later in Section 4.7.

3.3 Degree of innovation S-curves

With another re-statement of the phases, S-curves can be used to describe degrees of innovation and their impact on an economy. Figure 3.7 provides an illustration comparing incremental, evolutionary, and disruptive stages of innovation.

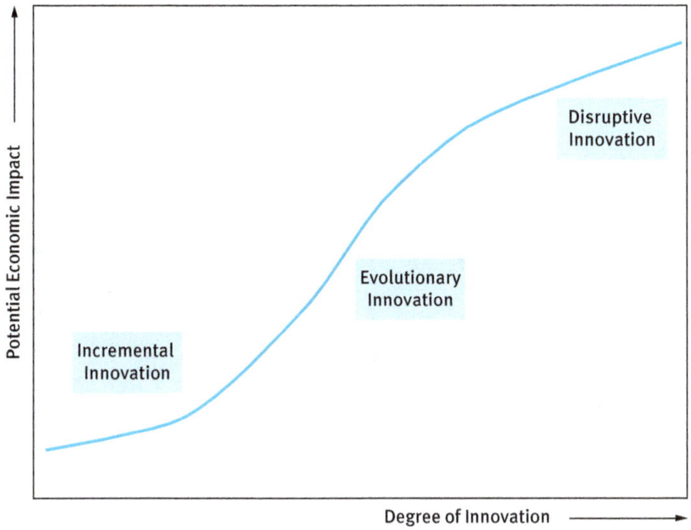

Figure 3.7: Illustration of an innovation S-curve.

3.4 Product lifecycle S-curves

Although the S-curve plateau can be caused by technology maturation and diminishing technical opportunities for further improvement, Schmookler [18] found many examples for which the plateau and/or subsequent decline was instead due to a shift in the trade-off between rising value and rising cost of technological advances, with the latter ultimately dominating. The metal horseshoe, one of the case studies described by Schmookler, provides a vivid example:

> *The metal horseshoe* was developed in the second century BCE (Before the Common Era), that is, some 2,200 years ago. One might think that the opportunities for technological advances in the horseshoe would have substantially diminished long ago and at least by around 1800, by which time two millennia had elapsed. However, the patenting activity related to horseshoe improvement inventions rose strongly throughout the 1800s and early 1900s, as illustrated by the U.S. patent statistics shown in Figure 3.8. It seems that changing agricultural conditions and practices led to new technological challenges and opportunities during the 1800s, such that there continued to be opportunities for new commercially applied inventions until the steam-tractor engine began to displace the horse as a means of motive power. Schmookler concludes that following about 1900, the number of horseshoe patents declined, not because the horseshoe had finally been perfected, but because the cost of developing improvements began to exceed the commercial value of such improvements.

For a new commercial product, process, or service, the typical market stages are (see Figure 3.8)

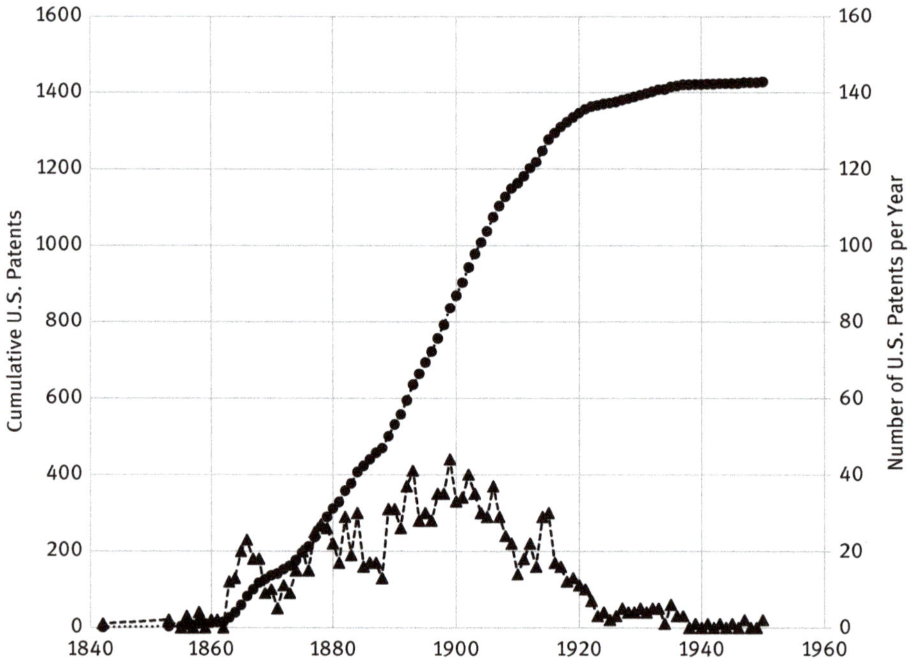

Figure 3.8: A technology S-curve from the US horseshoe industry. Drawn from data provided by Schmookler [18].

1. The introduction of a new product to the marketplace, with associated business development and early sales;
2. A phase of rapid sales growth as the product finds a successful market niche;
3. A sales plateau as the market niche becomes saturated; and
4. A phase of declining sales as the market-pull declines and/or competing products displace it in the marketplace.

Each of the above stages can be considered in terms of factors supporting the product's development and factors acting against the product's development [131, 133]. Some authors use different terminology to describe the stages. For example, the first two phases described above are sometimes referred to as the "*Era of Ferment,*" with the latter two phases being termed the "*Era of Incremental Change*" [138]. Here, again, a key concept is that all products and services eventually reach some kind of natural limitation and that they are eventually either completely or mostly[31] supplanted by new (disruptive) products and services.

31 There are often limited, niche markets for the older products/services, but not at their previous scale (e.g. sailing ships, and vacuum tubes).

Here's another way of looking at product lifecycles. A new product lifecycle concept, attributed to Windermere Associates (San Francisco, USA), is called the "*buying hierarchy*." In this model, a new product, process, or service goes through an evolutionary sequence comprising four phases, in which only the last phase involves competition based mostly on price (see also Figure 3.9):

1. *Functionality*: at this stage, the new product/process/service satisfies a customer need and competes based on a functionality not satisfied by anyone else (there are no competitors).
2. *Reliability*: at this stage, there are some competitors, but the product/process/service is able to stay ahead of those competitors based on being the most reliable.
3. *Convenience*: at this stage, the competitors are offering similar functionality and reliability, but the product/process/service is still able to compete by offering the most convenience (at this stage, however, competitors have a significant share of the market).
4. *Price*: at this stage, the competitors are offering similar functionality, reliability, and convenience, and the product/process/service has become a commodity; competition is now based on price (at this stage, the end of the product lifecycle is approaching; see Figures 3.9–3.11).

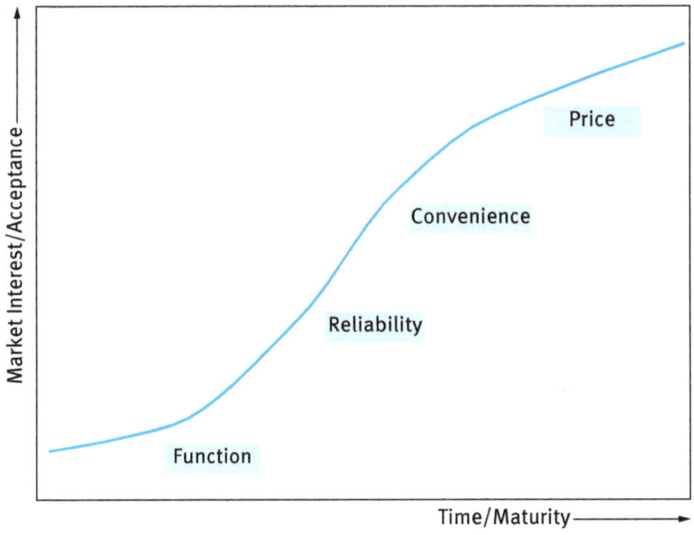

Figure 3.9: Illustration of a buying hierarchy lifecycle curve.

Some examples of technology lifecycles for solar PV and wind power technologies are given by Huenteler et al. [138].

Foster has coined the term "*caveat innovator*" to refer to the perils of ignoring the implications of the plateauing nature of S-curves, concluding that the evolutionary ap-

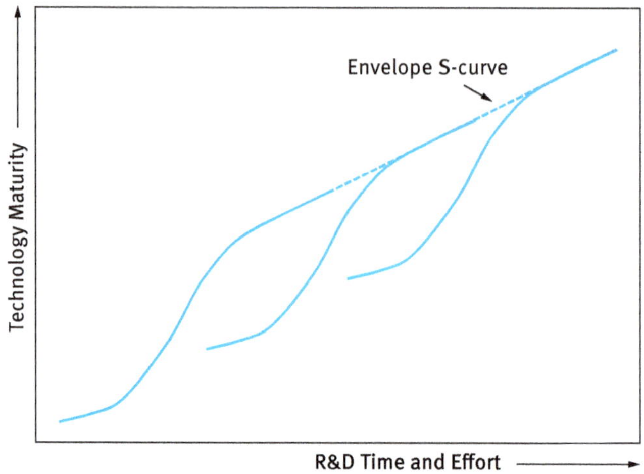

Figure 3.10: Illustration of a family of technology S-curves.

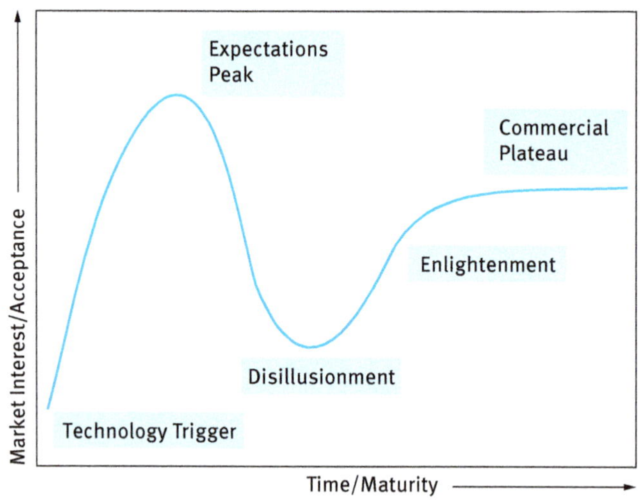

Figure 3.11: Illustration of a hype-cycle curve.

proach "is doomed to fail" in the face of discontinuities brought about by competing technologies or products/services [131].

A key concept in product life cycle S-curves is that all products, processes, and services eventually reach some kind of sales limitation and/or decline as they are eventually either completely or mostly supplanted by new products, processes, and services.

3.5 Families of S-curves and non-S-curves

In both the technology and product lifecycle cases, the displacement of older technologies by newer ones, or of older products by newer ones, can be represented by families of S-curves (see Figure 3.10). There is generally a gap, or transition phase, called a discontinuity (or technological discontinuity) between the plateauing of one S-curve and the rapid rise of a successive S-curve. Following this period of discontinuity, the newer technology or product has overtaken the earlier one. Within a given type of product, process, or service, a family of two or more S-curves, taken together, is sometimes referred to as an "*envelope S-curve*" (see Figure 3.10). Some examples of families of S-curves are given in references [4, 129, 131, 139]. According to Foster [131], it usually takes 5–15 years for a new technology to overtake an old one.

As illustrated in Figure 3.10, the ideal time to deploy or launch a new technology product is near the inflection point on the S-curve of the previous one. This is somewhat counterintuitive for many managers because this is exactly the point at which the previous technology or product is really "hitting its stride" [131, 139]. Two of the reasons companies do not purposefully "jump" from one product S-curve to another are:

- Unwillingness to intentionally displace a currently successful product.
- Concerns about customer acceptance (see the "technology readiness" and "technology acceptance" models in Section 4.7).

Foster has observed that:

> "*. . . CEOs will be roundly criticized by outsiders for venturing into new areas where they lack skills and for forsaking the tried and true. But in order to manage a technical discontinuity, that is exactly what they must do – forsake the past by abandoning a technology that, more often than not, has just entered the most productive phase of its S-curve*" [131].

Some examples of companies that intentionally strive to continuously introduce new products before their older ones have reached their full market potential, thus rendering their own products obsolete (such as IBM, Intel, and 3M), are described in references [131, 139] and in Section 7.7.

In considering families of overlapping S-curves, multiple processes and/or steps become parallel, and the two-dimensional, non-linear models become multi-dimensional. Some examples of multi-dimensional models of innovation are discussed below, particularly in the context of ecosystem models (see Section 5.1).

Before leaving this section, it may be worthwhile to explicitly mention that, although the S-curve model is extremely powerful, not all new products, processes, or services need follow these vaguely S-shaped patterns. Obviously, a new technological innovation that is initially promising in the market may not ultimately be very successful, or may later fail outright, for any of a number of reasons (see Section 2.5). The

Gartner Hype Cycle[32] provides an illustration of non-S-curve behaviour (see Figure 3.11) [140]. The essential stages of this cycle are as follows:
- *There is a technology trigger*. A new offering is announced to the market. Some venture capital may already be invested, but at this stage, the actual product, process, or service may not yet be available. Advertising and media interest, however, create interest in the market.
- *Expectations peak*. The product, process, or service launches in the market; the innovators and early adopters have made purchases, early reports appear promising, expectations are high, and sales surge upward.
- *Disillusionment sets in*. Some level of market disillusionment sets in as user reports are less positive, some expectations are not met, and sales fall. A second round of venture capital may have been invested by this stage, but some manufacturers may drop out or fail. Other manufacturers respond by improving the product/process/service in response.
- *Enlightenment arises*. A more realistic understanding of the product/process/service settles into the market, and the improvements made in the previous phase start to rebuild interest in the market. Sales begin to rise again. New-generation versions may be launched to further fuel sales.
- *A commercial plateau is reached*. Sales grow slowly, at best, as the product/process/service nears the end of its lifecycle.

Finally, the S-curve model implies a time frame for the evolution of a technology into a successful technological innovation. The amount of time required to develop an idea, concept, theory, or discovery into a commercialized innovation is termed the "*Knowledge Turn*" [8, 134, 141, 142].

The Knowledge Turn, or time for a new product/process/service to advance to the beginning of the commercial plateau stage described above, can be as short as a few years but is more commonly in the range from 20 to 50 years.

32 Gartner Inc. is an American information technology company.

4 Inside the innovation models

"If you can dream it, you can do it."

Attributed to Tom Fitzgerald, The Walt Disney Company
Written during development of the "Horizons" attraction at *EPCOT Center*.

4.1 Who actually does innovation?

The foregoing introduction of some linear and non-linear models of technological innovation provides a conceptual sense of the process steps, but does not completely describe how most technological innovation is actually accomplished. A common supplementary question to *"How is innovation accomplished?"* is *"Who actually does innovation?"* It is rarely a single person and seldom a single organization, although it can be. Some organizations have adopted the concept of *"whole product R&D"* by which the organization not only does its own R&D, but the R&D team is kept involved through all stages (Figure 4.1) of a product's launch, evolution, marketing, and life cycle in order to ensure that marketing and customer feedback is used to continuously improve, if possible, the product, process, or service – in order, in turn, to extend its life cycle. The value in keeping the R&D function involved is that it can be difficult for others to discern the difference between a technologically simple change and a substantial technological boundary. Moore provides a more in-depth discussion of this practice [143].

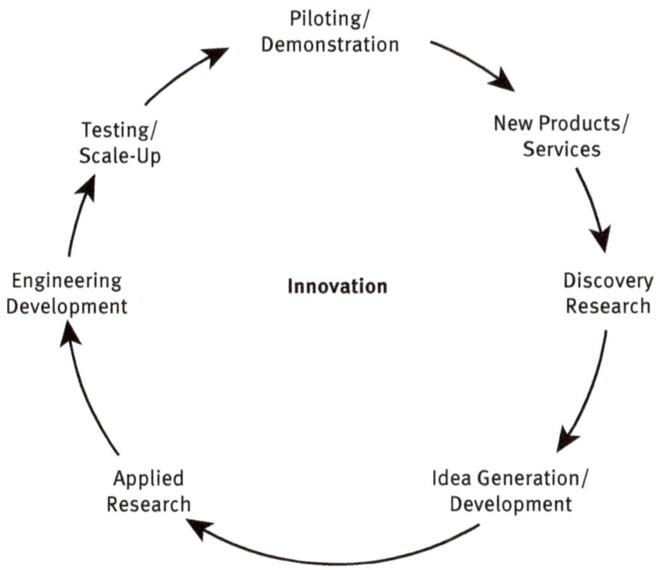

Figure 4.1: Elements of the process of technological innovation.

https://doi.org/10.1515/9783112206720-004

Universities, colleges, RTOs, and private-sector research and engineering companies seldom literally "do" technological innovation in a complete sense. Since technological innovation is the conversion of ideas and knowledge into new and commercially successful products, processes, and services, it hasn't fully occurred until the commercial success part has been accomplished, and that last part is almost always done by industries or other businesses.

In principle, industries can conduct virtually all of the steps needed to accomplish innovation on their own. Between the early 1900s and the 1980s, this was the practice of many large companies, including GE, Bell, Ford, General Motors, and Exxon, to name just a few. In the 1980s and 1990s, however, many companies changed their R&D strategies:

> Bell for example, had from the 1920s through the 1980s, operated large in-house R&D laboratories that made new discoveries, created new technologies, and developed and deployed into the marketplace some of the world's most advanced telecommunications networks. By the 1990s, however, Bell Labs had been restructured to focus on incremental technology improvements with shorter-term payoffs, relying on open innovation (*see* below) for everything else [144, 145].

In modern times, most companies can't afford the costs in time, money, patience, and risk that are required to do discovery research, idea generation and development, applied research, engineering development, entrepreneurship, and so on, all the way down the line to products in the marketplace. Instead, most of even the most innovative companies will rely on others to do components of the process for them.

Many organizations, however, are available to help with selected aspects:
- Universities, some colleges, and some large government labs specialize in discovery science and in elucidating engineering principles and processes.
- RTOs specialize in applied science, development engineering, testing and analysis, and sometimes pilot testing, scale-up engineering and even full-scale plant or field testing and demonstration [146].
- Service companies generally specialize in technical services, plant or field support, and sometimes process troubleshooting.
- Many manufacturing companies now specialize in just the manufacturing component.

Most companies will want to do their own marketing, distribution, and sales, but any or all of the components of the innovation process can be outsourced, even the idea generation component.

4.2 Closed and open innovation

The terms *"open innovation"* and *"closed innovation"* refer to two extremes in approaches that an organization can make to developing and commercializing technologies into innovations.

In closed innovation, a company conducts essentially all of its innovation activities internally, from idea generation through R&D and the subsequent activities such as experimental development, prototyping, pilot testing, demonstration, and preproduction. This doesn't mean that all the work is necessarily done by one organizational unit, however. Closed innovation can still be participative in the sense that others within an organization (beyond those officially in R&D) are engaged in the innovation process(es). Such organizations may appoint a chief innovation officer, or the like, to lead and coordinate such processes on behalf of the organization. The main advantages of closed innovation are the ability to control the process and to potentially achieve dominating competitive advantage positions in the marketplace. The main disadvantages are cost and the fact that the company is limited by its internal idea generation and innovation capacity.

In open innovation, an organization uses external resources to conduct some or even most of its innovation activities. This doesn't mean that the organization ignores its internal capabilities, but rather that it leverages its internal capabilities with those of the external world. For example, in open innovation, the organization can draw on both internal and external sources of knowledge and new idea generation. The potential external sources for relevant knowledge are both numerous and diverse ranging, for example, from universities and RTOs, to technology suppliers, technical specialists, inventors, entrepreneurs, and start-ups, among others [147]. Open innovation also enables the in-sourcing of proprietary technologies from others (licence-in and spin-in) and also the transfer out of technologies to others (licence-out and spin-off) as the R&D process is pursued. For more on definitions and examples of open innovation, see Kozioł-Nadolna and Świadek [148] and Gabison and Pesole [149]. Some case studies are discussed by Håkansson and Waluszewski [150], Jaekel et al. [151], and Mäkimattila et al. [152].

A key opportunity brought about by open innovation is *"adopt and adapt"* technology development, which refers to the practice of adopting an existing technology from another business, industry, or market and then adapting it to the innovator's needs.

> One example of this is called *"Reverse Innovation,"* which is the process by which innovations that have been developed to meet the needs of customers in developing nations are adapted (sometimes mostly by repackaging) for marketing and sales as low-cost innovations in developed nations. These are usually products rather than processes or services. The aspects developed in the developing nation(s) could be any or all of the product idea conception, the product R&D, the product design, or the initial market targeting. An example is GE's portable electronic medical instruments. [153–155]

In a study of over 500 product and process innovations, spanning five industrial sectors, between 1959 and about 1965, Marquis found that about 25% came from the adopt and adapt approach [102].

In ***closed innovation***, an organization conducts essentially all of its innovation activities internally.

In ***open innovation***, an organization uses external resources to conduct some or even most of its innovation activities.

The main advantages of open innovation are that a company is not limited by its internal idea generation and innovation capacity. Another option is to purchase, or barter for, externally derived ideas and knowledge. In addition to casting a broader net for ideas and knowledge, it could also be less expensive to adopt and adapt ideas and technologies from others than to develop them internally. In principle, open innovation could be as simple as making use of openly, readily available information such as published papers and patents, without the need for significant interaction with the sources of the information, but the real power of open innovation is that it lends itself to partnering with other organizations in the innovation ecosystem. Thus, open innovation during the new product development (NPD) process can involve companies in not just external technology in-sourcing, but a variety of practises along the way that are often more conventionally thought of as isolated activities, such as licence-ins and licence-outs, spin-ins and spin-outs, acquisitions, and divestitures [156].

In a 2006 presentation, Chesbrough describes the example of IBM's "open business model" in which applications, hardware, and even integrations are developed both internally (at a cost to IBM of about $100 million per year) and externally (at a cost to IBM's prospective suppliers of about $800 million per year) [156]. This example illustrates the value, or perhaps the need, to consider technological progress and business model progress together, as an organization pursuing technological innovation will ideally have the ability to acquire technology, profit from the technology, scale the technology, and build on or evolve the technology further [156]. A more recent example is the case study by Wolcott and Lippitz [157], which describes BP's open innovation evolution in the 2000s.

In "*collaborative innovation*," the open innovation concept is taken one step further in that two or more organizations work together on the development of innovation(s). Such organizations can include multiple customers, suppliers, and technology developer/providers. Innovation 2.0, introduced in Section 3.1, is an example. The advantages can include efficiency, and cost and risk sharing, plus the ability to leverage each other's technological and inventive capacities. However, the disadvantages can include difficulty maintaining cohesion and focus, and/or IP issues. Organizations that practise open innovation sometimes appoint a chief innovation officer, or the like, to lead and coordinate the collaborating and/or partnering processes and to manage the IP and other issues that arise as a result of the external interactions.

Finally, yet another form of open innovation is "permissionless innovation," in which an organization can invite the public to conceive and submit new ideas, solutions to problems, analyses of data, and/or solutions to opportunities in return for some kind of incentive. Chesbrough and Van Alstyne refer to permissionless innovation "as a complement to traditional research and development" [158]. One of the examples related by Chesbrough and Van Alstyne concerns Canadian company Goldcorp, which successfully offered the public access to decades-worth of proprietary geological data along with a challenge to find new ore-grade gold on their Red Lake property in return for financial incentives.

There are now many initiatives around the world that seek to incent permissionless innovations aimed at specific challenges or opportunities. Often bearing the label "*grand challenge*," such initiatives may be funded by individual organizations, industry groups, not-for-profit and/or charitable groups, or governments. In the case of industry, the challenges usually relate to business challenges, although sometimes with a socio-environmental aspect. In the other cases, the challenges usually relate to social and/or environmental challenges, frequently on a national or international scale. An example is the "Global Grand Challenges," which are aimed at incenting people to develop solutions to "global health and development problems for those most in need" [159]. A common theme to such grand challenge innovation initiatives is the search for disruptive, rather than incremental or evolutionary, innovations.

Although the activities associated with open innovation were being practiced since the 1980s, they became much better known with the publication of Henry Chesbrough's 2003 book titled *Open Innovation* [160, 161] and its successors, such as reference [162]. Chesbrough's work helped coalesce the previous open innovation activities into an approach that could be incorporated into an organization's strategy [161]. According to OECD, the open innovation approach is now more common than closed innovation [163]. The main disadvantages of open innovation are that competitor companies can usually access the same external capabilities and may be able to exploit then in ways that are better, faster, and or cheaper, making it difficult to achieve a competitive advantage in the marketplace. A case study involving open innovation at General Electric is given by Chesbrough [164].

An illustration of the evolution from closed innovation, to open innovation, to Innovation 2.0 is given in Table 4.1.

Table 4.1: Illustration of the evolution of closed and open innovation.

Closed innovation	Open innovation	Innovation 2.0
Organization goes solo	Bilateral	Part of an ecosystem
Self-dependency	Independency	Interdependency
External input by subcontract	Bilateral inputs-outputs	Quadruple helix
Planning	Validation, pilots	Experimentation

Table 4.1 (continued)

Closed innovation	Open innovation	Innovation 2.0
Organizational control	Active management	Orchestration
Win-lose relationships	Win-win relationships	Win more-win more relationships
In-the-box thinking	Out-of-the-box thinking	No boxes!
Single organization	Single discipline	Interdisciplinary
Value chain	Value network	Value constellation
Licensing by subcontract	Cross-licensing	Cross-fertilization

Adapted from reference [112].

The foregoing characterizations of open and closed innovation represent extreme positions. Many companies actually use a hybrid strategy and leverage internal innovation capabilities with external ideas, technologies, and other capabilities. For example, an organization might bring technology or other information in from the outside ("*inbound open innovation*") while at the same time sell or licence out technology or other information into the marketplace ("*outbound open innovation*"). The more advanced non-linear models of innovation assume that there is at least some degree of open innovation in the process, such as the "*integrated innovation model*" (Figure 3.3) and the "*system integration and networking model*" (SIN innovation model, Figure 3.4) discussed in Chapter 3, and this is also implicit in some of the innovation ecosystem models described in Section 5.2.

In previous chapters, when discussing innovation models, reference has been made to the relatively recent concept of generations of innovation models. Open innovation, and/or the functioning of regional clusters of innovation (Chapter 5) have been bundled into the category of 6th-generation models of technological innovation, which is illustrated along with the predecessor models in Table 4.2.

4.3 Creative thinking models

Another common innovation question is "*Where do the creative ideas come from?*" Henry Mintzberg has referred to traditional strategic planning in businesses as an oxymoron [168], meaning that what is really needed is creative, "*out-of-the-box*" thinking rather than numbers-oriented, linear extrapolation-based approaches. There is clearly a need for creative, out-of-the-box thinking and inventiveness in the world of technological innovation.

First, to get to technological innovation, there is the need for new products, processes, or services to introduce to the marketplace. Then there is the statistical reality that it takes, on average, about 3,000 new ideas to get to one commercial success. Finally, there is the question of how to filter through, and/or develop, those thousands

Table 4.2: Illustration of the generations of technological innovation [98, 126, 165–167].

	1920s–1960s	1960s–1970s	1970s–1980s	1980s–1990s	1990s–2000s	2000s–2010s
Generations	**1st-generation models**	**2nd-generation models**	**3rd-generation models**	**4th-generation models**	**5th-generation models**	**6th-generation models**
Features	Linear and sequential, "technology-push," "supply-side" approach	Linear and sequential, "market-pull," "demand-side" approach; may include Stage-Gate processes	Interactive: coupling market needs with technological opportunities, sequential with feedback loops	Parallel processes, vertically integrated across organization + customers + suppliers, with feedback loops, e.g. the chain-linked model	Vertical + parallel linkages + "lean innovation" time/cost-efficiency optimization, e.g. the SIN model	Adds "open innovation," flexibly leveraging internal with external capabilities, in-sourcing, "adopt and adapt" and/or regional innovation clusters
Risks	Risk(s): oversimplified; market information comes late in the process	Risk(s): oversimplified; market focus can lead to overweighting of short-term, incremental improvements	Risk(s): technological innovation focus can lead to underweighting organizational improvements	Risk(s): complexity and communication management	Risk(s): 4th-generation risks plus fast-tracking innovation increases development costs	Risk(s): 5th-generation risks plus IP + competitive advantage control; external relationship management
Example references	[57, 58, 60, 62, 95, 96, 104, 123]	[18, 22, 23, 98]	[100, 121]	[62]	[97, 125]	[154, 158, 160]

of ideas. Creative work is needed, and not just for the idea generation, but throughout the entire R&D process.

Richard Florida, in his book *The Rise of the Creative Class* [169], takes the importance of creativity and imagination to a new level, beyond that of corporate strategies to entire industries and our world economy. He writes:

> *Many say we now live in an "information" economy or a "knowledge" economy. But what's more fundamentally true is that we now have an economy powered by human creativity. Creativity . . . is now the decisive source of competitive advantage. In virtually every industry, from automobiles to fashion, food products and information technology itself, the winners in the long run are those who can create and keep creating* [169].

Ideation is the process of idea generation, whether through divergent thinking, brainstorming, Camelot Scenarios[33], or any other approach. Organized and/or facilitated ideation is sometimes used to originate ideas for new theories, applications, solutions, processes, or products. Stereotypically, one thinks of engaging highly creative people in such processes, and indeed some work has been done on assessing people's natural abilities in this area. For example, the "*rainmaker index*" refers to a creative-potential scale that is based on the Myers-Briggs type indicator (MBTI) personality-type indicator system [119] and that highly ranks people who score highly in the personality preferences intuitive (N) and thinking (T) in the MBTI system [118]. However, if you think that creative thinking is just for the few "naturally creative people," then you may want to reconsider. The idea that creative thinking skills can be learned and developed seems to have originated in the 1940s with thinking about the creative process itself, while the idea that inventiveness can also be learned and developed seems to have come a bit later, in the 1960s.

Brainstorming. In his popular 1949 and 1953 books, Alex Osborn developed the concept and process of "*think up*" – later known as "*brainstorming*," which stands for "*storming a problem in a commando fashion*" [170–172]. Osborn's brainstorming concept was aimed at groups and was designed to help get lots of ideas out into the open, avoid killing ideas with early criticism, and to enable multiple ideas to be combined and/or lead to new ones. This is sometimes termed "*horizontal thinking*." Osborn also originated (in 1953 [171]), and Sidney Parnes further developed (in 1967 [172, 173]), what is now known as the *Osborn-Parnes model*, or *creative problem-solving* approach, which is famous for its use of "*divergent thinking*" and "*convergent thinking*." Divergent thinking is open-ended, wide-ranging thinking used to create a broad set of options. Convergent thinking is solution-oriented thinking used to formulate a specific solution or approach to a problem using the results of divergent thinking.

33 An imaginary scenario in which some kind of substantial barrier that had been holding back progress in business and/or innovation does not exist. The creation of Camelot Scenarios is a brainstorming tool that is sometimes used in an attempt to find a solution or breakthrough.

Lateral thinking. In 1967, Edward de Bono introduced his creative thinking concepts of "*parallel thinking*" and "*lateral thinking*" [174, 175]. De Bono felt that most people think linearly, or "*vertically*," in attempting to solve problems, but that for problems requiring a creative solution, one should apply what he called parallel thinking, or lateral thinking, either of which could be considered, at first glance, to be illogical. Parallel thinking has to do with avoiding adversarial approaches in team-based creative thinking, focussing instead on more co-operative and constructive approaches. Parallel in this sense means having everyone on the team thinking in, broadly, the same direction. Lateral thinking refers to avoiding thinking about a problem head-on and/or in logical step-by-step fashion and instead thinking about it in indirect and/or nonlinear ways (sometimes referred to as thinking out of the box).

Six thinking hats. Another Edward de Bono approach, this is an example of a parallel thinking approach, in which a problem or opportunity is viewed from different points of view. In this case, individuals, or entire groups, consider the problem/opportunity from the perspective of a well-defined function or role. The metaphor, for each different role-play exercise, is that of wearing a specific kind of "hat." There are six different such hats, and they are each adopted in turn, but only one at a time. The six hats are as follows [176, 177]:

- The "*White Hat*," demands information, both the information that is currently available and the information that is needed.
- The "*Yellow Hat*," is bright and optimistic and wants to focus on the positive aspects and look for benefits and the value.
- The "*Black Hat*," is judgmental and wants to focus on the negative aspects, particularly why an idea may not work. This hat will play the "Devil's Advocate" role and try to identify difficulties, hazards, and reasons why something could (or will) go wrong.
- The "*Red Hat*," is intuitive, and relies on hunches and feelings. This hat will express emotional responses to ideas, both positive and negative.
- The "*Green Hat*," is creative and will focus on alternative ideas and the many possibilities inherent in any and all ideas.
- The "Blue Hat," is controlling and will want to have a process and to ensure that the process is managed so that it gets followed.

Synectics. William J.J. Gordon developed *Synectics* in 1961 [178]. The name comes from the Greek word *synektiktein*, referring to the joining together of different, and possibly seemingly irrelevant, ideas in the process of creative problem solving by a group of people. Synectics originally referred to the study of problem-solving and invention by groups, but later evolved into a set of techniques that include what are now known as *creative problem-solving, brainstorming,* and *lateral thinking* [179].

TRIZ. In the 1960s, Soviet engineer and inventor Genrich Altshuller developed a systematic approach to invention based on a comprehensive analysis of the patented solutions to hundreds of thousands of previously solved inventive problems. He identified a series of approaches that, taken together, were key to the solution of the majority of these problems. Altshuller called his approach the "*theory of inventive problem solving*" (TIPS, or TRIZ the Russian acronym) [72]. The TRIZ approach involves several groups of methods that can be used to look at a problem in ways that, either individually or in combination, frequently lead to finding an inventive solution.

Spider diagrams. These are diagrams that use text and drawings and/or pictures to visually organize information. The spider diagram, or at least its forerunner, seems to have originated with the famous notebooks of Leonardo Da Vinci. There are other variations of spider diagrams, including "*Idea Sun Bursts,*" and "*Mind Maps.*"

Medici effect.[34] This is when exposure to a range of fields, disciplines, or cultures leads to a combination of existing concepts that sparks a creative new idea [180]. The term "*intersection*" is used in the Medici effect, with reference to the making of multiple connections among the various fields of knowledge, philosophies, and approaches. Creating environments that facilitate explorations of such intersections among diverse disciplines has been advocated as a method for producing breakthrough innovation ideas [112].

The above approaches, brainstorming, lateral thinking, synectics, TRIZ, spider diagrams, and the Medici effect, are only some of the ways to get new ideas. There are many articles and books in the literature containing advice on nurturing creativity and on ways to get new ideas, e.g. references [181–184].

4.4 The product development process and the valley of death

Discussions of how innovation actually gets done inevitably lead to discussions about the "*valley of death*" (or "*technological valley of death*"; see Figure 4.2). The valley of death is the realm where the technical, financial, and market risks can be the greatest, and it therefore encompasses not only some of the most difficult technical challenges but also the greatest difficulties in securing the financial support to do the necessary work.[35]

The process of technological innovation is *not* linear, but just for now, let's pretend that it is linear. Imagine a pathway along which discovery science leads to new discoveries and/or understandings, and engineering research leads to practical new principles and processes. Now imagine that someone comes up with an idea that could use such advances to create a new technology that could in turn be developed into a practical product or service that others might be willing to buy.

34 The name Medici refers to a family in 15th-century Italy that funded creators from a wide range of disciplines, drew them to a single location (Florence), and helped trigger the European Renaissance in such fields as art, science, politics, literature, and architecture.

35 For this reason, it is sometimes termed the "*innovation progression gap*" or "*funding gap.*"

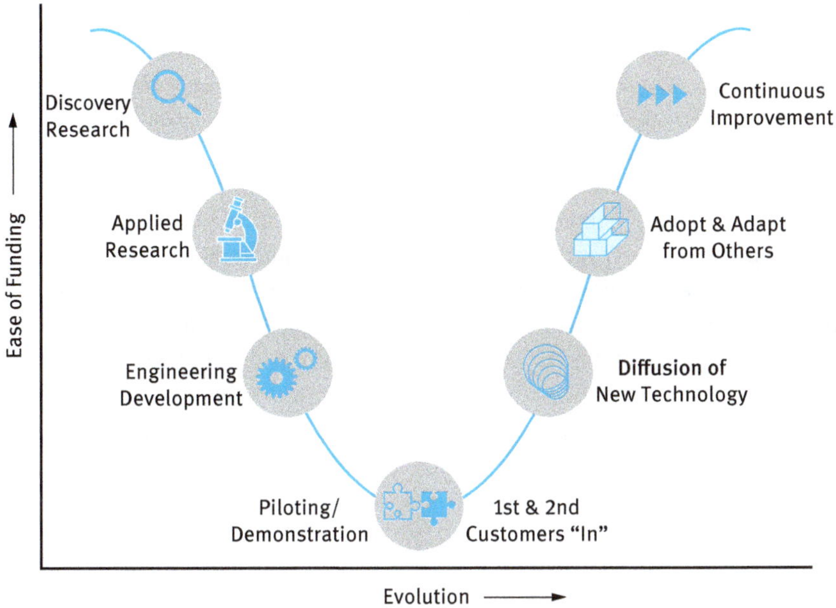

Figure 4.2: A simplified illustration of the "valley of death.".

Every one of the steps described so far is expensive and time-consuming and has great potential for failure, so this process is inherently very risky. The problem is not with any one of the steps but with the need to have enough of each of them, to be able to find the right ones, to be able to link a series together (in just the right way) to accomplish the early steps of the innovation process.

But it's worse than that. Great ideas backed by great science, great engineering, and great market opportunities are not nearly enough. The left-hand side of Figure 4.2 is sometimes referred to as "*upstream innovation*," the process of originating, evaluating, and developing concepts for new business products, processes, or services.[36] Even with all of those things in-hand, and after the new technology has been demonstrated "in principle" at "laboratory scale" (i.e. from technology readiness levels[37] [TRL] 1 through 3; see Section 4.5), the process is not yet complete. Unfortunately, as technology development proceeds, there is a tendency for financing, including government financing, to become increasingly difficult to obtain, as the technology is developed from basic and applied research through to speculative concept, engineering development, and proof of concept stages.

36 This can include the process of discovery.

37 Technology readiness and commercial readiness levels are described in Section 4.5.

Once one or more discoveries have been made, leading to one or more inventions, it may be possible to identify and develop a truly significant technology advance. The right-hand side of Figure 4.2 is sometimes referred to as "*downstream innovation*," the process of converting such concepts into market-ready products, processes, or services and then introducing them into the marketplace.[38] If the new IP is worth protecting, so much the better, but this is not the time for the innovator to relax – there are still many steps to fulfil in order to "*get to innovation.*" The next steps represent a transition from further technology development into the business side of innovation. These steps reside in the bottom of the valley of death [185], for which financing is the most difficult to obtain. These include technology development steps such as prototype development and testing, scale-up, and pilot testing (i.e. from TRL 4 through 8). This is where most potential innovations fail, where investors may become faint-hearted, and where product/service developers may give up.

Possibly the most difficult of the tasks involved in developing technological innovation is that of successfully navigating through the bottom of the valley of death: proceeding past proof of concept and beyond piloting and demonstration.

The NPD process. There will be a stage at which a number of potentially good ideas have been identified and some will be selected for further development (see Section 4.7). Some of the principal steps in product, process, or service development to marketability are [76, 186] as follows:

- *Product definition*: a basic summary of a new product, process, or service including a description of what it is, a preliminary design, and a description of how it would be used.
- *Mockup*: a model built to accompany a product definition and used to illustrate a product or process, and sometimes also aspects of the environment in which it would be used. For example, a mockup for a process concept might include a model of some of the process vessels, pumps, and piping in which it would be used. Mockups are usually built to scale, but usually are not working models.
- "*Works-like*" *model*: a model that illustrates the function of a product, process, or service without necessarily working exactly like the final product will. As such, a "*works-like*" model is more advanced than a mockup but less representative of the final product than a working model, engineering prototype, or production prototype. The benefits of a "works-like" model are the ability to quickly and inexpensively produce a model that can be shown to prospective customers in order to get feedback before finalizing the product design. Similarly, a "*looks-like*" model illustrates the appearance of a product, process, or service without functioning exactly like the final product.

38 This can include the process of invention.

- *Working model*: a physical model built to demonstrate that a product or process has been reduced to practice and basically works (proof of concept). Working models are expected to function but are usually not built to scale and their functioning has usually not been optimized.
- *Engineering prototype*: a working model of a product or process that has been built carefully enough for use in testing and demonstration of important design parameters. An engineering prototype may not be built to full scale but will be designed to be scaled-up. Once a prospective new product or process has been shown to work at a reduced scale, a next step in its development is to design and test (scale-up) a full-scale or intermediate-scale model. Depending on the nature of the product or process, a working model at larger scale might need to look and behave quite differently from the original. For example, in the case of a mechanical device, some components might have to be made with different dimensions or from different materials in order to function properly at larger scale. In the case of a chemical process, the nature and geometry of the vessels, mixers, and other equipment might have to be quite different from their smaller-scale counterparts in order to properly scale up critical process variables, such as shear and residence time.
- *Production prototype (or design model)*: a full-scale, completely operational model of a product that has been built to mirror a mass-produced unit but which has been custom built as a one-off or in very limited quantity. A production prototype is used to demonstrate performance requirements, including operation, safety, and durability.

In practice, depending on the specific technology and circumstances, some of these stages may be unnecessary or inappropriate, in which case they may be skipped over. Also, companies differ in how painstakingly each stage is pursued before deciding whether, and when, to move further forward in the NPD process (see Section 4.8).

Some increasingly common approaches are to engage multi-disciplinary teams and have them work together through the entire NPD process, to identify key stakeholders (including customers), suppliers, etc., all of whom are referred to as customers in this case, and to solicit their input, if not full engagement, throughout the entire NPD process. An example of the former is design for excellence[39] (DFX), in which customer needs are identified at an early stage in the process [187, 188]. An example of the latter approach is "*customer value-chain analysis*,"[40] in which the customer input helps design teams to recognize diverse product requirements at an early stage of the development process [189]. These are not mutually exclusive, so both can be applied to the process, together.

---

39 Some specific DFX tools include "*design for assembly*," "*design for manufacture*," "*design for manufacture and assembly*," "*design for excellence*," and "*design for all*."
40 Also termed "*voice of the customer*."

There are several kinds of pre-release testing, that is, testing conducted prior to commercial production and release. In manufacturing, a "*qualified production prototype*[41]" is a full-scale, completely operational sample of a product that has been built using a limited production run to demonstrate that it meets design standards. It's testing, "*alpha testing*" is pre-release testing, usually conducted internally, once all of the intended features have been built in, and usually includes testing for all aspects of integration and performance [190, 191]. If this version passes the testing, it is referred to as an "*alpha release*" of the product. Qualified production prototypes are also used to demonstrate that they meet industry or regulatory standards. In process industries, it is more common to refer to a "*full-scale demonstration,*"[42] which is a demonstration and/or test of a production prototype of a new or improved product or process, conducted at full-scale and under conditions of actual field, industrial plant, or market operation.

In the biomedical field, it is more common to refer to a "*clinical trial*" of a new medical or veterinary medical treatment, conducted with live human or animal subjects. There are four standard phases of clinical trial: Phase 1 refers to testing for safety, phase 2 refers to testing for the possibility of efficacy, phase 3 refers to testing for the degree of efficacy, and phase 4 refers to testing for long-term effects.

Another form of pre-release testing, "*beta testing*" (or pilot testing), is conducted by a selected number of external end-users. This is considered to be pre-release testing and provides a mechanism for obtaining end-user feedback while providing the marketplace with a preview of the intended product. This version is referred to as a "*beta release*" of the product. Another form, "*gamma testing,*" is sometimes conducted with customers in a limited market with a product that is almost but not quite ready for full market release. This is somewhat like a next level of Beta Testing and may be aimed at testing a specific feature of the product, such as an aspect of product safety, before the full release. This version is referred to as a "*gamma release*" of the product.

Beyond the bottom of the valley of death, different kinds of financing such as venture capital, private equity, debt financing tend to become available,[43] as the technology continues to mature, from gamma testing, field demonstration and first deployment, found to be practical and effective by a first customer and generally several others, and then reach commercial maturity in competitive markets (i.e. from TRL 9 through commercial readiness index [CRI] level 6).

---

41 Also termed "*product qualification prototype.*"

42 Also termed "*commercial-scale demonstration,*" "*field-scale demonstration,*" "*plant demonstration,*" or "*market demonstration.*"

43 In some literature, reference is made to a second, "*commercialization valley of death,*" which refers to later-stage technology development phases such as full field- or plant-scale commercial demonstrations and first and second customer sales and deployments. There is also an entrepreneurship valley of death, see Section 4.9.

In general, the elapsed time between origination of the concept for a new product, process, or service and the point at which it is available for sale in the marketplace is referred to as the "*time to market*"[44] [192]:

> *Time to Market and the "Valley of Death" in 1955: "Fifteen years is about the average period of probation, and during that time the inventor, the promoter and the investor, who see a great future for the invention, generally lose their shirts . . . This is why the wise capitalist keeps out of exploiting new inventions and comes in only when the public is ready for mass demand"* – Owen Young, quoted by Rupert Maclaurin in 1955 [105].

Marketing and sales. For any products, processes, or services that successfully emerge from the NPD process, there will be issues relating to supply chain management, manufacturing, and distribution, but these will only be needed if supported by the right marketing and sales.

"*Marketing*," in the context of products, processes, and services (as opposed to the marketing of an organization and its overall brand), represents the processes of identifying a product/process/service's main differentiating features, developing the key messages that explain these differentiators and their relevance, and developing and implementing strategies to position the product/process/service in its appropriate marketplace(s). Such strategies may include such things as branding, stakeholder relations, advertising, events, and communications (print, internet, and social media). The principal components of marketing are:

- *Market research and planning*, the evaluation of the market and the sales potential of a product, process, or service; identification of market barriers; planning for distribution channels; and identification of the target customers
- *Market management*, including advertising, promotion, and customer service related to a product, process, or service.

Marketing helps to create visibility and profile (brand awareness) for a product/process/service, and it sets the stage for business development and sales. See, for example, reference [193] and Section 7.3.

A company's sales function builds on the stage set by marketing and on the strategies developed by business development to create exchanges of value for a company's products, processes, or services. The sales function helps identify, create, and maintain relationships with clients, customers, and/or users in order to maximize sales. Sales also includes the closing, or securing, of sales deals, which may be through exchanges of value or the signing of contracts. The process of identifying and nurturing customer relationships is usually conducted in partnership with the business development function.

44 The "*product launch*" phase of the new product development process has been termed the "*fuzzy back end*" [192].

The valley of death is a challenge for innovative organizations of all kinds, but they particularly plague RTOs and start-up companies. It has been estimated that over 95% percent of start-ups fail to deliver their projected returns and that some 30–40% consume all of their up-front capital and fail [78]. However, the Valley-of-Death-challenge must be met in order for technological innovation to occur. What are required are people and organizations that do not give up, because it is out of the valley of death that market changing technological innovations and "creative disruption" emerge, and these are the big contributors to business and societal economic health and growth.

4.5 The first face of technology readiness: the technology itself

The term "*technology readiness*" is arising more and more frequently in discussions about technological innovation, especially as technologies are developed through the valley of death. When this occurs, one has to be a bit careful because there are two very different kinds of readiness, and both are very important. The first and possibly most obvious kind is the readiness of a technology for commercial release and deployment into the marketplace. The second kind of technology readiness is psychological and refers to the readiness of customers to accept and/or pay for the new technology, or indeed any new technology. This second kind of technology readiness will be discussed in Section 4.7.

As already noted several times above, technological innovation is NOT linear, but for thinking and discussion purposes, it can be helpful to categorize steps in the innovation process as if it were linear. Using this approach, tools like the TRL (technology readiness level) index enable the tracking of technology evolution from discovery ("blue sky") research up to actual demonstration in realistic conditions, while tools like the CRI (commercial readiness index) enable tracking of technology evolution as a commercial product, process, or service.

The TRL concept and system was first introduced by the US National Aeronautics and Space Administration (NASA) in the 1970s [194]. It evolved over time and by the 1990s its use had spread across multiple US government organizations, after which it spread to countries around the world. TRLs represent assessments of the maturity of an evolving technology. Different sectors and different countries use somewhat different TRL scales, but the general philosophy is the same. For convenience, they are each based on a linear model of innovation while at the same time recognizing that innovation is seldom linear, not all development cycles are the same, and the development of any particular technology may skip some readiness levels (see [195]).

Possibly the simplest TRL scale, with only four levels, is as follows:

TRL 1: Discovery
TRL 2: Development
TRL 3: Demonstration
TRL 4: Deployment

The United States, Canada, Australia, and the European Commission, among others, use similar but expanded scales, with TRLs ranging from 1 to 9 (see Table 4.3 and Figure 4.3) [195–198]. RTOs work across the full TRL span but generally focus on developing technologies from about TRL 2 or 3 through to about TRL 7 or 8. Both four- and seven-point scales are also in common use.

A few comments related to terminology may be appropriate here. During the process of invention, various concepts will be developed that could link knowledge and/or discoveries to some kind of application. Such *"speculative concepts"* may or may not be feasible or even physically possible. At this stage of development, the technology does not yet have to be shown to be feasible. In terms of TRLs, a documented speculative concept satisfies TRL 2. Also, as knowledge increases and discoveries occur during the development process, there frequently occur ideas, concepts, models, and analyses that, when viewed with hindsight, are perceived to have had a major impact on the development of a particular technological innovation and which led to significant improvements over the preceding technology. Such key advances are sometimes referred to as *"critical technology events"* [199].

A technology concept that is beyond the speculative concept stage and is either being developed into something feasible or has been demonstrated to be feasible is an *"early-stage technology."* The term *"proof of concept"* refers to a physical demonstration that a technology concept can be made to work and is no longer speculative. At this stage of development, the technology does not have to be practical, efficient, or cost-effective. In terms of TRLs, a demonstrated proof of concept satisfies TRL 3.

As a technology continues to be developed through the innovation continuum, attention will start to be paid to the gap between a technology that is at an early stage of development (i.e. at the discovery, technology concept, or invention stage) and the product development stage (from which a prototype could be developed). In terms of TRLs, this would be the gap between a technology at TRL 2–3 and at TRL 5–6, and it is referred to as the *"technology-translation gap."*

Table 4.3: A generalized description of technology readiness levels (TRLs).

Technology readiness level	Description
TRL 1	Basic principles have been observed and reported and are becoming translated into applied research and development.
TRL 2	Practical applications and inventions are being identified.
TRL 3	Applied research and development are underway at laboratory scale, including proof of concept.
TRL 4	Multiple technological components, if applicable, are integrated and demonstrated to work together, again at laboratory scale.

Table 4.3 (continued)

Technology readiness level	Description
TRL 5	The technological components are integrated for testing and validation in a simulated and/or realistic environment beyond the laboratory.
TRL 6	The integrated technological components in a model or prototype are tested and validated in a simulated and/or realistic environment beyond the laboratory.
TRL 7	A complete prototype, at or near full-scale, is ready for demonstration and/or demonstrated in a realistic operational environment.
TRL 8	A complete technology has been tested and demonstrated to work in its final form and under realistic operational conditions.
TRL 9	A complete technology, in its final form and including any final "fixes," has been proven through deployment in actual operational environments and conditions.

The commercial readiness index (CRI) is a representation of the readiness of a mature technology, and an extension of the TRL scale, transcending TRL 9. The original Australian CRI system uses a scale ranging from 1 to 6 (see Table 4.4 and Figure 4.4) [200].

The TRL and CRI scales are not mutually exclusive (see Figures 4.4 and 4.5). The CRI scale overlaps the TRL scale, starting at the proof-of-concept level and culminating at the fully commercially deployed and mature level. Regardless of the scale(s), this kind of technology readiness is measured and used when following the progress of innovations. Users of tools like TRL and CRI range from companies that dedicate significant resources to managing commercial innovation, to venture capital firms that fund prospective new commercial innovations as they are being nurtured past the "valley of death" by start-up and small- and medium-sized enterprises (SMEs),[45] to governments and agencies, like NASA, that buy and deploy new innovations.

Finally, although the foregoing discussion has focussed on technology and commercial readiness levels, TRL and CRI, other methodologies have been proposed for evaluating technology readiness. For example, Yeo et al. propose a bibliographic (scientometric) method that also produces maturity results that can be displayed in technology S-curves [201].

45 Business enterprises that are smaller than a specified number of employees (e.g. 500) and/or have annual revenues of less than a specified value (e.g. $75 million), as distinguished from large-size enterprises.

Figure 4.3: Illustration of a Canadian government TRL scale. From information in reference [195].

Table 4.4: A generalized description of commercial readiness index (CRI) levels. Based on [200].

Commercial readiness level	Description
CRI 1	The technology is commercially prospective but has not yet been commercially tested or proven.
CRI 2	The technology has passed a first, small-scale, commercial trial.
CRI 3	The technology has been commercially deployed, driven at least partly by market-pull.

Table 4.4 (continued)

Commercial readiness level	Description
CRI 4	Multiple commercial deployments have occurred, possibly with some level of government support.
CRI 5	Market-pull is driving broad commercial deployment and competition is emerging in the marketplace.
CRI 6	Mature commercial deployment with established standards, regulatory acceptance, and performance track-record.

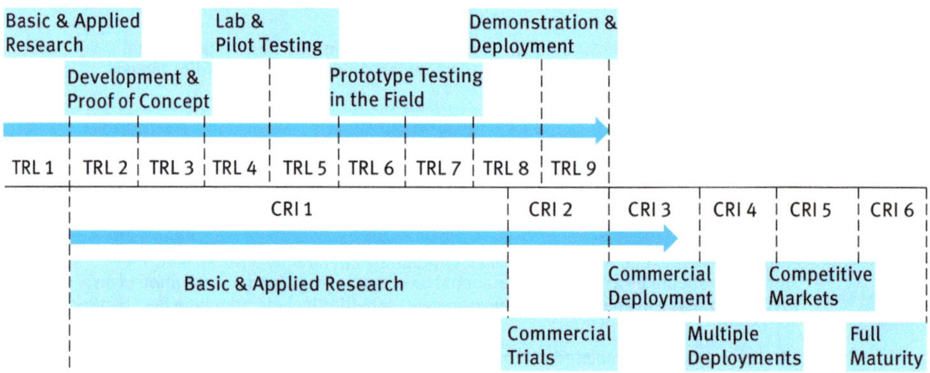

Figure 4.4: Illustration of technology readiness levels and commercial readiness index levels.

4.6 Market analyses, business plans, and financing

Along the technology development pathways outlined above, a parallel process will also have to be followed in which the associated business concept is developed. The "*business concept*" involves expanding the idea for commercializing a new product, process, or service into a description of the nature of the new product, process, or service; what it does; how it is different from and better than what is already in the marketplace; how it would be delivered; and who would be the customers for it. Thus, a business concept is often referred to as being "*a bridge between an idea and a business plan.*"

A "*feasibility study*" and a "*market analysis*" will need to be conducted and refined in order to make sure that continued development steps are worth undertaking, and also to provide the basis for an eventual commercialization and/or business plan.

A feasibility study is simply an evaluation of a proposed initiative or business concept that is aimed at identifying strengths, weaknesses, opportunities, and threats

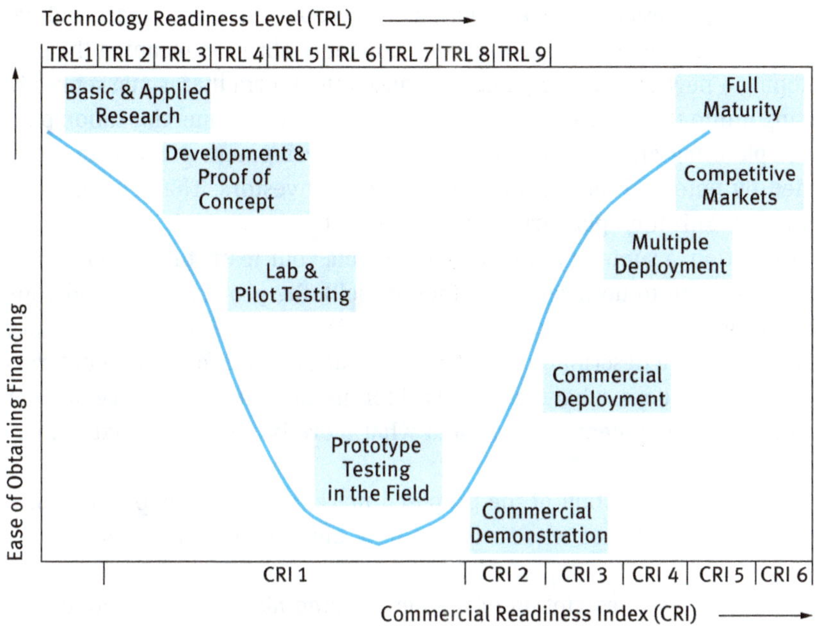

Figure 4.5: Illustration of technology development stages and the valley of death.

(SWOT); the potential for success; the resources needed to further develop the initiative or business concept; and whether its cost and selling price are likely to yield enough profit for it to be commercially feasible.[46]

The principal elements of a market analysis are as follows [76]:

- *Market identification:* a summary of the customer needs that will be satisfied; the specific customers who will buy the product, process, or service; the reason(s) they will buy it; and whether the timing is right in the marketplace
- *Market size:* a description of the target market and/or market segments and the size of those markets in numbers of units that can be sold
- *Customers:* a description of the target purchasers and also of the ultimate end-users
- *Distribution channels:* a summary of the pathways available to get the product to the customers and/or end-users
- *Competitors:* a description of the competitors in the marketplace[47] and how the product is or will be differentiated from them.

46 In this context, the classical terms "*Does it pencil?*" and "*back of the envelope calculations*" generally refer to feasibility studies.

47 There are almost always competitors. If there are truly no competitors, then there is a good chance there is no market.

As the development process and market analysis evolve, at some point, these will become integrated into a "*commercialization plan*," commercialization meaning the process of developing a new product, process, or service into a form that is market-ready and introducing it into the marketplace (product launch). The commercialization plan helps with the planning and conduct of the product development process and may also be needed by potential partners, funders, and/or investors. The principal elements of a commercialization plan are the following [76]:

- *Project description:* a summary of who you are (you, your team, or your organization), what you plan to do (license, partner, or sell), the project status, and your value proposition
- *Product description:* a description of the new product, process, or service in terms of what it is, what it does, what it can be used for, its current state of development (in the product development process), and what more is needed to make it market ready
- *Market analysis:* a description of the proposed market for the new product, process, or service in terms of what it is and how big it is (based on the market analysis)
- *Intellectual property:* a description of the new IP (and also the background IP if appropriate) and how it is being protected
- *Risk analysis:* a description of the principal risks involved in the commercialization process and how they are being managed
- *Next steps:* a description of the next steps in the product/process/service development plan

If the goal is to actually produce and sell the new product, process, or service, then the commercialization plan needs to be expanded into a "*business plan*," the principal elements of which are the following [76, 81]:

- *The project:* a summary of who you are (you, your team, or your organization), what you plan to do (license, partner, or sell), and your value proposition
- *The product:* a description of the new product, process, or service in terms of what it is, what it does, what it can be used for, its current state of development (in the product development process), the IP, and what more is needed to make it market ready
- *The market:* a description of the proposed market for the new product, process, or service in terms of what it is and how big it is (based on the market analysis)
- A *marketing strategy*
- An *operations plan*
- A *management plan*
- A *financial plan* that includes a risk analysis.

At this level of planning (the business plan), more sophisticated forecasts of marketing and sales may be needed. As an illustration, the ATAR (awareness, trial, availability, repeat) model can be used for such forecasting. The ATAR model is a sales- or profit-forecasting model that is based on advertising and brand-awareness data. It was developed for consumer product sales but could be used for process or service sales as well. A version of the model is [202]

$$\text{Sales} = (\text{purchasing unit no.}) \times (\%\,\text{aware}) \times (\%\,\text{available}) \times (\%\,\text{training}) \times (\%\,\text{repeating}),$$

where purchasing unit no. is the number of purchase participants (people, households, companies, or departments) in a sales region, % aware refers to the fraction of purchasing units that have heard about the new product, % available refers to the probability that the product is available to the purchasing unit, % trialling refers to the fraction of purchasing units that have purchased at least once for testing, and % repeating refers to the fraction of purchasing units that have purchased more than once. Based on such sales projections, profitability can be calculated as the number of units sold multiplied by (revenue per unit – cost per unit).

One more parallel process that generally has to be followed relates to *financing* of the process. Technological innovation can be expensive, and the costs typically rise as a new technology matures and begins to get developed into a product, process, or service. As a result, innovators tend to always be in need of additional capital. Depending on the manner in which the technological innovation is advanced, this may take the form of internal organizational funds, debt capital (some form of interest-bearing loan, often from a financial institution), grant capital (some form of government grant assistance), or equity capital (some form of partial ownership in return for financial investment).

Three of the more common forms of funding are "*bootstrapping*," in which the funds are generated internally from other income; "*debt financing*," in which funds are borrowed; or "*equity financing*," in which funds are obtained through the sale of a share in ownership. For entrepreneurial companies and/or specific commercialization initiatives, some other kinds and sources of financing may be available at different stages of company or product maturity (see Table 4.5) [203, 204].

Early-stage (usually modest) capital for a company and/or commercialization process is termed "*seed capital*" and is typically raised through informal investors, internal organizational funds, or government grants (Chapter 9 in reference [204] provides an introduction). Informal investors include "*angel investors*,"[48] or "angels," who are any or all of family, friends, and entrepreneurs. "*Sweat equity*" represents unpaid work that is contributed to the commercialization process, usually by the principal(s) involved. A combination of sweat equity and seed capital is sometimes enough to de-

48 Some angel investors contribute their personal expertise and experience, as well as money, into the venture.

velop a technology from a mock-up or working model stage to the engineering prototype or production prototype stage of product development and to develop a formal market analysis and a formal commercialization plan. Another, relatively recent, method of raising early-stage capital is through *crowdfunding*, in which a website serves as a platform that enables entrepreneurs to pitch their project(s) and interact with potential funders whose contributions could be any combination of donations, loans, equity, or future-product purchases (see, e.g. [205]).

Intermediate-stage (usually moderate-level) capital for a new young company and/or commercialization process is termed "*start-up capital*" or "*pre-venture capital.*" This kind of capital is typically raised through informal or formal investors, internal organizational funds, or government grants or loans. Pre-venture capital is sometimes enough to develop a technology from an engineering prototype or production prototype to the qualified production prototype stage of product development, possibly even limited production and a first introduction to the marketplace. By this point, the market analysis and commercialization plan have probably evolved into a market strategy and business plan, but most of the development costs still lie ahead. Two stages are sometimes distinguished. "*early stage*" (or formative stage) refers to companies beginning operations, and probably having a product or service in testing or pilot production, but which are not yet at the stage of commercial manufacturing and sales. "*Later stage*" refers to companies beginning commercial manufacturing and sales, but before any initial public offering of shares.

Two kinds of valley of death were described in Section 4.4, but there is a third. In start-up companies and the practise of entrepreneurship, there is an "*entrepreneurship valley of death,*" which refers to the gap, or period of time, between the initial funding and launch of a start-up company and the point at which it starts generating revenue from sales. During this interval, start-up companies tend to have a high risk of requiring more operating capital than they can afford, which can cause the enterprise to fail. An illustration showing steps in a linearized process leading from the initial business concept through to a successful, sustainable company is sometimes termed a "*death valley curve*" (see Section 4.4).

Final-stage, usually substantial capital for a commercialization process is termed "*venture capital.*" This kind of capital is typically raised through formal, institutional investment firms (hence "*venture capital firms,*" or simply "*vencaps*"). Chapter 10 in reference [204] provides an introduction. Venture capital is usually used to take a product, process, or service that has already been introduced into the marketplace and to rapidly drive it "up" past the marginal or break-even level and into profitability. A venture capital "*rule of thumb*" is the "*10/5 Rule,*" which refers to a goal of having a commercialized product, process, or service yield a 10-times ROI within a period of 5 years.

As a business struggles to gain sales and market share, if successful, there is ultimately reached a tipping point beyond which sales revenues offset costs and, ultimately, total revenues become positive and growing. This has been referred to as the "*J-curve*" effect, which was named for a graph in which plotted values first decrease for some time but ultimately reach a minimum and rise thereafter (see Figure 8.3). Once a sustained growth phase has been reached, the company or commercialized product, process, or service will usually be well established and growing, at which point the early investors are usually moving to realize the profit from their investments and move on to other opportunities.

Some case study illustrations of entrepreneurship and enterprise hubs in India are given in reference [206 630 661], and some international mini-case-studies in [207]. Additional case studies can be found in Section 10.2.

Table 4.5: Financing and technological maturity. Adapted from references [43, 203].

Maturity level	Description
Seed capital	– Still just a potential company or a potential product, process, or service. – Spans idea definition, initial research, mock-up or working models, market analysis, and scoping-out of commercial viability. – The economic potential at this stage is highly uncertain. – Heavily dependent on informal investors, internal organizational funds, or government grants. – Typically under $500k.
Start-up capital and pre-venture capital	– A new, young company (start-up capital) or product, process, or service (pre-venture capital). – There will be a *market strategy* and *business plan*, but most of the development costs still lie ahead. – Informal or formal investors, angel investors, internal organizational funds, or government grants or loans. – Typically under $2 M.
Venture capital	– A company or product, process, or service has already been introduced into the marketplace; has ownership of the key proprietary technology; has revenues and cash flow; and has a better knowledge of the market. – The goal at this phase is to rapidly drive it "up" past the marginal or break-even level and into profitability. – Formal, institutional investment firms. – For small businesses, typically $250k to $5 M. For medium and large businesses, typically $5 M–$25 M or more.

4.7 The second face of technology readiness: The customers

In Section 4.5, the first face (kind) of technology readiness was discussed: the readiness of a technology for commercial release and deployment into the marketplace. However, once a new product, process, or service is launched in the marketplace, that isn't the end of the innovation story.

The second face, or kind, of technology readiness (Figure 4.6) is psychological and refers to the readiness of customers to accept and/or pay for the new technology, or indeed any new technology. This is as important, or even more important, than the maturity of the technology itself because customers' personal views about technology influence the extent to which any new technology will "*diffuse*"[49] and be successful in the marketplace, if at all. American sociologist Everett Rogers distinguished several kinds of customer behaviours in his book *Diffusion of Innovations*, which outlines a model of the same name [77]. Rogers wrote: "*One reason why there is so much interest in the diffusion of innovations is because getting a new idea adopted, even when it has obvious advantages, is often very difficult*" [77].

Figure 4.6: Illustration of Janus, the two-faced Roman god of beginnings, gates, and transitions. Public domain image courtesy Wikimedia Commons; attributed to a book by Giovanni Battista Palumba (Rome, *circa*. 1517).

It should possibly not be surprising then, that there have been many studies of the diffusion of innovation. Rogers and Shoemaker's 1971 book *Communication of Innovations*

49 Technology diffusion refers to the transfer of technology from a person or organization to another where either party may be unaware of the identity of the other and occurs when technology is becoming widely disseminated. This is different from "technology transfer," which refers to the transfer of technology from a specific person or organization to another, and occurs for initial and/or limited technology dissemination.

was based on about 2,000 diffusion-of-innovation studies [208, 209], and Rogers commented in 1990 that he'd found about 4,000 publications on the diffusion of innovations [209]. Mahajan and Peterson [210] provide a summary of innovation diffusion models. Only a brief introduction to diffusion, and some illustrations, will be provided here.

MacVaugh and Schiavone begin two of their reviews of the diffusion of innovation as follows:

> *One of the least understood areas of innovation diffusion is the non-adoption of new technology . . . In some cases, individuals or groups eschew the functionality of technology, regardless of when it was developed . . . In others, existing technology users do not choose to purchase categorically similar . . . newer products when they become available. Further, some who have used a new technology may later become non-users . . .* [211, 212].

It turns out that major innovations may require decades, sometimes more than a century beyond their point of public or market introduction, to reach full implementation. Rogers provides a classic example from a very important non-commercial innovation: "*In the early days of long sea voyages, scurvy was the worst killer of the world's sailors, worse than warfare, accidents, and all other causes of death . . . In 1601, an English sea captain, James Lancaster, conducted a kind of experiment to evaluate the ability of lemon juice to prevent scurvy*" [77]. Lancaster found that on one of his ships, having the sailors take lemon juice every day kept most of the crew healthy, whereas on three other ships (all of which took the same voyage together), having none of the sailors take lemon juice resulted in over a third of the sailors dying due to scurvy. Although the results were made known to the British Navy, it was only about 150 years later that this innovation to begin to spread [77].

In the technological innovation realm, there are numerous examples of non-acceptance, or at least long delays in acceptance. Here are just two more:

- In the merchant shipping industry, some companies didn't replace their sailing ships with steam ships when the latter became available in the 19th century, nor even with diesel-powered ships when these became available in the 20th century [211].
- Some audiophiles remain convinced that music amplification and capture are best conducted with vacuum tube technology despite the fact that digital technology has extended to specifically recreating vacuum tube sound qualities. These consumers retain negative stereotypes about digital audio technologies that were introduced to the market more than 35 years ago [211, 212].

Consider the following two scenarios:

1. If people are optimistic about technology in general and tend to be personally inclined to try new technologies, they still may not accept a new technology if they do not perceive it to be useful and/or easy to use.

2. If people perceive a new technology to be useful and/or easy to use, they still may not accept a new technology if they are not comfortable with technology in general and especially if they mistrust new technologies.

The value of a product, process, or service (or brand) perceived by prospective customers has been termed *"perceptual equity."* For example, a high degree of perceptual equity for a particular product can contribute to a willingness to try the product, pay a premium price for it, and/or remain loyal to its brand versus those of competing products. In the same vein, *"net perceptual equity"* refers to the difference between the positive perceptions and any negative such perceptions on the part of the customers. Of the many characteristics of technological innovations, as perceived by individuals, that can influence the rate of their adoption, Rogers found that the five most important perceptions on the part of potential customers are (Figure 4.7)

1. *Relative advantage*, the degree to which a new product, process, or service is perceived to be better than the status quo or the alternatives;
2. *Compatibility*, the degree to which it is perceived to be consistent with their personal values, experiences, and needs;
3. *Complexity*, the degree to which it is perceived to be difficult to understand and/or use;
4. *Trialability*, the degree to which it is perceived to be something that they can try (test) in advance of committing to a purchase; and
5. *Observability*, the degree to which it, and its results and/or benefits, are perceived to be readily visible.

Rogers found that the greater the perceived relative advantage, compatibility, and trialability of a technological innovation, and the less its perceived complexity, then the more likely potential customers are to adopt it, and the more rapid will be its rate of adoption [77]. All of the above factors can be important in the innovation-decision process, that is, the process by which a potential customer evolves from first becoming aware of a technological innovation, to forming an opinion on it, to reaching a decision on whether or not to purchase it, to acting on the decision, to confirming the purchase decision.[50] Rogers has summarized these as (1) knowledge, (2) persuasion, (3) decision, (4) implementation, and (5) confirmation [77].

Of course, any given marketplace is populated by many different people who may have quite different views on the desirability of a new product, process, or service and whose innovation-decision processes may also be quite different. Rogers' *"technology adoption life cycle"* is a stereotypical pattern of technology diffusion, in which a new

50 Confirming the purchase decision is the process by which a customer considers the purchase that has been already made. This is important because the customer's early experiences with it may be negative and/or they may later learn of negative attributes from others, either or both of which may lead to discontinuing use and passing negative perceptions on to others.

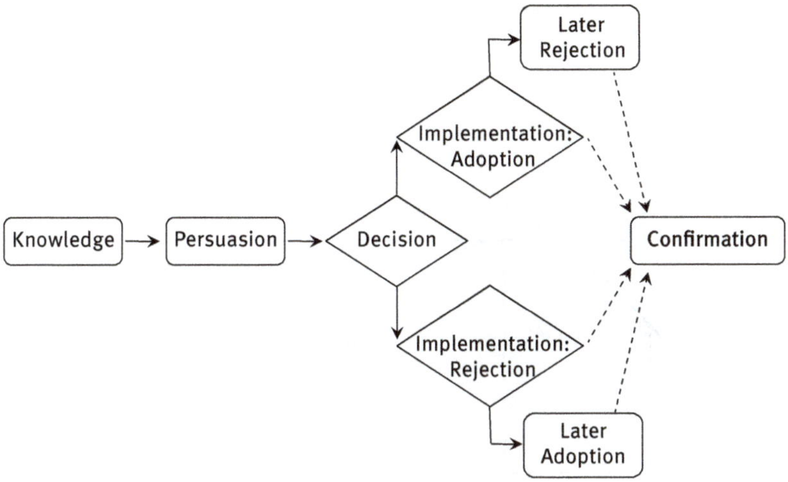

Figure 4.7: Simplified illustration of Rogers' five innovation decision process steps.

product, process, or service becomes adopted and spreads according to the sociological characteristics of different adopters. In this model, there will typically be a very limited number of initial adopters, then at least two distinct waves of subsequent adopters, followed by a limited number of very late adopters. Rogers found that only about 20% of the ultimate value of an innovation is created when an innovation is first established in the market, with 80% being realized in the later waves, leading to widespread adoption [77]. Rogers originally illustrated this using a normal distribution (i.e. a bell curve), as illustrated in Figure 4.8, to describe the different adopter types that he had identified [77]:

- *innovators* (the first few adopters, about 2.5% of market share);
- *early adopters* (the next few adopters, about 13.5% of market share);
- *early majority* (the wave of adopters, about 34% of market share);
- *late majority* (the wave of adopters, about 34% of market share); and
- *laggards* (the last few adopters, about 16% of market share).

More detailed descriptions of the characteristics of these different categories of adopters can be found in [77, 143].

Although Rogers developed his technology adoption life cycle and the normal distribution of adopters based on his studies of a single type of adopters (farmers), it has been found to apply quite well to other situations, including technological innovations in the coal, steel and iron, brewing, pharmaceutical, food product, and rail industries, among others [213–216].

There is an interesting transition between the first approximately 16% of adoptions (characterized by the innovators and the early adopters) and the next two large

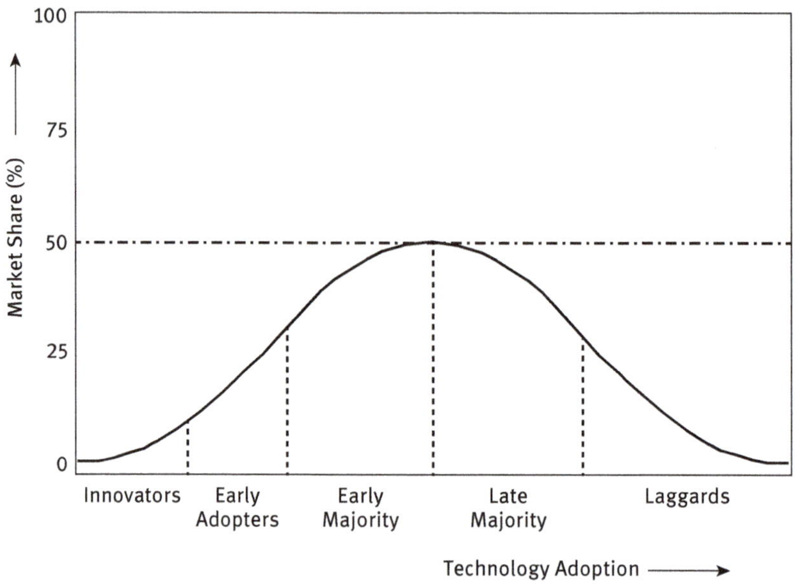

Figure 4.8: Illustration of Rogers' model of technology diffusion (based on data in reference [77]).

waves of adoptions, taking the total adoption levels to about 50% and 84% (characterized by the early and late majorities). This transition has been called the tipping point[51] [217], although Geoffrey Moore and others have referred to this as the "*Chasm*" or "*16% rule*" [143, 218] (see Figure 4.9). In this distribution, the first 16% of adopters, the innovators and early adopters, are often characterized as the customers who are most interested in the new technology and its performance. Conversely, the last 84% of adopters, the early and late majority especially, tend to be more interested in usefulness and convenience. This has huge implications for the marketing strategies that could be used for a new product, process, or service, as opposed to one that has already been embraced by the early adopters.

Another way of mapping the progress of technology diffusion in the marketplace is via "*technology diffusion S-curves*" (see also Section 3.2). If the technology adoption statistics described above are plotted cumulatively, then an S-curve usually results. Such a representation demonstrates the plateauing of technology adoption by the time a technology is adopted by what Rogers [77] termed the laggards. *See* Figure 4.10.

51 In general, a "*tipping point*" refers to a transition point beyond which some kind of "*critical mass*" has been reached, and an idea, trend, social behaviour, or product sales level spreads and/or increases very dramatically. For example, in epidemiology, the tipping point is when an infectious disease spreads beyond the capacity of any localized efforts to bring it under control.

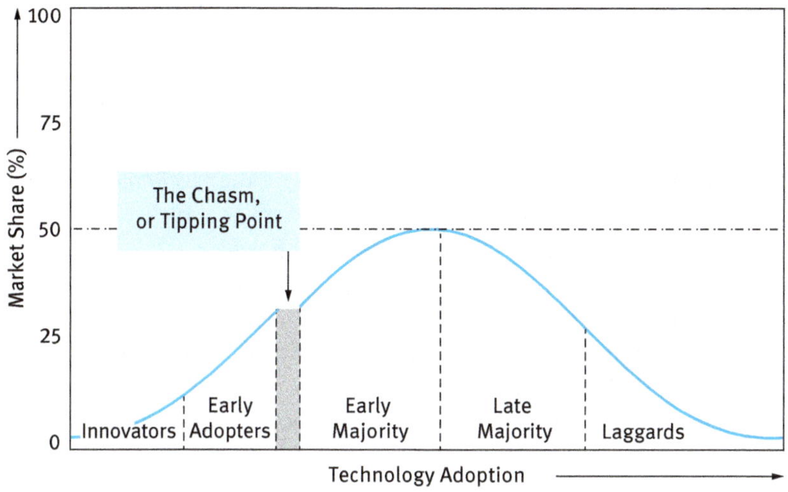

Figure 4.9: Rogers' technology diffusion model modified to illustrate the "*Chasm*" or tipping point.

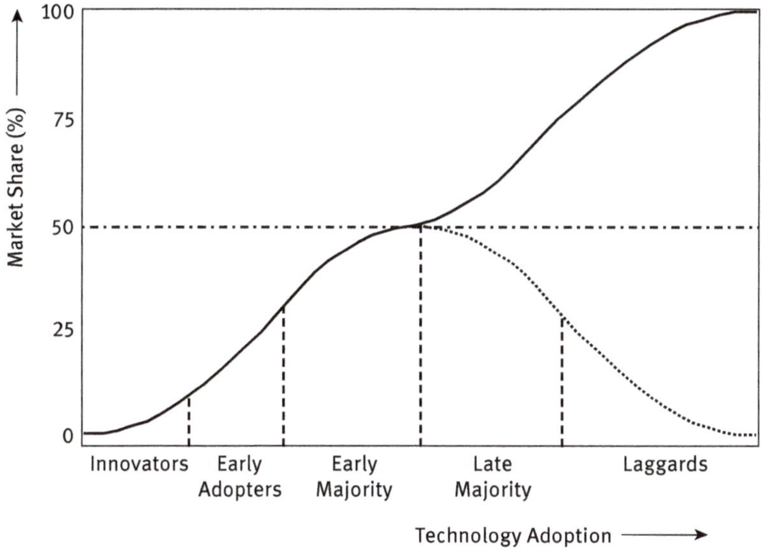

Figure 4.10: Rogers' technology diffusion model modified to illustrate the "logistic function" or cumulative market share as technology adoption increases.

The elapsed time between initial commercial launch and the point where adoption rates stagnate or start to decline can be quite variable, and some research suggests that, on average, the time span is shrinking. In 2002, for example, it was 36 years for televisions, 27 years for videocassette recorders (VCRs), and 18 years for compact discs [219]. By 2025, of course, VCRs and CDs had become virtually obsolete.

The S-curve shape is quite typical (see also Section 3.2), but for some technological innovations, the acceptance curve is fairly steep, while for others, it is fairly shallow. In attempting to explain such variations, more sophisticated tools and models have been developed. Two of the more recent tools are the *"technology readiness index"* (TRI), which enables assessments of prospective customers' personality preferences with regard to acceptance and use of a new technology, and the *"technology acceptance model"* (TAM), which enables assessments of the extent to which prospective customers perceive usefulness and ease of use in a new technology.

As just noted, the TRI is a psychological model developed to assess and describe how people's personalities influence the degree of acceptance and use of a new technology. The TRI represents an attempt to balance factors that would contribute to technology acceptance (optimism about the technology and personal association with innovativeness) versus factors that would inhibit technology acceptance (discomfort caused by the technology and insecurity caused by mistrust of the technology). Such views influence the extent to which a technology will be used. The original TRI model comprises 36 elements, while a more recent "refined" TRI model comprises 16 elements [220, 221].

A related, complementary model is the TAM, which is a psychological model developed to assess and describe how the features of a technology influence people's views of that technology in terms of perceived usefulness (PU) and perceived ease of use (PEU). Such views influence the extent to which a technology will be used by customers, if at all. There is also an integrated technology readiness and acceptance model (TRAM), which combines the TAM and TRI models to provide a single means of categorizing customer readiness to accept a new technology. The TRAM approach meshes the personality-related features of TRI with the system-related features of TAM.

In terms of psychological readiness of customers to accept and/or embrace new technologies, some literature makes reference to market segmentation in this context, a simple version of which would be to distinguish among:
- *explorers* (the first few adopters);
- *pioneers* (the next few adopters);
- *skeptics*;
- *hesitators*; and finally
- *avoiders* or *Luddites*[52] (the non-adopters).

52 The term *"Luddites"* has been used to describe people who are strongly opposed to mechanization, industrialization, automation, or even new technologies of any kind. The original Luddites were textile workers who, fearing job losses, destroyed textile machines in England in the Industrial Revolution of the early 1800s. Modern usage refers to people who are strongly opposed, or resistant, to the adoption of new technologies of any kind (hence "Neo-Luddites"). The "Luddite fallacy" refers to the hope that avoiding technological advances can ensure economic success and/or sustainability.

This has some analogy to the categorizations in Rogers' technology adoption life cycle above. Similarly, in the TAM, Rogers' five most important product/process/service perceptions (relative advantage, compatibility, complexity, trialability, and observability) have been incorporated to summary measures such as perceived usefulness (PU), perceived ease of use (PEU), and behavioural intention to use (BI). For a comparison between the two, see reference [222].

Armed with an awareness of the importance of prospective customers' personality preferences on new technology acceptance, plus tools with which to make assessments, it is perhaps not surprising that people in the innovation field sometimes categorize and "label" customer types (a nicer term for this is "*market segmentation*"). So, for any new technology that is introduced into the marketplace, one can imagine *explorers* (the first few adopters), *pioneers* (the next few adopters), *skeptics*, *hesitators*, and finally the *avoiders* (the non-adopters). Users of tools like TRI and TAM are generally the larger companies that dedicate significant resources to managing commercial innovation.

The word "*technology*" conjures up images of technical things, especially new technical things, but personality preferences can have a powerful influence on the way prospective customers view and make choices about new technologies. The most elegant and inventive product, process, or service, even if based on wonderful new discoveries, will fail to become innovation in the face of a marketplace full of "avoiders." On the contrary, it will have a chance of being adopted if it is perceived by potential customers to provide a better alternative than whatever it seeks to replace, is compatible with existing norms and needs, is easy to understand and use, can be tried and tested easily, and for which beneficial results are readily observed [77]. This adds another meaning to the adage "*it's all about the people*."

Some case study examples of diffusion of innovations are provided in Sections 6.1 and 10.2.

Before leaving this section, we can connect the innovation diffusion concept with the product development life cycle concept discussed earlier in Section 3.4. Knowing the innovation diffusion pattern, one could attempt to manage the development of new products so that by the time a first product (Product #1 in Figure 4.11) is in the hands of the innovators, early adopters, and early majority customers, it could continue to be sold to the late majority customers while launching a new, replacement product. This might not work out very well if the new product is simply an incremental innovation, as the late majority customers would probably just buy it instead. However, if the replacement product is a disruptive innovation (Product #2 in Figure 4.11), then it might appeal to the innovators and early adopters, who by this time are interested in something new, while the original product is just beginning to appeal to the late majority customers. Meanwhile, a company that is managing technological innovation really well would have a third disruptive product under development in hopes of catching "the next wave."

For example, "Product #1" could have been the manual typewriter. "Product #2" in this case could be the electric typewriter, which was initially sold to the innovators and early adopters, while the late majority customers were still buying manual typewriters. In this example, a "Product #3" would be the personal computer with word processing software and a printer, which again followed the same pattern in terms of diffusion, while the electric typewriter continued to be sold until the number of late majority adopters ran out.

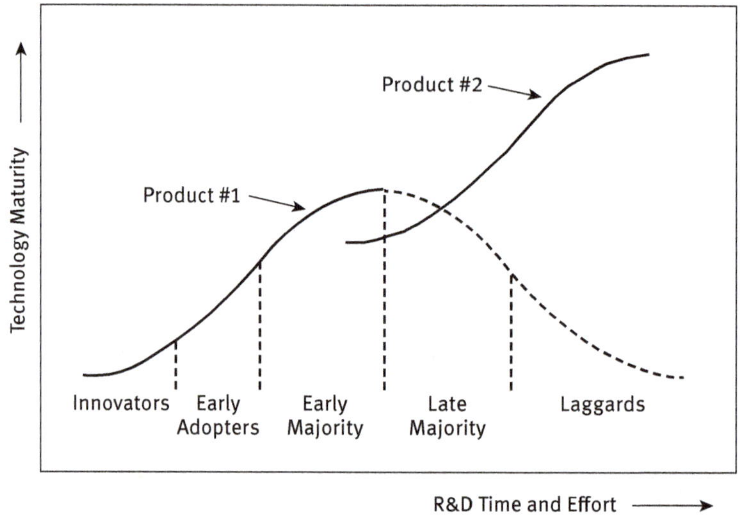

Figure 4.11: Illustration of managing product life cycles to match innovation diffusion patterns.

4.8 Managing the new product development (NPD) process

"No decision in business provides greater potential for the creation of wealth (or its destruction, come to think of it) than the choice of which innovation to back."

Robert Heller, British Business Author
In *The Decision Makers*, Hodder and Stoughton (1989).

Regardless of an organization's innovation development model, there will be a stage at which a number of potentially good ideas have been identified, but this number needs to be reduced in order to devote the most resources to the ideas with the best chance of being successful.

In Section 2.5 on success rates, it was noted that the literature suggests that it can take as many as 3,000 original/raw ideas to get to a workable, practical, and commercializable product, process, or service. While the challenge of finding this many ideas has been discussed above, another challenge is working through the large body of ideas generated in order to determine which ones have the most potential.

According to Gerry Katz [192], the concept of a funnel through which the number of ideas are winnowed down dates back to the 1980s, with (initially) a flowchart-type process for managing the process of making the selection and further development decisions. Many organizations use the idea of an NPD funnel,[53] hopper, or pipeline, in which many new ideas are fed in and there is a process for deciding which and how many of them should be further explored and/or developed (Figure 4.12). A case study example is provided by Barbieri and Álvares [127].

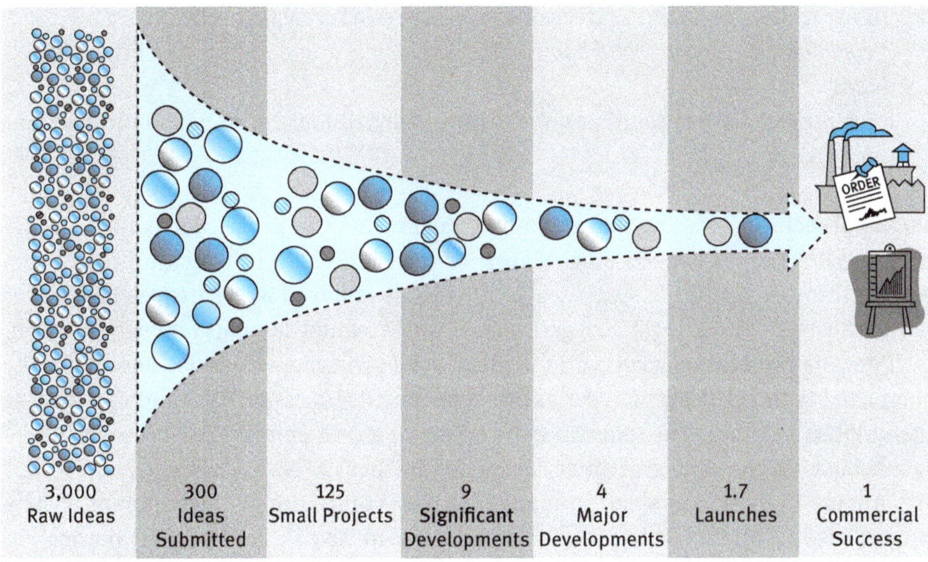

Figure 4.12: Illustration of an innovation funnel.

Since the 1980s, a number of "idea-to-launch processes" have been developed to guide the managed evolution of an idea into a commercial enterprise, product, process, or service [81]. An example is Whirlpool's "*I-pipe*" innovation pipeline (see Section 7.7).

The idea selection process can be done over several stages, rather than just picking one from the funnel, and an organization may develop specific criteria for assessing the ideas at any given stage and deciding which ones to advance and which to either discard or move back to the front-end of the funnel. Ultimately, only a few, or even just one (in a small organization), may be selected for final development and commercialization. An example of the sequence of stages is as follows:

1. *Identification of the problem or opportunity*, possibly using internal information, market data, customer input, or other information

53 Sometimes referred to as an "*innovation funnel*" or an "*idea funnel.*"

2. *Investigation, idea generation, and solution identification*, possibly using market knowledge, creative thinking approaches (see Section 4.3), or customer focus groups to develop a basket of product, process, or service ideas
3. *Further investigation and proposal development*, developing some of the ideas further together with plans for how they could be further developed
4. *Solution development and proof of concept*, possibly developing mock-ups or models and designing prototypes
5. *Prototype testing, demonstration, and marketability testing*, possibly involving customer trials, and at this stage, some re-design work may be needed
6. *Production, marketing, and commercial deployment*, of at least one of the original ideas
7. *Post-project review*, searching for lessons-learned and continuous improvement opportunities

One approach to assessing and potentially advancing the project ideas through the stages is Robert G. Cooper's Stage-Gate® process [223–225], in which a project progresses through a series of stages, each of which is followed by a gate (see Figure 4.13 for an illustration). Each gate triggers a review of accomplishments and decision as to whether the project may proceed to the next stage. The gate evaluations are normally completed with reference to a set of pre-determined success criteria and can involve many kinds of input. The tool can be used as an aid to objectively and consistently managing efficiency and risk (this is described further in reference [226]).

Among the risks involved in managing the evolution and development of new project ideas is the tendency to over-emphasize lower-risk, incremental innovation projects at the expense of higher-risk, breakthrough innovation projects. This risk is heightened if an organization over-weights the use of financial tools, such as net present value or/and economic value-added calculations, which are much easier to do for small, short-term projects than they are for large, long-term ones. As the use of such tools became more commonly used in the 1990s, Cooper [227] found from several surveys that between the 1990s and 2000s, there was an 80% increase in projects aimed at improvements and modifications to existing products, at the expense of projects aimed at developing new products and new product lines, which decreased by about 44% and 30%, respectively. In other words, organizations were increasingly abandoning all but the "sure thing" projects, which guarantees mediocre, at best, technological innovation performance. In order to avoid this trap, Cooper suggests specifically allocating resources to categories of projects having different degrees of projected innovation, hence risk, in accordance with an organization's overall strategy [227]. As an illustration, Cooper et al. [227, 228] show how the innovation funnel and Stage-Gate processes can be managed in concert, using a portfolio management approach. A case study example is provided by Barbieri and Álvares [127].

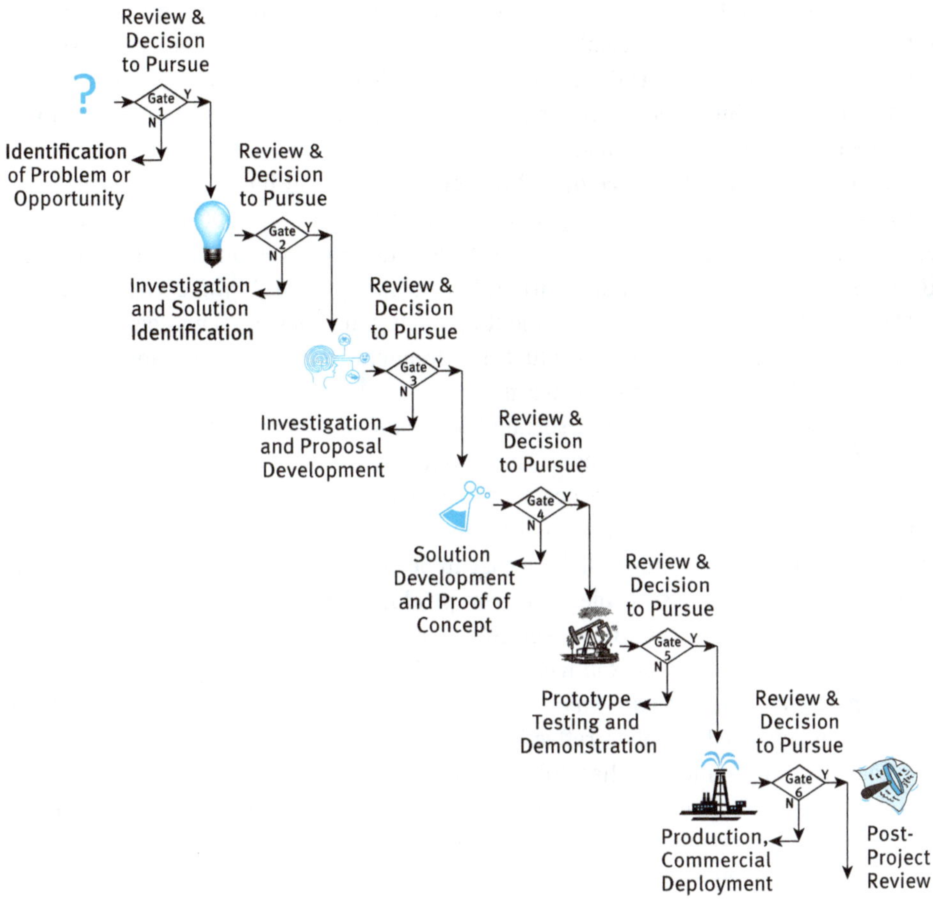

Figure 4.13: Illustration of a Stage-Gate® process for the development of a new petroleum industry process.

A related approach is the "*Technology Stage-Gate™ Process*" (TechSG process) [229], which is like the traditional Stage-Gate® process but designed specifically for technology development programmes and having more emphasis on managing through the development steps lying within the "*fuzzy front-end*" of the process – a reference to the early phase of new product/process/service development, where there is a very high degree of uncertainty about the nature of the future product, whether it will be technically successful, what it will cost in terms of time and resources, and whether it will be likely to be commercially successful.

The Stage-Gate® and Technology Stage-Gate™ processes are examples, but there are other approaches in the literature as well, including the "*waterfall*" method, Microsoft's "*5E*" framework, Bosch's "*innovation performance management*," and the "*agile*" method, among others [81, 183, 192, 193, 230–233]. Regardless of the methodology selected, when teams are established to take the concept from each commercialization step to the next, it is usually very important to ensure that some key personnel remain engaged throughout the entire process. This helps ensure a consistent line of sight from beginning to end, while ensuring transfer and consistency of information and linkages among different teams.

Failing forward. The pace of technological innovation has been increasing. According to Lewis [234], technology adoption rates in the year 2000 were already an order of magnitude faster than they were several decades previously, and as much as 10 times faster than 50–100 years earlier. Terms like *"fail fast," "fail cheap,"* and *"fail early"* are often used in discussions about managing the pace of technological innovation, especially with regard to product development, entrepreneurs, and start-up[54] companies. My favourite is *"fail forward."*

There are two key concepts associated with *"failing forward"* [235]. The first is not to fear failures but learn from them. The second key concept is that it's better to learn quickly if a new idea, product, process, or service isn't going to be successful so that it can either be improved or abandoned in favour of a new one. People aren't usually encouraged to fail at anything, and it's easy to become demoralized by a failure, but a failed experiment, process, project, or even product can present a great learning opportunity. In my own career, I've taken solace in a reflection by Danish physicist Niels Bohr, that *"an expert is a person who has found out by [their] own painful experience all the mistakes that one can make in a very narrow field."*[55] Thomas Edison would surely have approved of Bohr's reflection.

Learning why something has failed can be a critical part of the inventive process, and it has led shrewd researchers to develop some amazing inventions and technological innovations. For example, in 1953 at the Rocket Chemical Company in the United States, Norm Larsen and two colleagues were trying to develop a water-displacing rust-prevention coating for use in the aerospace industry. Their attempts failed 39 times, but because they were persistent and learned from their failures, their 40th attempt was successful and also gave them the name for the new product: WD-40 (i.e. water displacement product, "40th try") [65, 236]. Comedy and Grama provide additional examples [237].

The American industrialist Henry Ford wrote: *"Failure is only the opportunity to begin again, this time more intelligently. There is no disgrace in honest failure; there is disgrace in fearing to fail"* [238]. More recently, Steve Jobs was quoted as saying, *"You've got to be willing to fail. You've got to be willing to crash and burn"* [237]. It's not that a failure of any kind is some kind of goal – of course it isn't – the point is to

54 A start-up company is a new, or relatively new, company that is still in an early stage of development and may not yet have well-developed products, services, markets, or sales.

55 Quoted by Edward Teller in "Dr. Edward Teller's Magnificent Obsession," by Coughlan, R., *LIFE Magazine*, 6 September 1954, p. 62.

determine as early as possible if something isn't going to work, accept it, learn from it, come up with a new or revised approach, and move forward.

In product development, testing a new product/process/service idea and finding that it doesn't resonate with the market should be viewed as a valuable step forward. The learnings from such an experiment can be taken into account when designing and developing the next product/process/service, which can then be taken back to the marketplace for re-testing. Viewed as a technological innovation process, such early prototypes don't need to be viewed as failures so much as learning events along the pathway to an ultimate success (see the Thomas Edison quote related to this, in Section 2.5).

Like all tools, this one needs to be used with care. Experimenting at any stage of R&D costs time and money, and too many failures can make the total development cost-prohibitive, destroy projects and even entire companies. So, while there can be advantages to "fail fast" and "fail often," one might hasten to add "fail small" and "fail cheap."

Failing forward can mean designing and implementing small tests in order to develop resilience to failures, learning from them, being adaptable, and overcoming them without risking an entire initiative (or enterprise) on any of them. In product development, for example, an alternative to developing full product/process/services at an early stage is to develop "looks-like" or "works-like" prototypes that can be shown to prospective customers in order to get feedback. This can be a way to reduce the costs associated with each development cycle.

It can actually be helpful to experience one or more small failures from which useful learnings can be obtained, increasing the probability of achieving a significant technological success, rather than risking a single large failure that might be disastrous.

"*Failing forward*:" Don't fear failures, recognize them and learn from them, preferably at as early a stage as possible. It's better to learn quickly if a new idea, product, process, or service isn't going to be successful, so that it can either be improved or abandoned in favour of a new one.

Beyond the NPD process. As new products, processes, and/or services are developed within an organization, it sometimes makes sense to create an enterprise unit as a way to nucleate a mini-internal-company[56] that, when mature can either be operated as a subsidiary, spun-off, or spun-out. The enterprise unit might be used to focus on a single prospective innovation, or a bundle of associated prospective innovations. The process of creating and nurturing enterprise units is sometimes termed internal corporate venturing (ICV). An Amazon case study is described in reference [239]. See also Section 8.3.

---

56 Enterprise units usually have independent operations including strategy, business planning, management, financial system, culture, marketing, and sales functions.

4.9 Risks and paradoxes of innovation

The NPD process involves assessing a number of what might be called conventional risks, such as technological risk, legal risk, financial risk, commercial and market risk, and regulatory risk [193, 226]. There are also quite a few concepts of risk (or paradox) that are specific to innovation. These generally refer to either the pros and cons of pursuing technological innovation in the face of organizations or products/processes/services that are already successful ("*innovation system paradoxes*"), or else the challenges that may be inherent within the process of innovation itself ("*innovation process paradoxes*"). As such, they are usually more about risks than paradoxes in a literal sense, but the literature references are most often to "paradoxes."

Some examples of innovation system paradoxes include the following:

- The "*success/failure paradox,*" in which an organization's single-minded focus on "success" (and lack of tolerance for "failure") can blind it to the fact that it actually needs to be able to tolerate, and even embrace, failures in order to achieve success through technological innovation. This is termed the "*innovation paradox*" by Farson and Keyes [240], who argue that it could be limiting or counterproductive to even think in terms of successes and failures.
- The "*success paradox,*" in which an organization becomes so successful, through technological innovation, that it becomes blinded to the need for more or new innovations, causing it to miss out on breakthrough innovations that, to its ultimate detriment, are left to other, more nimble companies. An example is given by Nokia, which famously killed its smart phone because they were so deeply invested in "dumb phones." This is termed the "*innovation paradox*" by Davila and Epstein [241]. Another version of this is the paradox an organization faces when it recognizes that developing a breakthrough innovation could lead it to future success but that the pathway could involve destroying its currently successful products, processes, or services. This is termed the "*innovator's dilemma*" by Christensen [79].

Some examples of innovation process paradoxes include the following:

- The "*cultification paradox,*" in which the single-minded pursuit of innovation is taken to such extremes that it over-rides focus on other key activities such as sales and customer focus (i.e. too many pilots, too many specification changes, and too many new products) or that it causes changes to be made "for the sake of innovation" where there is no need for innovation at all (i.e. changing processes that are already effective and efficient).
- The "*collaborative compromise paradox,*" in which multiple groups are brought together to collaborate on a complex innovation challenge but the competing

needs and priorities of such groups lead to compromises that reduce their combined effectiveness, making the whole less than the sum of its parts, rather than more.

- The "*invention within convention paradox*," in which inventions that could lead to innovations are developed but then cannot be developed further because the organization's operational processes are not flexible enough to accommodate changes. This can create what Khan terms "innovation dissonance"[57] [242].
- The "*novelty paradox*," in which novelty is needed in product/process/service innovation, but a novel new product or a novel change to an existing product may not be perceived as an improvement by customers. Even worse, the introduction of a new or changed product is often accompanied by discontinuation of the previous one, which could lead to the loss of existing customers who are unable to purchase the old product and do not want the new one [8].
- The "*productivity paradox*," in which the problem that investing in the development of a new product, process, or service with an expectation of increased productivity may lead to very positive results in other areas but either no increase, or even a loss, in productivity itself. An example is the "*Solow computer paradox*": that increased investment in information technology could cause labour productivity to decrease instead of increase.[58]
- The "*organizational paradox*," in which an attempted technological innovation fails or performs poorly in the marketplace, or an attempted organizational innovation fails, and the impacts on the organization, such as supply chain problems or reduced spending budgets leading to reduced quality control (or reduced customer service), for example, lead to the loss of existing customers [8]. This could be considered to be "*negative innovation*."

57 The tensions that can arise among people and teams when an organization is trying to encourage and promote technological innovation and encounters any of the "innovation paradoxes" by which the drivers of innovation come up against elements of excess, competition, organizational momentum, or even strategy. Depending on the nature and degree, such tensions could be considered to be productive or unproductive.

58 In the early 1970s, many organizations began investing in computer systems with the expectation that they would enable labour productivity improvements (or even eliminate labour completely in some areas). However, although computers and information technology enabled many companies to achieve competitive advantage and market share, the desired labour productivity increases were largely unrealized, and in some cases, decreases were experienced.

5 Innovation ecosystems

Beyond the people and the organizations involved in technological innovation, there are also communities and clusters. There is evidence to suggest that, although not necessary for innovation, businesses can improve their innovation performance by collaborating with other organizations [112, 243–245]. A Conference Board of Canada study concluded that the share of new products in overall sales is significantly greater for companies engaged in collaborations than that for companies that are not, and that the former companies are more likely to introduce disruptive innovations [243]. An OECD study concluded that the difference can be as much as a factor of two [244]. Furthermore, several studies suggest that collaboration improves innovation performance, regardless of the nature of the collaborating partner (whether it is another company, university, government lab, RTO, other intermediary, or service company) [112, 243–245].

The term "*innovation ecosystem*" was coined by analogy with certain biological ecosystems, such as coral reef ecosystems, that provide a habitat conducive to fish, plants, and other marine animals. An innovation ecosystem, then, should enable organizations to practise open innovation, pool technical resources, achieve economies of scale, and gain synergies. Ideally, a healthy innovation ecosystem would provide an environment conducive to the development of technological innovations on the part of small-, medium-, large-, and multi-national companies. Depending on the circumstances, an effective innovation ecosystem could comprise a region, country, continent, or the globe (Figure 5.1).

The rate of technological innovation in a region or country is thought to be primarily related to seven major factors, all of which relate to the broader community within which such innovation takes place [43]. Governments seeking to improve the rate of technological innovation in their region or country typically focus on improving some or all of the following:

- *A competitive environment with demanding customers and/or clients*, without which there is little need for technological innovation. Although technological innovation is firmly rooted in the culture of some companies, others only innovate because they have to in order to survive. Furthermore, higher rates of technological innovation tend to be associated with higher levels of competition in an industry or region [43].
- *R&D infrastructure*, including academic research programmes and infrastructure, RTOs, industrial R&D organizations, linkages among research institutions and industry, and access to technology and technological expertise (since most technologies that are adopted in a region or country were not developed in that region or country).
- *Access to capital*, including informal investors, angel investors, government grants or loans, debt financing, institutional investors (venture capital), and pub-

https://doi.org/10.1515/9783112206720-005

Figure 5.1: Illustration of a global innovation ecosystem. (Courtesy pxhere, image #1575603, CC0 public domain.).

lic equity financing. Such sources of capital can provide the knowledgeable patience needed to span the many years involved in bringing a technological innovation through to commercialization.

- *Access to skilled workers* throughout the innovation continuum, from scientists, engineers, and technologists; to production, marketing, and sales staff; to entrepreneurs, business managers and leaders; and so on.
- *Access to information* of all kinds, not the least of which includes information about the science, technology, and engineering knowledge base, including previous technological ideas, inventions, and products, processes, and services. Lack of access to such information can obviously pose a hindrance or barrier to innovation or lead to *"reinventing the wheel"* – recreating things that have previously been created by others. In the context of technological innovation, the term "information" can be somewhat too general and even misleading. As Pavitt notes, "*Most technical knowledge turns out not to be 'information' that is generally applicable and easily reproducible, but specific to firms and applications, cumulative in development, and varied amongst sectors in source and direction*" [19].
- *Supply chains*, including technology suppliers, can be accelerators of technological innovation. For example, the presence of large companies in a region, whether or not they are part of regional clusters, can drive not only the creation of supply chains but also innovation within those chains. It has been observed

that *"major companies within a supply chain often drive innovation by requiring their suppliers to reduce costs, improve quality, or adopt specific supply chain management processes and systems"* [43].

- *The general environment and other supporting infrastructure.* This refers to the extent to which a political, legal, financial, and social environment exists that is conducive to innovation. Some factors that can either encourage or discourage business innovation, investment, and activities include: the general business environment, inflation, interest rates, taxation policy, the regulatory environment, the legal environment, trade barriers, sources of energy and raw materials, and transportation systems. Although none of these typically drive technological innovation, any or all of them can provide help, hindrance, or barriers.

5.1 Innovation ecosystem and multi-dimensional innovation models

There are two key aspects to innovation ecosystems: the nature of the components, described in Section 5.2, and the degree to which they interact with each other. The latter aspect, in a general sense, will be explored first and used as an introduction to the former aspect.

A sub-set of an innovation ecosystem is an *"innovation system,"* which is a network of innovation ecosystem people and/or entities that interact with each other to develop, enable, and/or produce innovations. It has been argued that such interactions are crucial to successful regional and national innovation systems [245, 246]:

> *"Within a firm, for example, successful innovation results from . . . a process of interactive knowledge-creation, in which – for example – skills in marketing are used to channel relevant information about user needs into the development processes which shape the technical and performance attributes of products. In this context, innovation is far from being simply the 'transfer' of knowledge which has been developed elsewhere. A second interactive dimension follows [in that] . . . successful innovative firms are usually those which are open to their environments. That is, they engage in interactive learning involving other institutions: partners, rivals, and a wide range of other knowledge-creating and knowledge-holding institutions."* Smith et al. (1995) [244].

A very narrow innovation system model involves a primary company and its suppliers. A slightly broader model would include its complementors as well (see Figure 5.2). Several examples of such a system are given by Adner and Kapoor [247] and Kerzner [122], including the following illustration from Adner and Kapoor:

> *"Consider, for example, Airbus's monumental investment in pioneering the super-jumbo passenger aircraft with its A380 offer. Airbus, as the focal firm, faces substantial challenges in designing and manufacturing the core airframe of the airplane. Beyond this internal challenge, it also relies on a host of suppliers for subassemblies and components. Some of these suppliers are themselves confronted with significant innovation challenges to deliver components that meet Airbus's require-*

ments (e.g., engine, navigation system), while others will not need to innovate at all (e.g., carpeting) . . . Airbus faces the additional challenge of integrating these components with the airframe in order to deliver a functioning aircraft to its airline customers. In order for the aircraft to be used productively by airlines, however, a number of other actors in the environment, outside of Airbus's direct supply chain, confront additional innovation challenges as well. Complementors such as airports need to invest and develop new infrastructure to accommodate the oversized aircraft, regulators need to specify new safety procedures, and training simulator manufacturers need to develop new simulators on which aircraft crews can be trained . . . The key point is that it is not enough to consider whether and how Airbus will successfully resolve its internal innovation challenges; in order for the A380 offer to create value, all of the other ecosystem partners need to resolve their own innovation challenges as well."

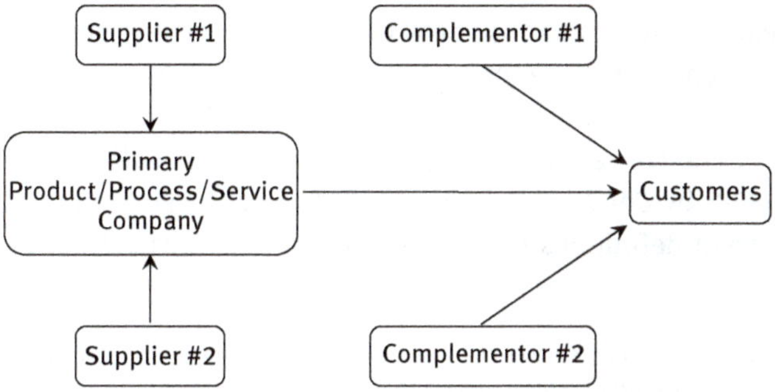

Figure 5.2: Illustration of a simple innovation system model involving a primary company and its suppliers and complementors. Adapted from information in Adner and Kapoor [247].

Viewed more broadly than simply primary companies and their suppliers and complementors, there is an "*innovation system theory*," which is a hypothesis that a nation's or region's innovation performance depends upon the breadth and depth of their innovation ecosystem and on the quality of the relationships and interactions among the innovation ecosystem entities. A corollary is that entities like government departments and agencies, RTOs, universities and colleges, businesses, and even entire industries in a country or region need to interact effectively with each other in order to stimulate and/or enhance their innovation system performance [248].

It was noted in Section 3.5 that in considering families of overlapping S-curves, multiple processes and/or steps become parallel and the two-dimensional, non-linear models become multi-dimensional. Some examples of multi-dimensional models of innovation are provided by "*innovation ecosystem models*," which focus on the interacting roles of multiple institutions in the process of creating technological innovations. These include the "*Holy Trinity*" *model*, *triple-helix model*, *quad-helix model*, and *N-tuple helix model*, among others. Of these, only the first three will be discussed further here.

One of the key concepts in ecosystem models is that a society has several broad kinds of resources involved in its development and growth: human resources, technology, capital, consumers and the marketplace for goods, processes, and services [64]. Another key concept is that society develops and supports several broad kinds of organizations that deal with aspects of the above kinds of resources: governments or regions, universities, intermediary organizations, and industry. The models have been developed in attempts to describe how discovery, knowledge-generation, invention, product development, market launch, and so on, depend on interactions among institutional players, involving elements of control and/or incentives, knowledge generation, and wealth generation.

There is also a broader concept: that of regional and national innovation environments, or "*systems of innovation*," which can involve institutions, society, governments, etc. Here again, the concept is that the innovation process involves flows and iterations of interactions, knowledge, and technology among people, businesses, and other organizations [245, 249]. The concept of a "*national system of innovation*" was developed in the 1980s and 1990s by Christopher Freeman [250] and Bengt-Åke Lundvall [120, 251, 252], and a range of definitions have been proposed [245, 246, 249–251] (see also Section 11), such as:

> "*A national or regional innovation system comprises the network of public- and private sector innovation-related institutions and their mutual interactions*" [2].

The concept of regional innovation system (RIS) appears to have emerged in the late 1980s and early 1990s [246, 248, 253–256]. The character of such RIS varies widely depending on the nature of the region and the depth of its pool of public and private innovation resources.[59] In general, an RIS will aim to achieve an optimal mix of collaboration and competition within its key existing or emerging industries, which requires identifying which activities are best conducted collaboratively and which are not. A wide range of approaches has been used worldwide. China, for example, has used the RIS approach as part of the catalyst for an innovation economy and the rebuilding of their economic power [257], which has risen exponentially since the 1980s. A simplified taxonomy for RIS is illustrated in Table 5.1 [253].

59 Clusters could be considered to be a special kind of regional innovation system [163] (see also Section 5.3).

Table 5.1: A simplified taxonomy for regional innovation systems. Adapted from information in reference [253].

Regional innovation system type	Technology transfer initiation	Funding	R&D competence	Technical specialization	Degree of system coordination
Grassroots	Local	Diffuse, mostly local	Mostly very applied R&D	Low, geared to problem-solving	Low
Network	Multi-level: local, regional, external	Guided by companies, banks, government	Discovery research plus applied R&D aligned with companies	Flexible, wide ranging	High, with many stakeholders
Dirigiste	External to the region	Centrally determined by government	Mostly discovery research	High	High, by government

Regional innovation system type	Innovation drivers	R&D depth	Degree of associations
Localist	Mostly driven by small companies	Limited, few public R&D resources	High, involving entrepreneurs and regional government
Interactive	Balance of drivers from large and small companies	Mix of public and large company resources	High, involving regional industry networks and associations
Globalized	Dominated by global companies	Determined by large company resources	Low, unless led (and directed) by large companies

The "Holy Trinity" model. The *"Holy Trinity" Model* is an RIS model advanced by Michael Storper in 1997 to describe the roles and intersections of territories, technologies, and organizations in linking regions, technologies, and production capabilities into innovation systems that generate new products and economic development [258]. Storper termed this way of thinking about economic development (and innovation) the *"heterodox paradigm."*[60] The first-order interactions in this model would produce economic technological systems, regional technology systems, and regional productions systems, all of which, combined, should lead to a stream of new products entering the marketplace (see Figure 5.3).

60 Heterodox economics refers to approaches to economics that go beyond, or at least fall outside of classical or conventional economics, such as socio-economics or eco-economics.

Triple-helix model. The "*triple-helix model*"[61] was developed by sociologists Henry Etzkowitz and Loet Leydesdorff in 1994 to describe the roles and intersections of governments, universities, and industry in advancing knowledge-based economies[62] (KBEs) [259–262]. This model has been used to explain (or construct) the inner workings of RIS and clusters (see Section 5.3). By extension, the model has also been applied to the advancement of economies based on technological innovation (the "*innovation-related triple-helix*"),[63] in which case:

- Governments provide a regulatory framework, and S&T policies, programmes (including funding programmes), and infrastructure (such as research parks or incubators).
- Universities provide[64] new knowledge, understanding, and technology.
- Industry provides production and deployment of the innovations into the marketplace.

In the triple helix model, the innovation process is considered to be more like a spiral than a straight line or S-curve, and within the spiral, there are multiple points of intersection and interaction among the three kinds of organizations; hence the imagery of a triple helix (Figure 5.4). For a case-study example, see Chapter 5 in reference [263].

Etzkowitz coined the term "*innovation in innovation*" to describe the broadening of the term "*innovation*" to include improvements in the way organizations interact in order to create the conditions for innovation, and/or to enable innovation to occur [262].

The triple helix image is intended to reflect several features:

- Each of the three kinds of organizations is independent but interacts with the others, both individually and collectively.
- Each has a certain degree of dependence on the others.
- Each has the potential for conflict as well as cooperation with the others.
- Innovation can be enabled, advanced, and/or created at the multiple points for which the activities of all three intersect.

As others have pointed out, the triple-helix model is an oversimplification and has some basic shortcomings [264]. Among the shortcomings are:

---

61 Also termed "*Triple-Helix Field Theory*," "*TH Theory*," or the "*ABG Model*," where ABG refers to academia, business, and government.

62 In this context, knowledge-based economy (KBE) is distinguished from information-based economy and refers to the development of knowledge workers, embodied knowledge, and also codified knowledge. Two other principal economic models are the market-based and political economies.

63 There is also a sustainability-related triple-helix in which the public-at-large is substituted for industry, to represent societal concerns about technologies, industry, the environment, and sustainability.

64 In this connection, greater participation by universities in working with governments and industry to enable economic growth (via innovation) and social progress has been referred to as the "third mission" (after teaching and research) of an "Entrepreneurial University."

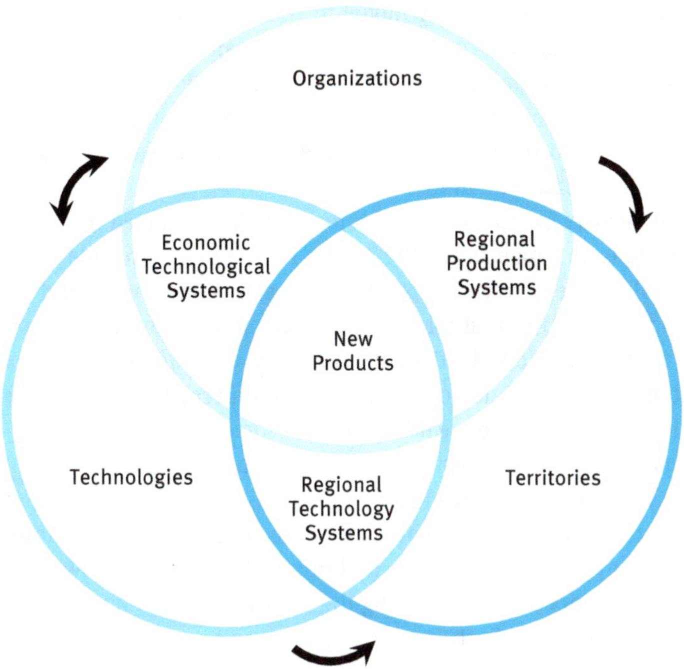

Figure 5.3: Illustration of the Holy Trinity model of a regional innovation system. The arrows are drawn to illustrate primary linkages [258].

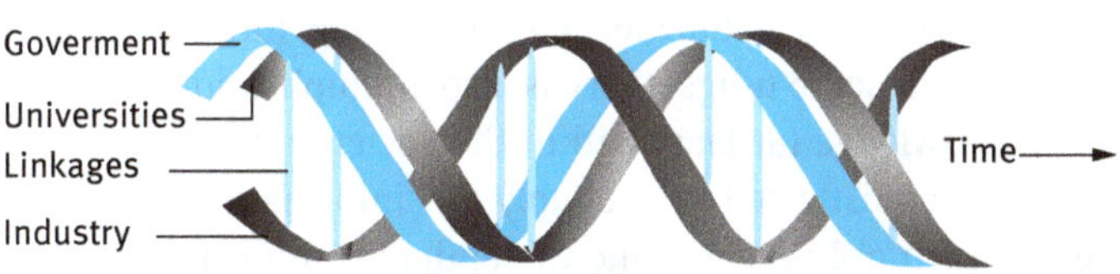

Figure 5.4: Illustration of the triple-helix model of a regional innovation system. The vertical bars are drawn to illustrate the existence of continuing linkages along the development pathway.

- Real-world innovation is actually a much more complex process.
- The model does not consider the role of intermediary organizations, particularly RTOs in the innovation continuum.
- The roles of entrepreneurs, start-ups, and SMEs are much more often fulfilled by independent organizations within the innovation continuum than they are by sub-units within universities.

Quad Models. The addition of a fourth kind of institution to any of the three-component models can improve their sophistication and explanatory power. This is the same concept that underlies the characterization of *Fourth-Pillar Organizations* (see Section 5.2).

Wilson's "*quad model*" is like a Holy Trinity model to which a fourth segment has been added, comprising resourceful organizations[65] that can act as catalysts, intermediaries, and/or entrepreneurs (hence the term "*quad leaders*" [264, 265]). Wilson's four kinds of institutions are governments, research institutions (universities and RTOs), non-government organizations, and industry. An adaptation of Wilson's model is governments, universities, intermediaries, and industry.

Similarly, the "*quadruple-helix model*" (or "*quad-helix model*"), developed by Etzkowitz, Leydesdorff and others, is an expansion of the Triple-Helix Model to include intermediary organizations [266]; hence, governments, universities, intermediaries, and industry. In any of the quad models, the multiple points of intersection and interaction among the four kinds of organizations lead to the imagery of a quadruple helix (Figure 5.5). In terms of the three main functions of an innovation ecosystem – controls and/or incentives, knowledge creation, and wealth creation – some illustrations of first-order interactions among the four kinds of institutions are also shown in Figure 5.5.

From an innovation system point of view, knowledge creation is pointless without knowledge flow, and the helix metaphor encompasses the main knowledge flows among the entities in an innovation ecosystem, including diffusion of knowledge among the entities and movements of people among the entities. It is not an accident that the most sophisticated models for both innovation systems and innovation processes (Section 3.1) involve such parallel processing.

The "*Knowledge-Based Economy*"[66] (KBE) is an economic model that refers to a society whose economy is largely based on a combination of knowledge workers, the creation, dissemination, and utilization of knowledge (embodied knowledge and codified knowledge), and the translation of such knowledge into learning, innovation, and economic development [267, 268]. In some usage, there is also an implication of broad access to and sharing of the knowledge (at least within the society under consideration). A KBE is therefore more sophisticated than simply an "information-based" or "learning-based"[67] economy (which only creates and disseminates information) or a "market-based" economy and also different from a "political" economy.

Example. The manufacturing sector provides an example for which technology diffusion is critical. In this sector, technology diffusion traditionally occurs through the dissemination of new equipment and machinery – a process that can be extremely slow, as is the process of adopting and adapting technological innovations developed elsewhere. Participating in an innovation sys-

---

65 The original definition focussed on individuals [260]. It has been broadened here to include individuals and/or intermediary organizations.

66 Also termed "Knowledge Economy" or, more broadly, "Knowledge Society" (in which economic health and growth can be translated into social health and growth).

67 According to Lundvall [252]: "*The learning economy concept signals that the most important change is not the more intensive use of knowledge in the economy but rather that knowledge becomes obsolete more rapidly than before; therefore, it is imperative that firms engage in organizational learning and that workers constantly develop new competencies.*"

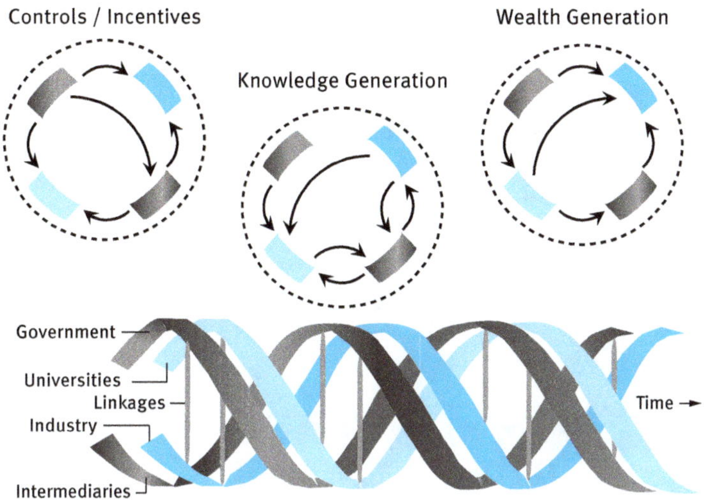

Figure 5.5: Illustration of the quad-helix model of a regional innovation system. The vertical bars are drawn to illustrate the existence of multiple and recurring linkages among the innovation system entities, along an evolutionary pathway.

> tem can both improve and speed-up such processes by virtue of constant interactions with customers, suppliers, competitors, and intermediaries. This is particularly important for companies that are not themselves R&D performers or purchasers.

The helix metaphor illustrates another critical flow in innovation systems, the flow of personal interactions and the knowledge that is exchanged via such interactions. Again, this applies to all of the entities in an innovation system, and beyond the specific, specialized kinds of knowledge that can be transferred are the more general kinds of knowledge, such as approaches to idea generation or problem solving, for example. The ability to access and engage with diverse networks of people is a highly valuable knowledge asset and can enhance any organization's ability to adopt and adapt technology.

In the quad-helix model, intermediary organizations would be the primary incubators for many technology-based start-ups and most SMEs. RTOs as intermediaries would be providers of applied R&D, piloting, demonstration, and commercialization assistance to organizations of all sizes, from SMEs to multi-national enterprises (MNEs). Figure 5.6 illustrates the positioning of RTOs as intermediaries in the quad-helix model.

The quad models illustrate some of the kinds of interactions that are inherent in a KBE. S-curves (see Section 3.2) have been used to describe phases within which a community's organizations and resources begin to both grow and interact supportively and synergistically, then evolve to a rapid growth phase, then a plateau phase,

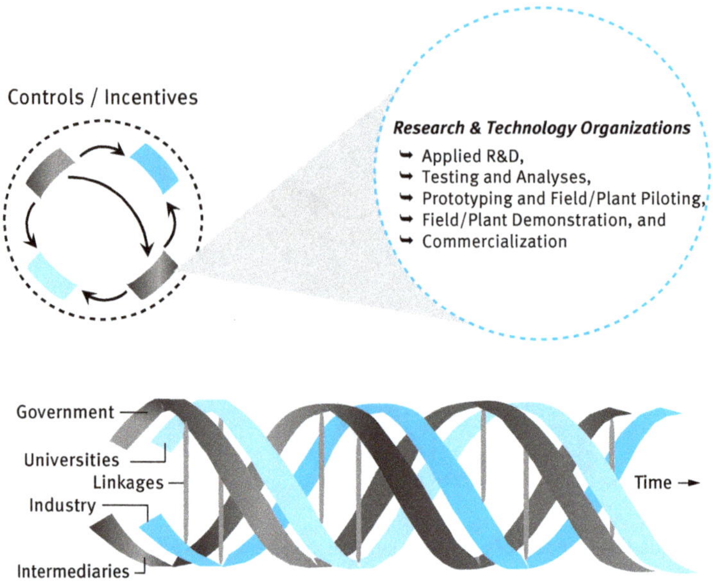

Figure 5.6: Illustration of the positioning of RTOs as intermediaries in the quad-helix model. The vertical bars are drawn to illustrate the existence of multiple and recurring linkages among the innovation system entities, along an evolutionary pathway.

and finally, possibly, failures leading to a regression [64]. Such a sequence illustrates the "*generative principle*" of KBEs.

The "*quad-helix model*" is an example of a sophisticated regional innovation system model, in which governments, universities, intermediaries, and industry dynamically interact to enable and support technological innovation processes.

Other Models. Although the quad-models represent an improvement over the triple-helix model, there are, of course, more than four possible kinds of contributors to a multi-dimensional, dynamic innovation system (or process). There are models for this as well. For example, Ahonen and Hämäläinen have suggested a penta-helix model in order to include a distinct role for communities (separate from government) [269]. In another example, the N-tuple helix model was developed by Etzkowitz, Leydesdorff, and others as an expansion of the triple-helix and quad-helix models to include additional players, beyond governments, universities, intermediaries, and industry [266, 268].

With the above models of innovations systems in mind, it can be worthwhile to revisit some of the more specific models of innovation itself, such as those of Kline and Rothwell, discussed in Section 3.1.

5.2 Innovation ecosystem entities

An innovation ecosystem includes all of a region's public and private sector people and organizations whose activities include any or all of developing, enabling, producing, or diffusing innovations. Some of these have been introduced in Section 5.1. Examples of innovation ecosystem entities and some examples of their roles in the system include the following:

- Academic organizations: universities and colleges that develop people's intellectual capacity and skills and which undertake discovery research aimed at new knowledge and understanding.
- Intermediary and fourth-pillar organizations that connect industry with knowledge, service providers, and partners and help industry to overcome barriers during the technological innovation process.
- RTOs that go beyond the above roles and participate with industry in applied research, development, demonstration, and commercialization.
- Government organizations providing funding and/or tax credit programmes for R&D, sometimes funds for infrastructure, and sometimes providing government research laboratories, which may themselves conduct discovery and/or applied research.
- Financial organizations providing sources of financing to assist the industry in the development of businesses and/or business products, processes, and services.
- Individual inventors and entrepreneurs.
- Companies and industries.

These people or organizations constitute innovation ecosystem entities because of their capabilities and their potential to contribute to processes of innovation, regardless of whether or not they actually make such contributions, and regardless of whether or not they interact with each other. Some of the benefits that can be realized through industrial collaboration with these entities are summarized in Table 5.2.

Table 5.2: Some benefits of industrial collaboration with other partners [120, 243, 249, 252].

Partner	Benefits
Other industry	– Cost reduction (sharing) – Risk reduction (sharing) – Access to new markets
Universities	– Strong discovery research capacity – Leading-edge, specialized expertise – Cost reduction (government leverage) – Access to specialized infrastructure – Access to new employees (graduates)

Table 5.2 (continued)

Partner	Benefits
Government labs	– Strong applied research capacity – Leading-edge, specialized expertise – Access to specialized infrastructure – Cost reduction (government leverage) – International science linkages
Research and technology organizations	– Strong applied research capacity – Leading-edge, specialized expertise – Access to specialized infrastructure – Cost reduction (government leverage) – Ability to manage large projects – Strong development engineering capacity – Prototype development, piloting, scale-up, and demonstration capacity – Business incubation capability
Service companies	– Strong specialized product/process expertise – Strong engineering design capacity – Pilot-testing and demonstration capacity

In industrial sectors, a "*business ecosystem*" (or "*virtual cluster*") is an organized network of organizations that collectively support a product, process, or service business. The anchor and/or leader of a business ecosystem is termed the "*ecosystem leader*" (or "*keystone*" or "*platform leader*") [270]. The other ecosystem organizations are termed "*niche players*" (or "*complementors*"), whose products and services contribute to the productivity and outputs of the greater ecosystem.

Wal-Mart, Apple, and Mozilla, for example, have been ecosystem leaders for their respective business ecosystems. A business ecosystem is sometimes referred to as an *innovation ecosystem*, but this term is better used to refer to the broader meaning of the important entities involved in technological innovation (i.e. industry, academia, intermediaries, and government).

Academic organizations. The "*first mission*" of a university is teaching, while the "*second mission*" is research. The evolution of universities to have a second mission has been referred to as the result of the "*first academic revolution*" from the teaching university. Much of the world's new knowledge and understanding come from the basic (discovery) research (also called "*mode 1 research*") conducted at academic institutions. Almost all of the knowledge developed by universities gets published in the public domain and is fairly readily accessible for anyone in the world to search, access, and use, but it appears that much, if not most, of these results are only used by other researchers. Nevertheless, discovery research is a necessary component of an innovation system because scientific and engineering knowledge and advancements form a base of knowledge and understanding from which other things can be developed. However, such knowledge does not itself drive technological innovation, as discussed in previous chapters.

Although universities do sometimes support applied R&D, especially in support of the potential commercialization of their own inventions, such activities represent a very small portion of that done by others in the innovation ecosystem. This is partly a question of strategy and focus, and partly because university researchers generally tend not to have the interest, the resources, or the broad understanding of industrial processes, practices, or needs to be able to conduct successful applied R&D. Nevertheless, as the principal performers of discovery research, the research-intensive universities produce not only an ever-expanding body of basic knowledge, but also new scientific and engineering methodologies and instrumentation, and, of course, highly qualified people.

Some universities, however, have begun to engage more fully in the process of technological innovation and in innovation ecosystems. The so-called *"Entrepreneurial University"* is a university that has incorporated a *"third mission,"* beyond those of teaching and research, which is to engage with other organizations in the development of practical uses of new knowledge (also called *"mode 2 research"*). Such an entrepreneurial university can play a larger role in any innovation ecosystem (see Section 5.1), and such universities are frequently associated with research and/or technology parks, incubators, and/or start-up and spin-off companies (see Section 5.3). For universities that have transitioned to a "third-mission" strategy, reference is sometimes made to a shift from an *"endless frontier"* of discovery (Mode 1) research as an end unto itself, to an *"endless transition,"* in which discovery research is translated into applied (Mode 2) R&D and thence into commercial deployment and use [261]. Some case studies are presented by Owen et al. [271].

In some definitions (including that of OECD), the entrepreneurial university is also itself associated with an entrepreneurial culture and can play a role in helping to develop entrepreneurs, such as by operating business incubators [272] (see also the discussion under the heading *"Intermediary and Fourth-Pillar Organizations"* further below in this section). The evolution of the entrepreneurial university with its third mission has also been referred to as the result of the *"second academic revolution"* [262, 273]. Any university, but most particularly the entrepreneurial universities, can be a vital part of any regional or national innovation system.

To facilitate the transfer of the knowledge gained through discovery research to those that would undertake applied R&D, many universities and polytechnics have established some form of technology transfer and/or industry liaison office (often termed university-industry liaison offices, UILOs, or simply industry liaison offices, ILOs). The principal focus of such offices is to build links between academic researchers and industry and to negotiate licencing agreements in order to earn revenue. With universities' increasing need (or desire) to diversify and increase revenues, what started out as an innovation-enabling concept has increasingly become a barrier however. As Lundvall has warned, *"Private companies might, in the short run, appre-*

ciate that universities become more profit-oriented but they will soon experience that the barriers around the knowledge accumulated will become higher and that access to the most relevant knowledge will become more difficult" [120].

Intermediary and fourth-pillar organizations. In general, an intermediary organization functions between an organization and some or all of its stakeholders by providing some kind of service, such as a programme function or technical assistance. For example, brokers and dealers are sometimes referred to as "*technology market intermediaries.*" In the world of technological innovation, an intermediary organization[68] is one that functions between industry and the marketplace in a way that supports technological innovation processes. Many intermediary organizations are for-profit, such as brokers, dealers, patent and trademark agents, and lawyers.

Expanding on this concept, a "*Fourth-Pillar Organization*" is an organization that works to enable and/or assist with the innovativeness and competitiveness of companies by working with industry, government, and academia (which represent the other three "pillars" – see the discussion of "*quad models*" in Section 5.1.3) [274, 275]. Fourth-Pillar Organizations are usually either government-owned corporations or not-for-profit corporations, such as RTOs, industry associations, economic development organizations, business incubators, or science (or technology or business) parks (see Sections 5.1 and 5.3) [275]. These are sometimes termed "*Technology and Innovation Centres.*"

All Fourth-Pillar Organizations help to link partner organizations together, and they use these linkages to help companies develop and deploy new commercial products, processes, and/or services. The specific functions provided by an intermediary or Fourth-Pillar Organization could be almost anything that enables, assists, or leverages an innovation process, also called "*knowledge-intensive business services*" (KIBS),[69] such as [276, 277]:
- business and technology intelligence;
- foresight;
- information, knowledge, and technology gathering, processing, and transfer;
- management assistance, such as providing project management, troubleshooting, risk assessment, and strategic planning assistance;
- linking together other entities and/or facilitating networks;
- applied R&D, including co-development;
- testing, evaluation, and accrediting;

---

68 Also termed "*Bridging Organization*" or "*Innovation Intermediary.*"

69 *Knowledge-intensive business services*" (KIBS) can encompass almost any kind of business service that involves or is strongly reliant on sophisticated technological and/or other professional knowledge. The former, technological KIBS, is sometimes referred to as T-KIBS (e.g. science, engineering, and R&D), while the latter, professional KIBS, is sometimes referred to as P-KIBS (e.g. accounting, law, and marketing).

- lobbying governments (especially on behalf of industry associations);
- branding and marketing; and
- commercializing.

Some case study examples are provided in Section 10.2.

The process of conducting such activities is termed *"Innovation Intermediation."* At a minimum, intermediaries help businesses to overcome barriers during the technological innovation process. This is especially important for start-ups and SMEs (see, e.g. Makerspace in Section 11). Some case studies related to the work of intermediaries are listed in Section 10.2.

In general, intermediary organizations specialize in particular stages of the innovation development process, whereas RTOs (Section 5.2.3) tend to work across the entire innovation development spectrum, and this is reflected in the nature and breadth of expertise in RTOs versus other intermediaries. Some case study examples are provided in Section 10.2.

Example: Business Accelerators and Incubators. Business accelerators and incubators comprise organizations, facilities, and/or services in which modest (accelerator) or significant (incubator) amounts of capital, mentorship, and/or other services are provided to a business start-up or early-stage business in return for small (accelerator) to significant (incubator) amounts of equity. In both cases, the goal is to provide enough support for entrepreneurs to survive the *valley of death* period between initial financing and significant sales of the new business' products, processes, or services. The services may span a wide range from standard business services, such as legal, financial, recruitment, office facilities, and marketing, to custom services, such as access to angel and venture financiers and successful entrepreneurs, research, development, access to specialized facilities, or prototyping. Depending on the nature and maturity of the business, the time spent in an accelerator can be as small as a few months, whereas the time spent in an incubator could be several months to several years. For more information, see [278]. For some case studies, see [205, 279–282].

Some examples of intermediary organizations include the following:

- *Knowledge integration communities* (KICs). These are intermediaries based at the University of Cambridge but created by the Cambridge-Massachusetts Institute of Technology (MIT) Institute, which is a strategic alliance between the University of Cambridge and the MIT. The KIC mission is to enhance knowledge exchange between academia and industry in order to advance research and accelerate the pace of technological innovation. To accomplish this, each KIC draws on research teams at Cambridge and MIT in a collaborative manner that engages outside participants as well. The KICs also play a coordinating role in facilitating the innovation development projects. The KICs' revenues derive mostly from government, with some funding from industry partners for specific projects [283].
- *Technology Advanced Metropolitan Area Association* (TAMA). This is based in Tokyo, Japan, and is a government-supported intermediary organization whose mission is to strengthen university-industry linkages. The principal industrial ac-

tivity areas supported are electronics, machinery, transportation, and precision instruments, and the industrial companies concerned range from SMEs to large firms. TAMA's main activities are facilitating knowledge exchange and coordinating research consortia. In some cases, TAMA is able to help facilitate, flow, or even supply project funding. TAMA's revenues derive mostly from government, with some funding from industry partners for specific projects [283].

- The *Holst Centre* is a joint initiative of the Dutch and Flemish governments, based in Eindhoven, the Netherlands. Holst's main role is to facilitate linkages and partnerships between academia and industry, mostly in pre-competitive R&D (up to the proof-of-concept level) in the area of advanced microelectronics. Unlike RTOs, Holst does not normally engage in the co-development with clients of technological innovation. Holst's revenues derive mostly from the government, but with the aim of achieving 50% from industry partners for specific projects. Reference [283].
- The Trade and Business Development Body, *InterTrade Ireland*, based in Newry, is a government intermediary organization that aims to build networks, research programmes, and partnerships to assist SMEs. Some of its intermediary roles involve knowledge exchange; identification and development of new markets, product, or process development advice; accelerator assistance through a "voucher" programme; and joint marketing initiatives. In some cases, InterTrade Ireland is able to help facilitate, flow, or even supply project funding. InterTrade's revenues derive mostly from government, with some funding from industry partners for specific projects. Reference [283].
- *Concierge*[70] was a Canadian intermediary programme operated by the National Research Council of Canada (NRC) in the 2010s, under its Industrial Research Assistance Programme and in collaboration with other federal and provincial organizations [284]. Aimed primarily at SMEs, "*concierge*" provided a single point of access for innovation-related resources and support programmes available through Canadian governments, particularly the federal government. The service offered "free, one-on-one assistance" from advisors who provide customized guidance [284].

Research and technology organizations (RTOs). Whereas most intermediary organizations specialize in particular stages of the innovation development process, RTOs tend to work across the entire innovation development spectrum. RTOs are either government-owned corporations or agencies or private not-for-profit companies that are primarily focussed on developing and deploying practical technologies that address commercial marketplace problems or opportunities and spanning multiple sectors of an economy. RTOs are unusual organizations. They differ from academia,

70 Formerly the Canadian Technology Network, CTN.

mainstream government, for-profit companies, and even from mainstream not-for-profit companies.

The principal differentiators of RTOs are that they:
- Have "public good" missions and are not primarily profit-driven
- Are either government-owned corporations/agencies or private not-for-profit companies
- Are focussed on developing and deploying practical technologies through companies and other organizations
- Work in the public interest, using a business-like approach, and with a social and environmental agenda

RTOs specialize in applied science, development engineering, testing and analysis, and sometimes pilot testing, scale-up engineering, and even full-scale plant or field testing and demonstration. Additionally, RTOs specialize in bridging the "valley of death."

In his 2003 article entitled *"The Missing Link in Canada's Innovation Chain,"* Joseph Wright wrote about the under-supported but critical activities involved in translating new knowledge and ideas into commercial products, whereas *"[t]oo much of the Innovation Agenda funding has been directed at academic research . . . Linkages into the needs of the resource-based and more traditional sectors, that still effectively drive the Canadian economy, have not been acknowledged"* [285].

One of the most important functions of an RTO is to help business enterprises access, absorb, adapt, deploy, and exploit new technologies in order to enhance industries' ability to innovate. This can involve the ability to share and decrease R&D and commercialization costs and to share and reduce risks. Being "at the hub" and connected to virtually all other innovation ecosystem entities makes RTOs uniquely positioned to forge linkages and enable people and technology flows among the entire ecosystem (Figure 5.7).

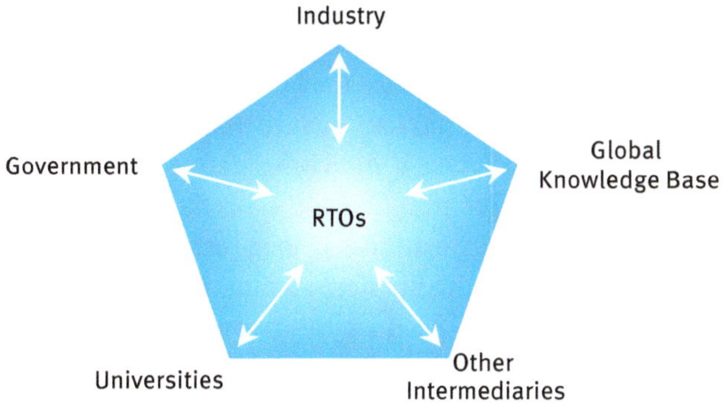

Figure 5.7: RTOs can work "at the hub" and connected to virtually all other innovation ecosystem entities, making them uniquely positioned to forge linkages and enable people and technology flows among the entire ecosystem. Adapted from EARTO [286].

The best RTOs focus their work in multiple sectors of the economy and provide research, development, demonstration, testing and analysis, and problem-solving services to small-, medium-, and large-sized enterprises. In doing so, they help businesses and industry access, adapt, deploy, and exploit new technologies, which helps them to achieve technological innovation [287] and, therefore, their competitiveness and sustainability. RTOs can help businesses to increase productivity, develop natural resources, develop and deliver new products/services, create and/or maintain jobs, achieve compliance with regulatory requirements, and improve socio/environmental sustainability. This is expensive, time-consuming, and risky work where failures are not uncommon and must be accommodated and overcome. As a result, the period of engagement of RTOs with businesses is generally longer than that of the non-RTO intermediaries discussed in the previous section. One of the ways that RTOs reduce their business risks is by listening closely to "the voice of the client," identifying industries' key challenges and opportunities, and using them to focus the RTOs' work.

RTOs help industries realize technological innovation in a number of ways. Some high-level examples include helping companies to improve safety; increase productivity; develop natural resources; develop and deliver new and improved products, processes, and services; create and/or maintain jobs; achieve compliance with regulatory requirements; and improve socio/environmental sustainability (Figure 5.8).

RTOs can also provide support for clusters, networks, and "networks of networks" [288]. Whereas intermediaries help businesses to overcome barriers during the technological innovation process, RTOs also help to decrease R&D and other innovation costs, reduce technological risks, and overcome internal constraints. SMEs, in particular, have reported [289] that the most important time for them to get help is once they get beyond the start-up stage, and that the two most important kinds of support they can gain are (1) accessing R&D resources and (2) helping the SME develop as a business.

RTOs generally earn some of their revenue from contract services, which are usually mostly provided to industries, and some in the form of "base" government funding. Regional and federal governments generally invest in RTOs' applied R&D, demonstration, and commercialization activities in part because they are direct users of the information and know-how developed, and in part to ensure that a reasonable breadth and depth of R&D will be conducted (even if industries cannot or will not support it on their own). As a result, RTOs have to balance the needs of industries with the needs of governments, whether short-, medium-, or long-term. A "balanced drivers" model tends to encourage and support the greatest amount of technological innovation and is, therefore, the one that tends to produce the greatest economic impacts and jobs (see, e.g. reference [290]).

An international benchmarking study of best practices for RTOs was conducted by the World Association of Industrial and Technological Research Organizations (WAITRO). They concluded that RTOs should receive unrestricted government "base" funding of 25% to 50% of total revenues, with a "best practice" level of about 35%

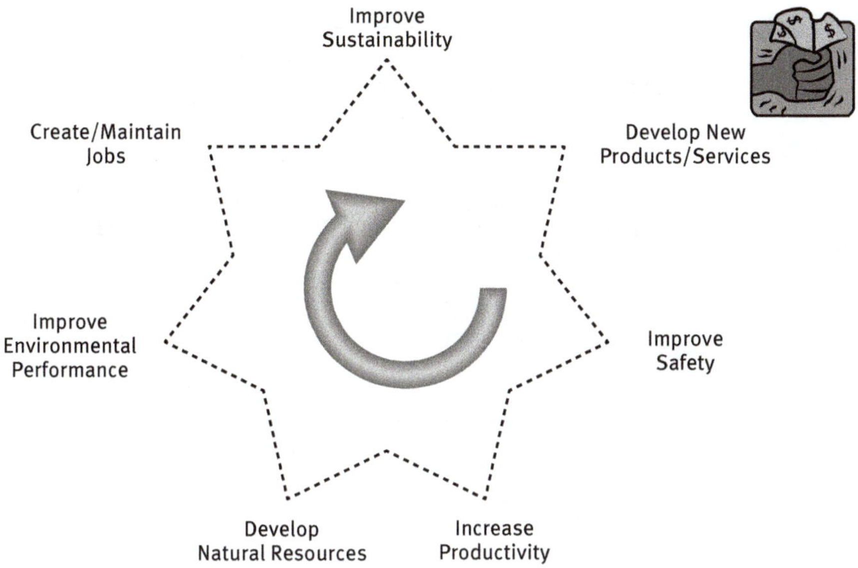

Figure 5.8: RTOs help industries realize technological innovation in many ways.

[291]. The WAITRO study found that when government investment falls too low, RTOs naturally tend to focus on industrial markets and their short-term needs at the expense of medium- to long-term needs. In this scenario, the amount of innovation enabled falls dramatically. Conversely, when government investment becomes too high, RTOs naturally tend to focus on governments' needs rather than industry's needs. In the extreme case of government funding (90%–100%), RTOs tend to focus on just "Blue-sky" discovery research. In these latter scenarios, the amount of innovation enabled falls dramatically. Essentially, the same results and implications were found in a subsequent study by Hauser [292]. These effects are illustrated in Figure 5.9.

There are hundreds of RTOs around the world. Some examples are the following (see also Section 10.1):

- *Fraunhofer-Gesellschaft* is a group of *Fraunhofer Institutes* in Germany. They have as their main mission "strategic and applied research for government and industry with a focus on the development of new technologies" [293]. The Fraunhofer Institutes mainly specialize in fields that are strategically important for Germany. They support companies of all sizes (including SMEs and MNEs). Some of the activities associated with the Fraunhofer Institutes are in knowledge brokering, applied R&D, the co-development of innovation, and facilitating commercialization. The Fraunhofer Institutes work with businesses, universities, other RTOs, banks, and government (see the description of the "*quad models*" in Section 5.1.3). The Fraunhofer's revenue mix is approximately 40% public and 60% contract research [283].

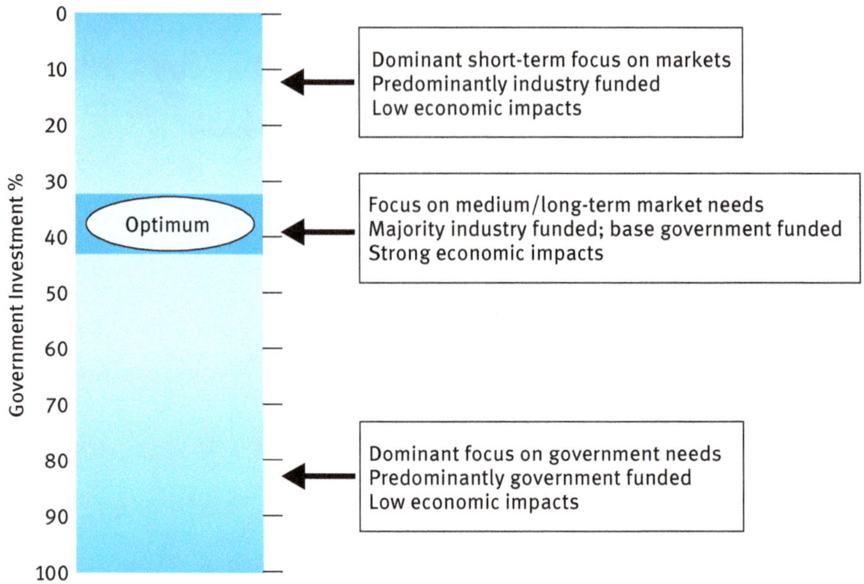

Figure 5.9: Illustration of the impacts of different levels of base government funding for RTOs.

- The not-for-profit *Inter-University Micro Electronics Centre* (IMEC) in Belgium focuses on micro- and nano-electronics, especially related to ICT. Initially modelled after the *Fraunhofer Institutes*, IMEC's mission is "*operating 3 to 10 years ahead of industrial needs and to foster the development of the local industrial base through spin-off creation, collaboration and training*" [283]. IMEC is multi-disciplinary and project-driven. Its intermediary roles include applied R&D and co-development of innovation, facilitating communications, brokerage, and network construction. IMEC's activities are focussed on the early stages of the development process rather than the entire innovation continuum, in contrast to most other RTOs. IMEC's revenue mix is approximately 20% public and 80% contract research.
- The non-profit *Industrial Technology Research Institute* (ITRI) is Taiwan's principal RTO. Its main activities span technology needs assessment, technology development, and technology commercialization (with a focus on international export markets) [248]. The main fields addressed have been electronics, machinery, chemical engineering, materials, energy, and refining. In recent years, ITRI has included an emphasis on providing services to assist in the creation of new enterprises via an "Incubation Center" [248].
- TNO, the *Netherlands Organization for Applied Scientific Research*, is an independent organization that seeks to enable business and government to apply knowledge in ways that create innovation [294]. TNO's main research and innovation activities are focussed in the areas of industry *per se*, healthy living, security and defence, low-carbon energy, and urbanization [294].

- *NRC* is an agency of the Government of Canada that provides innovation support, strategic research, and scientific and technical services. Among its functions is to provide industry with technical and commercialization services that enable lower-risk ways to develop innovative ideas, reduce start-up costs, and shorten time to market. An explicit part of NRC's role is to bridge the innovation gap between university-based discovery research and industrial commercialization of technologies [295].

There are also two principal multinational associations of RTOs: the European Association of Research and Technology Organizations (EARTO) and the World Association of Industrial and Technological Research Organizations (WAITRO) [296, 297].

- The *EARTO* is a not-for-profit association based in Brussels. EARTO represents about 350 RTOs from across the European Union and supports the functioning of European RTOs in nodal positions linking and enhancing the efforts of other innovation ecosystem entities, particularly European companies and industries [296].
- The *WAITRO* is a not-for-profit association established by the United Nations in 1970. Rather than a permanent head office, WAITRO has a "roving" Secretariat. WAITRO's activities typically focus on the RTOs of developing nations and helps link them to other RTOs and best practices from around the world. WAITRO represents about 180 RTOs from about 80 countries and facilitates co-operation and/or collaboration with many other technology-related institutes and agencies worldwide [297].

Government Organizations. Beyond the traditional entities (Figure 5.10), government laboratories form an important component of many regional and national innovation systems (see Section 5.3). In many cases, government laboratories are able to provide access to highly specialized, national science facilities and international science connections, plus an ability to connect and collaborate with any or all of the other innovation ecosystem entities. Government labs are also sometimes able to offer strategic funding programmes and even (technical) business incubation services.

The United States, for example, has a wide range of national laboratories, each with specific missions. Among these are the U.S. Department of Energy's 17 National Laboratories that address large-scale, complex R&D challenges in the energy field, with a multidisciplinary approach that places an emphasis on translating basic science into technological innovation [298]. These are principally funded by the government but are usually managed under contract by either private-sector companies or universities. As a system, they have been referred to as an *"iron triangle" of military, academia, and industry"* [299].

A relatively recent phenomenon is the emergence of *"Government Innovation Labs"* (GILs) around the world. GILs aren't actually laboratories at all but rather physical or virtual spaces in which government officials, sometimes with other stakeholders such as business or community people, meet to conceptualize, discuss, and propose innovative government practices, policies, and/or services [300]. These are usually conducted in the context of federal governments and their activities. Innovation in this context is not technological innovation but is a government form of organizational and/or social innovation.

5.3 Government innovation strategies, accelerators, clusters, and innovation parks

Regional or national innovation systems comprise the respective networks of public and private sector innovation-related institutions and their mutual interactions [254–256, 301–303], including:

- Industry: small, medium, and large companies;
- Intermediaries, including not-for-profits and RTOs;
- Education and research entities, including universities, polytechnics, colleges, and government research organizations; and
- Governments.

At the heart of these innovation systems are government policies and organizations. Governments tend to be interested in industrial development and economic health and growth. They are sometimes also interested in opportunities to link technological innovation with solutions to societal "grand challenges" such as, for example, solutions to climate change impacts or opportunities for alternative, renewable energy and power.

Taking into consideration the implications of the Schumpeter and Solow-Swan theories relating technological innovation to economic health and growth for regions and nations, many regional and national governments have developed "*research and innovation*" strategies aimed at boosting economic growth in their region or country. Such strategies are aimed at influencing their innovation systems. Governments have been doing this for many years now but have been slow to adapt to the wealth of new knowledge of how technological innovations are developed. This is a key to understanding governmental research and innovation strategies because, even in the present day, governments tend to base their policies and programmes on a linear and sequential model of innovation rather than a more realistic model (see the discussions in Chapter 2 vs. those in Chapter 3). This is despite the fact that the importance of communication and interactions among the innovation system components is becoming more widely recognized.

Another public policy issue is how to influence the innovation system in a way that produces the desired outcomes. In this connection, McFetridge has observed [95]:

> *"Any important innovation threatens existing interests and entitlements, and threatened interest groups might be able to forestall innovation politically. It is the degree to which the political pro-*

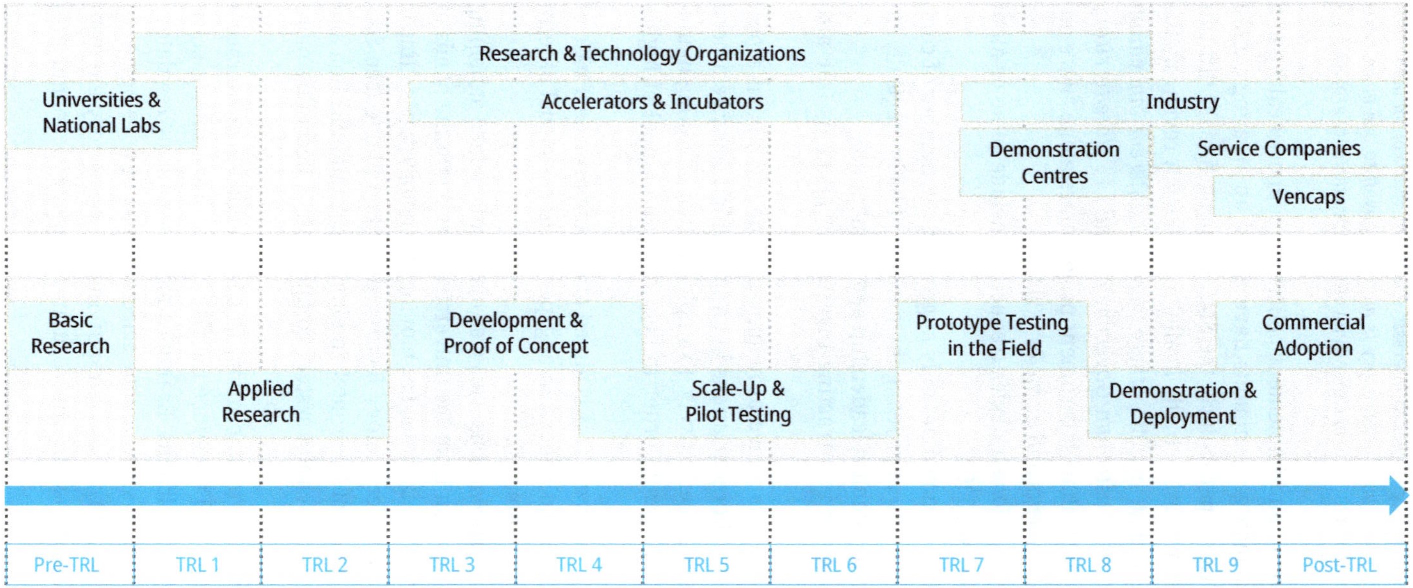

Figure 5.10: Illustration of typical spans of technology focus of innovation ecosystem entities (excepting governments). In this figure, the technology maturity increases from left to right.

> *cess insulates itself from the pressures of entrenched interests that is the mark of an innovative society. A political environment in which innovation policy is merely a payoff to one more lobby group . . . is unlikely to generate much in the way of either innovation or productivity growth."*

Finally, in their attempts to develop and implement policies to support healthy and growing innovation ecosystems, may governments have ended up creating such a myriad of programmes that industry, particularly the SME segment, finds it difficult to determine what kinds of assistance are actually available to them. A review by the Australian Parliament [304] found that, in 2003, "*there were 169 different innovation programs available . . . through the Australian Government, and . . . state/territory governments.*" Furthermore, the Australian government programmes alone "*were administered across 11 different departments and agencies*" [304].

Government *research and innovation* strategies frequently include some or all of the following key elements [43, 95, 305, 306]:

- *Attract leading companies.* This is aimed not only at growing the number of companies contributing to GDP but also at growing the number of technologically innovative companies in order to establish leadership and momentum.
- *Attract investment.* This is aimed at attracting investors of all kinds (see also Section 4.6).
- *Support advanced education.* This is aimed at helping provide the human aspect of the infrastructure base that all organizations in an innovation ecosystem need in order to function efficiently and effectively. Examples include investing in universities, polytechnics, and colleges in order to help maintain a pool of highly qualified people.
- *Provide funding for discovery research.* This is aimed at helping provide a base of ever-expanding new knowledge and understanding, per the linear model of technological innovation (see also Section 2.1).
- *Encourage technology-transfer mechanisms.* This was historically aimed at helping new ideas and knowledge transition from the university worlds into the hands of industry and/or intermediary organizations, i.e., moving technology "up" in linear fashion from the bottom of the Technology S-curve described above (see Section 3.2); however, the importance of horizontal and iterative communication and interactions among all of the innovation system components is becoming more widely recognized.
- *Remove or ease regulatory barriers to innovation.* This is intended to encourage a more competitive environment and to reduce the risk of key organizations starting up or relocating to other jurisdictions. Examples include reducing regulatory "*barriers to entry*" in selected industries and/or selected kinds of developments, assisting with product approval processes, simplifying the regulatory environment, reducing "*red tape,*" and making the regulatory regime clear and predictable.

- *Remove or ease financial barriers to innovation.* This is intended to encourage a more attractive environment for both start-up and established organizations. Examples include reducing tax and tariff *"barriers"* in selected industries, whether by special tax reductions or tax credits, providing job creation tax credits, training allowances, low-cost financing programmes, and export financing and guarantees.
- *Provide and/or improve supporting infrastructure.* This is aimed at helping provide the infrastructure base that all organizations in an innovation ecosystem need in order to function efficiently and effectively. Examples include investing in roads and communication networks.
- *Provide funding for applied R&D.* This is usually aimed at helping SMEs and intermediaries to develop pre-commercial technologies up to the beginning of the valley of death (see Section 4.4), whether through grants, R&D tax credits, or voucher programmes, for example. Innovation voucher programmes for industry generally provide some kind of credit note that can be used to purchase expertise and technical solutions from R&D or other intermediary service providers [307].
- *Encourage the formation of networks.* This is aimed at developing a healthy innovation ecosystem and recognizes the importance of linking industry, intermediaries, universities, and government, as illustrated by models such as the quad-helix model described above (see Section 5.1). The encouragement of cluster formation is one of the policy tools that governments use to accomplish this.
- *Provide funding for technology demonstration.* This is aimed at helping companies and intermediaries to develop technologies through the trough of the valley of death, past the prototype stage, and higher "up" the Technology S-Curve. This can help build both awareness of and confidence in technologies that are becoming market-ready and can help attract investment capital to a region or country.
- *Act as a first purchaser.* Governments themselves are large players in their own economies and can use their own purchasing power to promote technological innovation in their region or country by serving as first adopters or early adopters (see Rogers' diffusion model; Section 4.7).

As noted above, of this long list of strategic and policy areas, two recent trends have been to increasingly recognize (1) the importance of taking a global perspective and (2) the value of a systems approach. Both of these aspects are particularly important for smaller economies trying to find niches in which they can compete with larger, and sometimes more highly productive, economies.

For example, the Conference Board of Canada has found that "*. . . for smaller economies such as Canada's, which account for but a small share of the world innovation effort, technology diffusion contributes more to productivity growth than does the country's R&D effort*" [308].

On the value of a systems approach, according to the OECD, "*Today, most OECD countries are placing emphasis on developing networks and linkages (i.e. collabora-*

tions) in order to enhance the knowledge diffusion power of their national systems of innovation" [244]. Similarly, the business world has found that technological innovation has become increasingly dependent not only on internal capabilities and capacity but also on the strength of their external relationships, which can be leveraged and optimized through networks [309].

Some government approaches that have consistently failed include those by which governments try to "pick winners," whether in terms of choosing particular applied R&D pathways, particular products, processes, or services, or particular companies [95, 245, 310, 311]. Similarly, the approach of government-led ("top-down") creation of clusters has not worked well, partly because every successful innovation ecosystem cluster is so unique in terms of its internal relationships and interactions that it is a difficult model to replicate unless it is led by mutually cooperative and supportive companies themselves [245, 311–313].

However, governments can play an important role in creating and implementing policies that foster healthy business environments and supportive innovation ecosystems. For example, several approaches are increasingly being tested and used as mechanisms for encouraging interactive regional- or national-scope networks with enhanced communications, interactions, technology transfer, development, and commercialization.

Some regional examples include the following:

Research and technology parks are essentially real estate developments that provide infrastructure and a "like" environment for organizations with a strong interest in aspects of research, development, and/or commercialization of technology. Such parks are generally located adjacent, or in close proximity, to a major university. They generally comprise units of the local university itself, RTOs, other intermediaries, industry, government agencies, SMEs, and even start-up companies. Some research and technology parks provide business services and even business incubation programmes. The government policy aspects are discussed further in reference [306].

Clusters comprise groups of organizations in fairly close proximity that have similar or complementary business interests, are part of each other's value chains, and may include supporting or coordinating organizations.[71] These organizations normally share sufficient common interests that they become interconnected in formal and/or informal ways. In some cases, a cluster can achieve economies of scale that its individual constituents could not otherwise achieve. The cluster theory of economic development holds that clusters strengthen both entrepreneurship and innovation by providing a supportive, knowledge-rich, and resource-rich environment (including human resources through labour market pooling). For these reasons, a cluster environment can be particularly attractive to start-up enterprises and SMEs.

71 For this reason, clusters are often viewed as having a particularly important role to play in supporting the evolution and growth of SMEs [311].

Clusters need not be located in research and technology parks, but there are some obvious advantages to doing so. Some of the multi-dimensional innovation ecosystem models discussed in Section 5.1, such as the quad-helix model, apply particularly well to clusters, and the close proximity to one another can reduce the transaction costs of interacting, enabling more opportunities to share knowledge and develop productive working relationships. The term "*Ecosystem innovation*" is sometimes used to refer to technological innovation that is identified, developed, and commercialized through a multi-organization partnership. This is similar to an R&D Alliance except that ecosystem innovation is aimed at developing and commercializing large technological innovation opportunities in a short time period, whereas R&D alliances are more focussed on the R&D components and over moderate to long periods of time [314].

Successful clusters are normally led by industry but are frequently supported by government agencies and/or government funding [245, 256, 306, 312, 315, 316]. Some aspects of government policies or actions in support of cluster development include competition and regulatory reform policy, providing strategic information through technology foresight studies, supporting broker and network agencies, explicit cluster development programmes, and supporting joint industry-research centres of excellence [163, 256]. Potter and Miranda [263], Harrison et al. [317], Wolfe and Lucas [318], and Holbrook et al. [319] provide some case study examples. Some additional case study examples are provided in Section 10.2.

6 Eras and waves of innovation

In Chapter 4, it was noted that technological innovations, especially major innovations, may require decades, or even more than a century beyond their point of public or market introduction, to reach full implementation. Some of the technical and psychological reasons for this have already been discussed, but there is another.

Most people are familiar with the concept that, for various reasons, specific industries or even entire economies can cycle through periods of building, then rapid growth, then a plateauing and/or crisis, then a recession, and finally some kind of recovery or rebuilding. Several kinds of economic cycles have been identified, each with its own drivers, measures, and time frames (periods).

Probably the first business-economic cycle model was proposed in 1862 by French physician and statistician Clément Juglar. Based on his analyses of French, British, and US markets, he proposed [320] that there was a typical business-economic cycle of about 7 to 11 years, caused by oscillations in the demands on and levels of investments in production facilities. Within this cycle, businesses experienced first an expansion phase, then a crisis, followed by a recession, and ultimately a recovery phase.

Another major business-economic cycle model was proposed in 1923 by British businessperson and statistician Joseph Kitchin. Based on his analyses of British and US markets, he proposed [321] that there was a typical business-economic cycle having a period of about three to four years and caused by the influence of the time required for information to flow in companies' decision-making processes. This factor has, of course, become less important as the nature and pace of information flows have dramatically improved. In modern usage, the "*Kitchin cycle*" tends to refer more to the time to recognize that a market saturation has occurred, make decisions in response, adjust production and inventories, and/or wait for market demand to re-establish.

Similarly, various observers have noticed that innovations sometimes seem to be clustered in eras or waves, whether in certain sectors, regions, or both:

> "*The great inventive contributions of mankind had always come in sudden bursts: an era of electrification; another of automotive building [and yet another era of information building]. Each cluster of inventions had resulted in a spurt of investment, but when each had run its course, the hectic activity of building was succeeded by a period of quiescence.*" (From Heilbroner, R.L., The Worldly Philosophers, Simon & Schuster: New York, 1961, quoted by Courvisanos [52].)

https://doi.org/10.1515/9783112206720-006

6.1 Industry waves of innovation

Some eras or waves are specific to a particular industry, or even a specific industry in a particular region or country.[72] There have been several systematic case studies on technological innovation, some of which are quite general, some cover a specific industry, and some cover a specific innovation. An example is Robert Gordon's study of the role of technological innovation in driving the wave of economic growth and societal transformation experienced by the United States between about 1870 and 1970, roughly spanning the third and fourth Kondratieff waves of scientific and technical revolutions (which will be discussed in the next section) [322].

Some in-depth case studies focused on technological innovation in a specific industry and region include the following:[73]

- air conditioning, United States, *ca.* 1900–early 1930s (Nagengast [49, 323]);
- automobile industry, United States, *ca.* 1914–*ca.* 1933 (Schumpeter [15]);
- aviation industry, United States, *ca.* 1903–*ca.* 1945 (Launius [324]);
- aviation industry, United States, *ca.* 1984–*ca.* 2004 (Williams and Weiss [325]);
- computer industry, United States, *ca.* 1940s–*ca.* 2000s (Malerba et al. [326]);
- cotton textile industry, Britain and United States, *ca.* 1721–*ca.* 1929 (Schumpeter [15]);
- cutlery industry, United Kingdom, *ca.* 1918–*ca.* 1955 (Carter and Williams [22]);
- electric lighting industry, United States, *ca.* 1876–*ca.* 1926 (Reich [145]);
- electric power industry, United States, *ca.* 1840–*ca.* 1930 (Schumpeter [15]);
- farming industry, United States, *ca.* 1850–*ca.* 1950 (Schmookler [18]);
- jute industry, United Kingdom, *ca.* 1945–*ca.* 1955 (Carter and Williams [22]);
- mechanical refrigeration, United States, *ca.* 1900–early 1960s (Briley [48]; Nagengast [49, 323, 327]);
- military, United States, 1919–1987 (Hoffman [328]);
- military, United States, 1945–1963 (Sherwin and Isenson [89, 90]);
- military, United States, *ca.* 1967–*ca.* 1995 (Lyons et al. [92]; Chait et al. [93]);
- paper making industry, United Kingdom, *ca.* 1927–*ca.* 1955 (Carter and Williams [22]);
- paper making industry, United States, *ca.* 1850–*ca.* 1950 (Schmookler [18]);
- petroleum refining industry, United States, *ca.* 1850–*ca.* 1950 (Schmookler [18]);
- pharmaceutical industry, United States, *ca.* 1940s–*ca.* 2000s (Malerba et al. [326]);
- railroad industry, United States, *ca.* 1835–*ca.* 1910 (Schumpeter [15]);
- railroad industry, United States, *ca.* 1850–*ca.* 1950 (Schmookler [18]);
- steel industry, United States, *ca.* 1817–*ca.* 1900 (Schumpeter [15]);

72 Usher [124] provides a more general review of the origins and development of some early economically successful inventions (i.e. technological innovations), including water wheels and windmills, water and mechanical clocks, mechanical printing, textile industries, and electrical power.

73 Some more broadly focused, innovation-related case studies are provided in Section 10.2.

- telephone network and industry, United States, *ca.* 1876–*ca.* 1926 (Reich [145]); and
- textile machinery, United States and Germany, late 1800s–*ca.* 1970 (Sabel et al. [329]).

Some case studies focused on specific innovations and/or diffusion of innovations include the following:
- The locomotive engine, *ca.* 1804–*ca.* 1829 [97]
- Bessemer's iron to steel conversion process, *ca.* 1854–*ca.* 1878 [97]
- The introduction of the tunnel oven in the U.K. pottery industry, *ca.* 1912–*ca.* 1955 [22]
- The development of VCR products in the United States and Japan and the success of the latter and demise of the former, *ca.* 1951–*ca.* 1980s [330]
- The laser, *ca.* 1954–*ca.* 1960 [331]
- The (unsuccessful) development of the RCA videodisc, *ca.* 1967–*ca.* 1986 [332]
- Twelve distinct technological innovations in the coal, steel, iron, brewing, and rail industries from about 1890 to about 1960 [214]
- IT diffusion [333]
- Health-related technology diffusion in developing countries, from about 1960 to about 2010 [334]
- "Clean" technologies ("Clean-Tech") diffusion [335–337].

6.2 Societal waves of innovation

Socio-economic waves. In similar fashion to business-economic cycles and the early works of Juglar and Kitchin, several kinds of socio-economic cycles have been identified. Probably the first socio-economic cycle was advanced in 1925 by Russian economist Nikolai Kondratieff,[74] who proposed that economies tended to go through cycles of approximately 50 to 60 years' duration, in which an economic depression (of about 10 years) would be followed by a recovery and expansionary period of technical and economic advances (of about 30 years), followed by a crisis period of economic uncertainty and recession (of about 10 years) [338]. A recent statistical analysis suggests that the Kondratieff waves have continued with a period of about 50 years [339].

Also identified have been shorter cycles, such as the Juglar cycles, which are medium-term macroeconomic cycles with periods of about 7–11 years [340]. It has been observed that there are several – perhaps as many as five – Juglar cycles ("J-cycles") in each Kondratieff wave ("K-wave") [340].

74 Kondratieff is sometimes written as "Kondratiev." Kondriatieff's work fell out of favour with the Stalin-era Soviet government, and he was executed in 1938.

Probably, the next major socio-economic cycle was advanced in 1930 by American economist and statistician Simon Kuznets.[75] Kuznets proposed an economic cycle (often called "*Kuznets Swing*") having a period of about 15–25 years and reflecting demographic changes, usually due to immigration or emigration waves (which could also be within a country, such as between rural and urban areas), and their effects on building construction [341].

Cesare Marchetti distinguished "basic" innovations from innovations that start entirely new industries and found essentially the same "*long-wave*" cycles,[76] of diminishing period, spanning the previous 250 years [342]. Like Kondratieff, Marchetti found that many inventions underlie a wave of innovations and that there are significant lags between inventions and innovations and also between waves of inventions and waves of innovations.

Of the identified cycles,[77] the Kondratieff cycles have received the most attention and further study. Schumpeter coined the term "*Kondratieff waves*" for these cycles, although they are also often referred to as "*long-waves*" [15]. Kondratieff waves have at their heart periods in which a number of key technological (as well as social/political and economic) developments enable bursts of disruptive product or service innovations that fuel a sharp increase in industrial and, therefore, economic growth. The waves of change also carry with them the seeds of the next wave, so the cycles repeat, apparently with periods of approximately 40 to 60 years but in diminishing periods over time [343]. There is no universal agreement on the waves, but a summary view is provided here, and an illustration in Figure 6.1 [344, 345]:

- The *first Kondratieff wave*: spanning the Industrial Revolution era and lasting from about 1780 to about 1830. Key developments during this wave included the steam engine and general industrialization.
- The *second Kondratieff wave*: spanning the industrial production era and lasting from about 1830 to about 1880. Key developments during this wave included railways, steel, and heavy engineering.
- The *third Kondratieff wave*: spanning the scientific revolution era and lasting from about 1880 to about 1930. Key developments during this wave included electricity, chemistry, and the chemical industry.

75 Kuznets received the 1971 Nobel Prize in Economics "*for his empirically founded interpretation of economic growth, which has led to new and deepened insight into the economic and social structure and process of development.*"

76 In a previous section, it was pointed out that inventions are frequently based on multiple discoveries, and innovations are frequently based on multiple inventions. Industry-transforming innovations are frequently based on multiple smaller innovations.

77 There are additional macroeconomic cycles, including Akamatsu cycles, but they are beyond the scope of the present work.

- The *fourth Kondratieff wave*: spanning the technical revolution era and lasting from about 1930 to about 1970. Key developments during this wave included automobiles, mass production, and the petrochemical industry.
- The *fifth Kondratieff wave*: spanning the information and telecom revolution era and lasting from about 1970 to about 2010. Key developments during this wave included microcomputers, information technology, and telecommunications technology.
- The *sixth Kondratieff wave*: spanning the current era and predicted to last from about 2010 to about 2050. Key developments during this wave might include renewable energy, telehealth (especially post-COVID-19), genetic engineering, nanotechnology, robotics, smart grids, cloud computing, and artificial intelligence (AI).
- The predicted *seventh Kondratieff wave*: spanning a future era and predicted to last from about 2050 to about 2090. It has been speculated that the seventh wave may be driven by, for example, a merging of mechanical technology and AI to create autonomous robots, whether at nano-, micro-, or macro-scales, that are capable of independent action, self-repair, and replication (see also Section 9.3). It has also been predicted that the seventh wave will be influenced by ageing populations and, in particular, "above retirement-age" populations becoming greater than the corresponding "active working-age" populations [340].

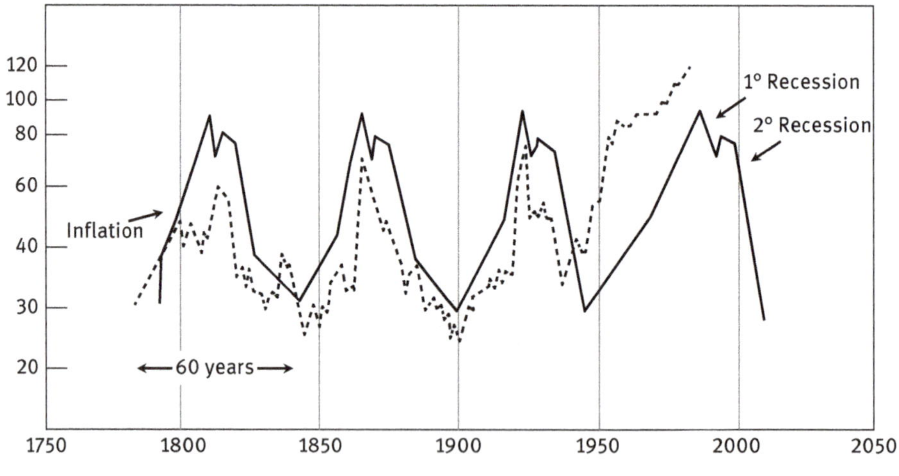

Figure 6.1: Illustration of a series of Kondratieff waves. The solid curve represents a series of idealized 60-year-long waves. The broken curve represents US wholesale prices. From data in reference [344].

It has been suggested that as economic growth fuels the "up" part of the cycle, the risk tolerances of investors decrease, eventually causing a decrease in the wave of innovations [343]. Later, as stagnation and/or recession occurs, the pool of low-risk investment opportunities disappears, and investment shifts back to increased technological innovation as investors' risk tolerances increase.

Historically, the development of inventions into technological innovations, and individual technological innovations into clusters of innovations, seems to require one or two Kondratieff waves. The reason for needing two waves is that, sometimes, innovations occur in the marketplace during one cycle but only become "mainstream" or "catch the wave" in the following cycle [343]. This means that the lag between invention and innovation, and between individual innovations and waves of innovations, can easily range from 40 to 120 years (note that this does not take into account the lag between discovery and invention).

Kondratieff and Marchetti found that there are significant time lags between inventions, many inventions underlie a wave of innovations, and there are significant lags between waves of inventions and waves of innovations.

Taken together, these time lags can easily amount to 40 to 120 years. This does not take into account the time lag between any needed discoveries and the inventions!

Technological ages. Technological ages (or revolutions) represent periods of major technological, economic, and societal changes. At least four Industrial Revolutions have been broadly identified (see also Figure 6.2):

The First Industrial Revolution (mid-1700s to mid-1800s). Some of the principal changes included transitions from wood to coal power and hand-to-machine production methods (particularly in the textile industry); the advent of chemical and iron manufacturing processes, then increasing use of steam power and the development of machine tools; and the transition to high-volume printing via the steam-powered printing press. The first Kondratieff wave lies within this period.

The Second Industrial Revolution (mid-1800s to mid-1900s). Some of the principal changes included the transition from coal to oil power; the advent of electricity, the telegraph, and the telephone; mass production; the internal combustion engine; and highways and rapid transportation. In a broader sense, this era marked the emergence of very large industries, with huge economies of scale. This period is also known as the "*Technological Revolution.*" Most of the second, and all of the third, Kondratieff waves lie within this period.

The Third Industrial Revolution and Post-Industrial Age (mid-1900s to mid-2020s). Some of the principal changes have included transitions from analog to digital technology (particularly digital computers and communications); nuclear energy; the advent of the Internet; and information technology, mass media, and globalization. This period is *also known as the "Digital Age," "New Media Age," "Information Age,"* or the "*Post-Modern Age.*" Most of the fourth, and all of the fifth, Kondratieff waves lie within this period.

Several other perspectives on post-Industrial Revolutions (ages or societies) have also been identified, beginning with the "*Scientific Revolution*" (with key developments including electricity, chemistry, and the chemical industry) and the "*Technical Revolution*" (with key developments including automobiles, mass production, and the petrochemical industry), followed by the "*Information Age,*" the "*Knowledge Age,*" and the "*Creative Age.*"

Of the latter ages, an "*Information Society*" creates and disseminates information. Although this is not new, the growth of ICT in recent years has massively increased data production and dissemination, and the Internet has made global connectivity possible and rapid. However, the production and communication of information alone do not necessarily lead to knowledge creation. A "*Knowledge Society*" develops, processes, shares, and uses knowledge to improve economic, social, and/or environmental conditions. Knowledge societies are by nature also "*Learning Societies*" and embrace the concept of lifelong learning [346]. A "*Creative Society*" comprises people who are developing their natural creative talents and energies and combining the power of information, knowledge, and creativity into an important economic force [347].

The Fourth Industrial Revolution (mid-2020s to the future). As of the time of writing of this book, it appears that the 3rd Industrial Revolution is in the process of transitioning into the 4th. Some of the principal changes predicted and beginning to emerge include advances in biotechnology; the Internet of Things (IoT [348]); cloud computing; artificial intelligence; big data analytics; and '*Smart Manufacturing*' (*Industry 4.0)* in which supply chains, logistics, production, and product lifecycle management may become inter-connected in systems that are much more intelligent, adaptable, and dynamic than they are at present, with the ability to improve through self-optimization and autonomous decision-making (*see*, for example reference [350]). The 6th and 7th Kondratieff waves are likely to lie within this period. A case study on the introduction of Industry 4.0 solutions to a manufacturing plant is provided in Chapter 7 of reference [349], and several other case studies related to the challenges and opportunities of implementing Industry 4.0 are cited in Chapter 4 of the same reference, and also in reference [350].

Waves:	1st Wave: Industrial Revolution	2nd Wave: Industrial Production	3rd Wave: Scientific Revolution	4th Wave: Scientific-Technical Revolution	5th Wave: Information and Telecom Revolution	6th Wave: ...
Date range:	~1780–1830	~1830–1880	~1880–1930	~1930–1970	~1970–2010	~2010–2050
Economic trend (US S&P 500)	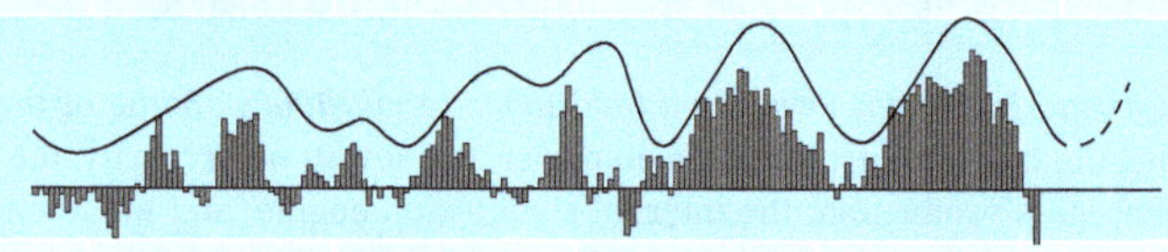					
Key developments	Steam engine, industrialization	Railways, steel, geavy engineering	Electricity, chemistry, chemical industry	Automobile, mass production, petrochemical industry	Microcomputers, information, telecom	
Events near the peaks	War of 1812	US Civil War	World War I	Vietnam War, Six-Day War		
Events near the troughs		1873/79 Depression	1929 Depression	1974/80 Oil crises	2007/09 Financial crisis	

Figure 6.2: Illustration of Kondratieff waves in the context of global industrial and scientific revolutions and other key events. Economic trend adapted from information in Wilenius and Kurki [343].

Figure 6.3 shows a comparison of Kondratieff waves and the industrial revolutions.

It has been fairly widely suggested that the world is presently in the "Sixth Wave" of Kondratieff cycles and that the technological drivers probably lie in carryovers from the previous (fifth) wave, such as digitization and computing power (Grinin et al., refer to a Cybernetic Revolution [340]), plus newer areas such as environmental technologies, biotechnology (including genetic engineering), nanotechnology, robotics,

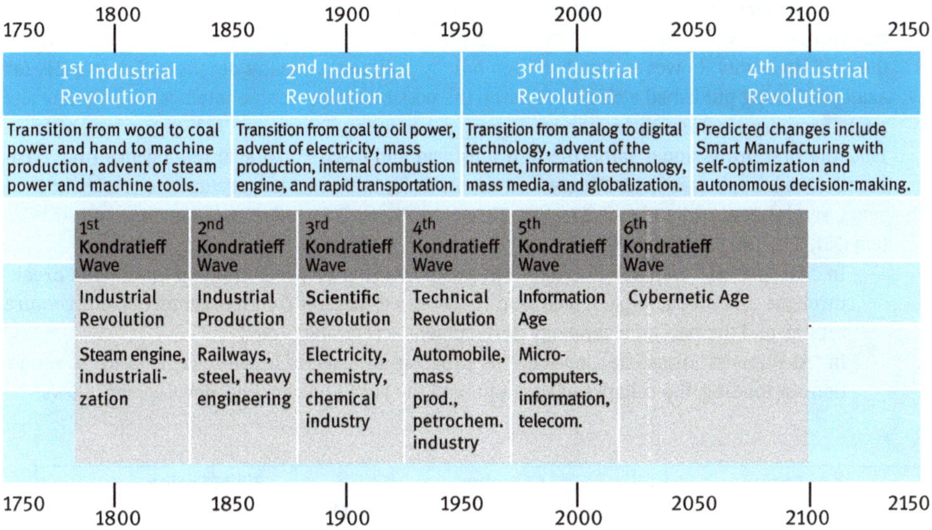

Figure 6.3: Illustration of Kondratieff waves in the context of global industrial revolutions.

and health technologies [53, 343]. Some of the associated social changes may be the ageing of the population, increases in the size of the global "middle class," and the numbers of workers displaced by automation (i.e. robots). Some of the associated environmental changes may be climate change and the diminishing availability of non-renewable natural resources.

At a more detailed level, Figure 6.4 provides a forest sector example of a possible sixth wave transformation [343]. An example could be the development of wood-based polymers for use in the manufacture of such products as toothpaste, drugs, containers, and clothing. This could represent a major (positive) disruption in the forest products industry into higher value-added manufacturing, which could dramatically improve the forest sector economy (and which could be negatively disruptive for the oil and gas sector economy).

5th Kondratieff K-Wave	6th Kondratieff K-Wave			
Raw Forest Materials	Forest-Based Value Creation			
Pulp and paper based	Fibre to replace oil-based products	Wood construction of houses and buildings	Wood product designs	Forest-based biofuels and bioenergy

Figure 6.4: A forest sector illustration of a possible 6th wave transformation. Based in part on reference [343].

Example: Artificial Intelligence (AI). Since the publication of the first edition of this book, AI applications have burst onto the scene, and it's tempting to think that we may be living in the dawn of the field. In reality, however, this has been quietly building for many decades [351]. In 1950, for example, Turing published a classic paper on the possibility of machine intelligence and how one might compare machine vs. human intelligences (hence the 'Turing Test') [352]. Not long after, in 1956, the first professional gathering to brainstorm and discuss artificial intelligence (the 'Dartmouth Workshop') was held. Progress in the AI field, however, has occurred in cycles of 5–15 years, reminiscent of Kondratieff waves. These have been referred to as AI Springs and AI Winters [351, 353, 354] (see Figure 6.5):

- In "AI Springs," new ideas create optimism, leading to predictions of imminent breakthroughs and media hype, which lead to increased funding from governments and venture capital, and the pace of technology development accelerates), and
- In "AI Winters," dissatisfaction with the progress and lack of breakthroughs leads to reductions in funding, the collapse of start-ups, and the pace of technology development slows.

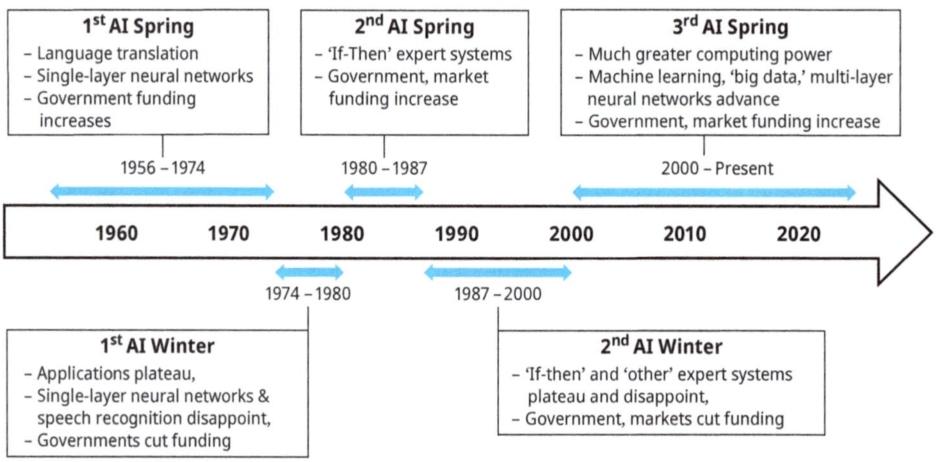

Figure 6.5: Illustration of AI waves (or cycles). The date ranges are approximate. Adapted from summaries in Mitchell [353] and Lutkevich [355].

At the time of writing, the 3rd AI Spring is underway, and machine learning, multi-layer neural networks, and 'big data' have made it possible to develop functional AI tools that have become widely available, such as the AI-driven internet search engines that are now almost unavoidable. On the other hand, machines that are conscious, self-aware, and sentient will likely have to wait for a future AI Spring and the seventh Kondratieff wave, if not later. There has been speculation that the 2020s may herald a 3rd AI Winter [356].

Example: The U.S. Some waves can have transformational effects on a society. In the period between about 1870 and 1970, roughly spanning the third and fourth Kondratieff waves and the scientific and technical revolutions, the United States experienced an unprecedented wave of economic growth and societal transformation. Robert Gordon calls this the U.S. "*Great Leap Forward*," and he traces the impact, during this period, of the advent of "*electric lighting, indoor plumbing, home appliances, motor vehicles, air travel, air conditioning, and television*," among others, on their post-Civil War society and economy [322]. This wave of productivity growth has been assessed as being mostly driven by a unique "*wave of waves*" of technological innovations. Figure 6.6 shows the U.S. productivity growth that spiked during this period of time. The main driver of this productivity growth was technological innovation, as represented by total factor productivity (TFP; *see* Section 8.1), and it drove an economic revolution [322, 357]. At least one of the causes of the decline has been attributed to the off-shore outsourcing of manufacturing [358].

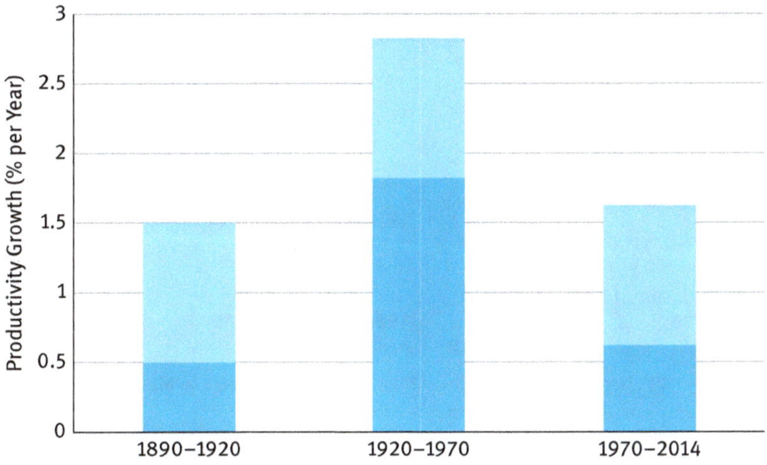

Figure 6.6: Productivity growth in the United States between 1890 and 2014. The contributors are shown as total factor productivity (dark shading) and the combination of educational attainment plus capital input (light shading). Drawn based on data from Gordon [322].

Example: China. Missing out on waves can be economically disastrous. For centuries leading up to the nineteenth century, China had had a dominating economy. By the end of the eighteenth century, China must have felt that they had everything they needed to remain dominant. In the nineteenth century, China not only missed the Industrial Revolution but failed to accept and adopt its technological developments. This was compounded by rebellions, lack of infrastructure maintenance, and war. As a result of this, China also missed the Industrial Production, Scientific Revolution, and Scientific-Technical Revolution entirely, and their economic power collapsed for about 140 years (Figure 6.7). Their recovery, however, has been underway for some years now [359, 360]. To 'catch up,' China has developed and applied a number of strategies. One of these, as noted in Section 5.1, has been the establishment and support of regional innovation systems (RIS) [257]. Another has been by deploying a 'catching up' innovation financing model, in which their "*large banks offer ample credit resources to big companies to promote incremental innovation,*" and by supporting a 'leading' financing model "*where mid-and small-sized companies use the money from [the] capital market to fund innovation*" [257]. This model may help China become a global frontrunner in technological advances. Another strategy, of course, has been the support for mass-scale, "rapid appropriation of a wide array of foreign inventions, whether devices or practices, from abroad decades and deployed in a newly receptive environment" [361]. As a result, China's economic (GDP) output has risen exponentially since the 1980s. As Figure 6.7 illustrates, as of the time of writing (2025), their recovery was well underway.

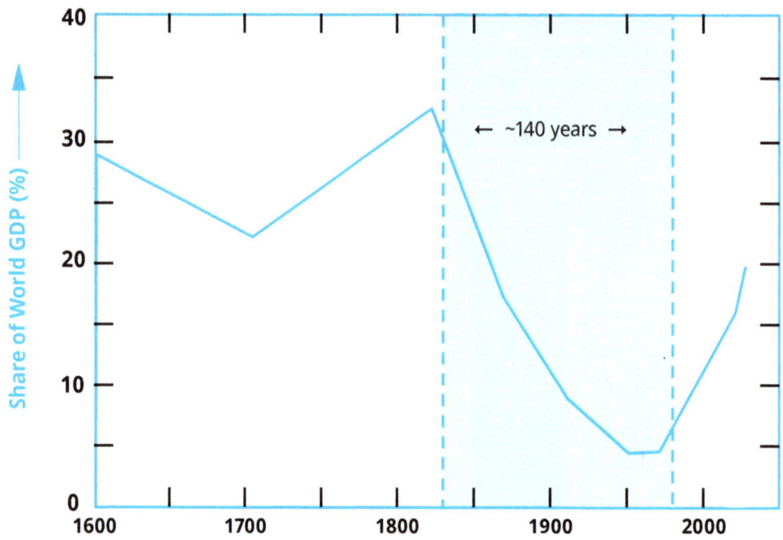

Figure 6.7: Illustration of China's share of global GDP over time. Adapted from data in Maddison [359, 360].

Future seventh Kondratieff wave and Fifth Industrial Revolution. Some features of a future Seventh Wave might include the merging of technology and intelligence to create autonomous robots, whether at nano-, micro-, or macro-scale, that are capable of independent action, self-repair, and replication. *The Fifth Industrial Revolution* is already being predicted to possibly involve a transition from Industry 4.0 to a new Industry 5.0, in which industry transitions from profit-driven to sustainability-driven, from efficiency-focused to resilience-focused, and from technology-centric to human-centric [362, 363]. Time will tell, but there is already a growing literature related to the Circular Business Model (CBM) and the Circular Economy (CE).

7 The management of innovation

> "You've got to think about big things while you're doing small things, so that all the small things go in the right direction."
>
> Alvin Toffler,
> American Writer and Futurist

Earlier sections have dealt with tactical approaches to managing technological innovation processes and intra-organizational approaches to enabling and accelerating technological innovation. In between these extremes lies another aspect, that of the strategic leadership and management of innovation within an organization in order to maintain and improve an organization's differentiation and competitiveness. Leadership and management of technological innovation in an organization are needed because it is not natural for most organizations. Govindarajan and Trimble write [364]:

> *"Why is it that innovation leaders so often feel that their biggest enemy is not the competition but their own company? There is a simple answer. Organizations are not designed for innovation. Quite the contrary, they are designed for ongoing operations."*

Can innovation be managed? Baker writes [365]:

> *"While some argue that innovation cannot be managed – that it just happens – most researchers and theorists agree that the organizations can be designed to have a structure, a culture, and processes that are conducive to innovation."*

In a related vein, Johnston and Grant conclude that "*The literature shows that most innovation projects fail because of poor innovation process management*" [230]. So, the pursuit of technological innovation may require organizational innovation. In this chapter, some starting points will be introduced.

7.1 How much innovation, if any, does an organization need?

Before discussing the management of innovation, it may be appropriate to pause and ask how much innovation an organization really needs. This will obviously have an impact on decisions around strategy.

It has become popular for governments and agencies to attempt to measure and compare their region's "innovation performance" with that of others, particularly among countries. This has created a huge flow of reports on relative innovation performance, hand-wringing, and attempts to assign blame for less than some kind of ideal performance, and a seemingly never-ending series of reports and articles on how to fix the situation. It is common for such reports to lay the blame on the shoulders of industry for not pursuing sufficient technological innovation, or at least not

https://doi.org/10.1515/9783112206720-007

pursuing it vigorously enough (see e.g. [95]). It is equally common for such reports to judge that the solution is for governments to make more funding available, usually for academia [95]. The problem with the latter advice is that the linear model of technological innovation has been demonstrated to be hugely inefficient (see Chapter 2), so that unless the academic institutions in a given region or country are significantly underfunded (which is not usually the case in the developed countries), this is not going to be an effective way to deal with the issue.

The earlier conclusion, that any perceived lacking in the amount of technological innovation and competitiveness on the part of a region or country must be the fault of industry, implies that companies are intentionally allowing themselves become less competitive. Geoffrey Moore wrote [143]: "*It has become very fashionable of late to talk about how high-tech companies can and should become market-driven organizations . . . however . . . All organizations are market-driven, whether they acknowledge it or not . . .*" Whether or not individual organizations need innovation, or more innovation, comes down to individual business strategies, which can be influenced by any or all of the following:

- *The individual character and culture of the organization.* Some companies operate in high-volume, low-margin businesses for which the products have changed little in many decades because there is neither an internal driver nor a market need for change. Such companies may not have, need, or want a culture or strategy of innovation.
- *The business strategy of the organization.* A company's strategy may be simply to maintain pace and market share, another's may be to grow, while another's may be to grow aggressively. Even where a company has a growth strategy, it may be via acquisitions rather than by increasing product/process/service competitiveness,
- *The nature of the business sector, especially in terms of competitiveness.* Many, if not most, small businesses survive and/or succeed based on a combination of great customer service and/or convenience [366]. Others serve customers who value "consistency and authenticity" over change [367]. Unless such businesses are driven by a desire or need to increase their scale, scope, or market reach, they may not perceive a need for technological innovation (although they would probably see the value of service or convenience innovations).
- *Monopolies are different.* Very large organizations that have a natural monopoly, for example due to a very high cost-of-entry to their markets, may not need innovation because their customers have no alternative providers [367]. Similarly, government policies, regulations, and programmes may hinder or even prevent competition, such as in the case of government-enabled monopolies, which may have little incentive or need to innovate.

In a global survey of companies, PricewaterhouseCoopers found that the majority (57%) did not view innovation as a "competitive necessity for their organization"

[367]. In a 2014 survey of Canadian companies, Grant found [366] that the number of companies that self-identified as "innovative" correlated with company size (as measured by annual revenue). Whereas all companies having annual revenues in excess of $500 million self-identified as being "innovative"; this was only the case for about half of companies having annual revenues of $1 million or less. It may be that the largest companies cannot compete on such benefits as service and convenience and therefore have to rely on other benefits, requiring technological innovation in order to differentiate themselves and remain competitive [366]. Figure 7.1. illustrates the difference that innovation can make to net cash flow for a company that invests in innovation as opposed to one that does not, where both are operating in the same competitive marketplace (competition drives prices and/or market share down, so the "no innovation" net cash flow projection is negative rather than zero).

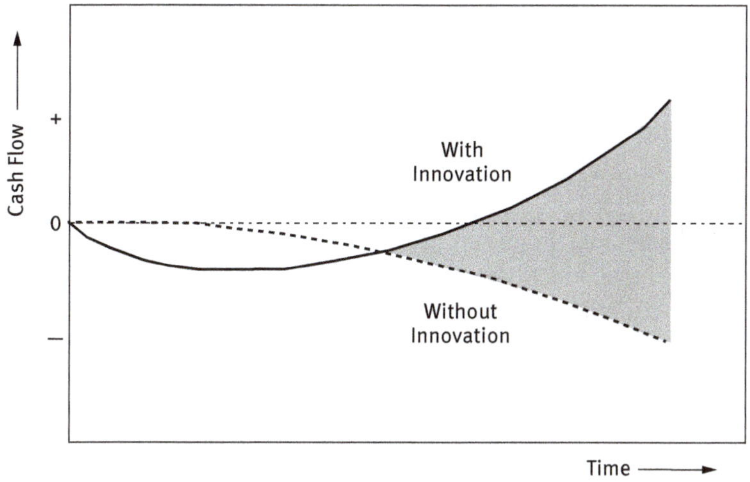

Figure 7.1: Illustration of cash flow with (solid curve) and without (broken curve) innovation in a competitive market. The shaded area highlights the difference over longer periods of time. Adapted from Johnston and Grant [230] and Christensen et al. [368].

Ultimately, any well-led and managed organization will decide for itself how much and what kind of technological innovation it needs, and the level of innovation that it needs may not fit well into the naïve view that all organizations must aggressively pursue technological innovation at "world-class" levels, all the time.

An organization can have too little innovation (*"hypo-innovation"*), such as when it has not produced significant technological innovations and is being outpaced by its competitors.

Conversely, an organization can have too much innovation (*"hyper-innovation"*), such as when it has been highly successful with one or more major technological innovations and it has become vulnerable to a culture shift towards hubris, undermining the kind of culture that originally enabled the innovation(s).

Furthermore, as discussed by Abernathy and Utterback [369], an organization's technological innovation needs and its innovation strategy and tactics may all change over time, particularly as an organization matures. A Conference Board of Canada survey asked Canadian companies, among other things, whether innovation has been a long-term priority for them: only 25% reported that it has been an objective for 20 years or more [370].

Some approaches to innovation strategy and tactics are discussed in the next few sections.

The pursuit of technological innovation may require organizational innovation.

7.2 Key success factors for innovation in organizations

Some of the most important factors thought to be involved in successful innovation in organizations are shown in Table 7.1.

At the top of the list is strategy. If long-term innovation is to be taken seriously in an organization, then leadership "from the top" will be needed, and innovation will have to be made a core component of corporate strategy. Otherwise, there are just too many other things that can, and will, get in the way. This is discussed further in Section 7.3. Originating and maintaining an organizational strategy is going to require leadership, but an innovation formula (f_i) that takes the form

$$f_i = (\text{leadership} + \text{strategy})$$

is incomplete.

Table 7.1: Some key factors and competencies in innovation management [98, 365, 371–373].

Strategy	Inclusion of innovation	A clear positioning of technological innovation in strategy, to ensure that innovation processes are embedded in the organization's long-term focus and that they are continuous and integral to operations, and including the nurturing of a creative and entrepreneurial culture and activities.
	Strategic approach	Promotion of a broad definition of business boundaries, fluid organizational boundaries, and an open market for ideas and talent.
	Leadership of the strategy	The organization's leaders consistently make decisions consistent with the organization's strategy.

Table 7.1 (continued)

	Foresight	Including regular or continuous environmental scanning for ideas, opportunities, and threats.
	Openness	Open innovation provides opportunities for new ideas and technology, plus opportunities for networking, collaboration, and partnership (including with customers).
Culture	Commitment	Visible senior management commitment and support for technological innovation, to ensure that internal innovation processes do not falter due to internal barriers or resource allocation limitations. There should be at least one senior executive champion for innovation.
	Innovation culture	An enterprise-wide culture and structure that promotes innovative thinking, experimentation, free flow of information, change management, and technological innovation.
	Openness	The organization's employees, managers, and leaders are open to new ideas and consider them seriously.
	Integrity	Employees need to trust their leaders and their organization in order to take the personal risks involved in innovating.
	Collaboration	There is a willingness to collaborate across organizational boundaries.
	The voice of the customer	The "voice of the customer" is actively sought out and seriously considered as part of the NPD process.
	Risk tolerance	Senior management appetite and ability to accept and manage risk and entrepreneurial behaviours, so the organization is flexible in its outlook and is capable of accepting failures and using them as learning opportunities.
Management (operations)	Consistency with strategy	Operational management is consistent with the organization's mission, core values, and vision.
	Flexibility	Organizational flexibility and adaptability, especially in its production processes, so that there is a built-in ability to introduce product/process/service improvements or even complete changes.
	Sharing	Continuously promoting knowledge sharing and communications across organizational boundaries.
	Challenging	Challenging the status quo models for business mission, market scope, products, processes, and services, target customers, and markets.
	NPD funnel	There is a good process for the ongoing classification, screening, and prioritization of new ideas and then managing the new product/process/service development process.

Table 7.1 (continued)

	Project management	Effective innovation project management, with decentralized decision-making.
	Mistakes are allowed	Managers and leaders recognize that mistakes are a necessary part of the innovation development process.
	Individual support	Providing time and resources for employees to develop and test new ideas, like 3M's "*15% Time.*"
	Employee recognition	Strategic thinking and innovation are motivated, rewarded, and recognized across the whole organization.
	Metrics	Innovation needs to be measured and tracked, and it should be on the organization's balanced scorecard.

Some organizations place their focus on searching for the next great new idea and/or on developing lots of new ideas. These are important elements, but if a culture of innovation does not already exist, then to accomplish anything really substantive and long-standing, such a culture will have to be created and developed. Even if a culture of innovation already exists within an organization, it will need nurturing and support in order to be sustained. This is discussed further in Section 7.4.

Originating and maintaining an organization-wide culture of innovation is going to help with the implementation of an innovation strategy, but an innovation formula (f_i) that takes the form

$$f_i = (\text{leadership} + \text{strategy} + \text{culture})$$

is still incomplete.

The function concerned with coordinating organizational efforts to accomplish innovation is "*innovation management.*" Such efforts could include planning, organizing, staffing, and/or leading processes involving, for example, research and development (R&D), product development, manufacturing, and marketing. Some technological innovation pathways and stages have already been discussed in earlier chapters, but mostly from the perspectives of new idea generation, technology development, and the technical management of new product, process, or service development processes. Under an organization's innovation strategy, there are other aspects that are related to organizational behaviour and structure.

With leadership support and an innovation strategy in place, the focus can shift to execution, including making plans and making investments in people, training, technologies, equipment, and facilities and then implementing the plans in order to deliver the desired results and, if necessary, realize positive change in practices and culture. This provides a better innovation formula:

$$f_i = (\text{leadership} + \text{strategy} + \text{culture} + \text{management}),$$

where management includes not just planning and resources but also measuring outputs and outcomes from the implemented strategy and plans, which enables an organization to manage for success and continuous improvement.

Our innovation formula now looks like this:

$$f_i = (\text{leadership} + \text{strategy} + \text{culture} + \text{management} + \text{monitoring} + \text{continuous improvement}).$$

These aspects are discussed further in Section 7.5.

7.3 Innovation strategy

"If you don't know where you're going, you might not get there."

Attributed to L.P. "Yogi" Berra,
American Professional Baseball Player, Manager, and Coach

Chapter 2 briefly introduced the dawning of organized research, development, and technological innovation in the 20th century in North America. As the first research and technology organizations (RTOs) were established, they began to organize themselves in ways that would enable them to conduct industrial research aimed at helping industry to advance. As this goal was being taken up by organizations, rather than lone wolf-type individuals, the RTOs evolved to actively manage teams of people engaged in systematic research. It was with this aim in mind that Maurice Holland, the first head of Engineering and Industrial Research at the US National Research Council (NRC), put forward his description of the *"cycle of research"* leading to technological innovation and industrial growth – which is the linear model of innovation described in Section 2.1.

As has been shown in earlier chapters, it turned out that the technological innovation process is only rarely linear, or even a single pathway, but by expanding on the linear model, it did prove to be a good guide for the kinds of processes that would be needed for successful industrial research and, therefore, the people, tools, and facilities that would be needed. This thinking found its way to major industries in North America shortly after their introduction at the US NRC. By the early 1900s, the concepts of organized industrial research had taken hold in the form of RTOs and industrial research laboratories in US major companies (like GE, Bell, and Du Pont), as had the notion of programmes of systematic R&D with teams of researchers and dedicated facilities.

Technological innovation at the organizational level needs leadership and management because it requires a forward-looking vision, a business commitment of the necessary resources, and a willingness to accept risk. Possibly the most important thing an organization's leader can do to enhance technological innovation is to set bold, stretch organizational goals and to demonstrate a willingness to accept (and manage) the commensurate risks to the organization.

By the 1970s, some companies began to develop and implement management practices for the process of technological innovation [374]. "*management innovation*"[78] can be considered to encompass new ways of organizing work or work-related functions, practices, or processes, including the management function, within an organization, in order to achieve organizational goals. This is a form of non-commercial innovation; it is broader in scope than administrative innovation and can be a component of organizational innovation. Management innovation can provide a means of focusing and leveraging an organization's technological skills and knowledge to improve its performance in terms of technological innovation, productivity, and competitiveness [375].

Why have an innovation strategy? Part of the answer is to clarify the organization's vision and R&D philosophy in the context of its overall strategy and indicate how committed management is to technological innovation and to the provision of resources in support of technological innovation. After all, the costs of pursuing technological innovation in terms of time, effort, and money have to fit within management's need to achieve overall responsible and reliable corporate performance [376].

It has been said that more than anything else, "*vision . . . drives successful innovation*" [377].

Another part of the answer to why have an innovation strategy is because technological innovation requires a lot more than just creative ideas. Certainly, creative new ideas will be needed, but, as discussed in earlier chapters, much work also has to go into deciding which ones to pursue, and the subsequent R&D process and pathways can take many different directions. It has also been noted that there is more involved than just R&D. According to OECD estimates, R&D expenditures account for only one-third to one-half of the total expenditures on technological innovation, the bulk being spent on associated expenditures on "equipment, training, licences, marketing and organizational change" [43].

It has also been noted that "*it is strategic oversight that prevents organizations [from] falling prey to the dangers of research and technology push when there is no market for its outcomes, and over-reliance on demand-pull processes where customers*

78 Also termed "*managerial innovation.*"

can be conservative and stifle potentially disruptive innovation" [378]. An appropriate technological innovation strategy should ensure that there is ongoing assessment of the external environment (in terms of such features as business, social, political, economic, and technological) and ongoing identification and assessment of technological opportunities, and it should identify the organization's approach to assigning and managing resources related to producing technological innovations.

If technological innovation is going to be pursued by an organization, then having a technological innovation strategy and aligning it to the organization's overall strategy are the secrets to investing wisely and to managing the risks inherent in conducting R&D. Without such overall direction ("*mission orientation*"), an organization's R&D efforts will be more expensive and time-consuming and less successful in yielding new commercial products, process, and services in the organization's intended business arenas than they could, or should, be. In the 1980s, Tom Peters wrote:

> *The course of innovation – idea generation, prototype development, contact with initial user, breakthrough to final market – is highly uncertain, to say the least. Moreover, it always will be, sloppy, and unpredictable, and this is the important point. It's important because we must learn to design organizations that take into account, explicitly, the irreducible sloppiness of the process and take advantage of it rather than attempt to fight it* [379].

Having an organizational innovation strategy enables management, at a high level, to pursue R&D in alignment with the corporate strategy and in a manner that is as focussed, efficient, and effective as possible. According to a 2005 American Management Association (AMA) survey "the ability to foster creativity and innovation is among the top competencies required of leaders today," and this ability is expected to become even more important in the future [380]. AMA's top recommended actions leaders should take are [380]:

- develop an organizational strategy for innovation;
- redesign organizational structure and/or workflow; and
- increase employee involvement.

However, as will be discussed in Section 7.4, this needs to be more than a top-down process. Jaruzelski and Dehoff [381] describe the critical importance of having bottom-up contributions from the individual business units, "*which are so much closer to the customer, must first see an opportunity, and begin to innovate.*" In addition, some organizations use a Technological Innovation Committee to oversee and/or manage the implementation of their "innovation" or "R&D" strategy [374]. Such a committee usually comprised people who are knowledgeable and experienced in areas such as research, development engineering, manufacturing, marketing, and finance. Regardless of how it is set up and labelled, the role of such a committee would usually be to integrate and coordinate the various activities involved in ideating, developing, and commercializing a technological innovation (such as have been introduced in Chapters 3 and 4).

There is a significant, and growing, body of literature on innovation strategy and management. Some examples include references [24, 79, 122, 374, 375, 377, 382, 383]. At a high level, there are two traditional approaches to strategy: "rationalist" and "incremental." The rationalist approach essentially consists of attempting to scan and understand the surrounding environment, develop a strategy taking the environmental scan into account, and implementing the strategy. This approach is similar to classical military strategy and builds upon the classical organizational SWOT (strengths, weaknesses, opportunities, and threats) approach to planning. In contrast, the incremental approach essentially consists of developing and implementing tactics intended to take the organization in the desired general strategic direction, evaluate their effectiveness, and adjust and repeat as necessary. This kind of approach has also been referred to as "trial and error." The relative strengths and weaknesses of these approaches are discussed elsewhere [384].

In recent years, the concepts of environmental, social, and governance principles, management, and metrics have become widely adopted, and therefore the concepts of sustainable innovation, or responsible innovation, have come into vogue [385, 386]. That is, innovation that improves an organization's sustainability performance, whether measured in terms of economic, environmental, or social aspects.[79] Some case studies are described in references [385, 387–389].

In the following sub-sections, several lenses on organizational innovation strategy that are specific to the world of technological innovation will be introduced.

Some approaches to organizational innovation strategy. A common categorization of organizational approaches to innovation strategy is based on an organization's intended positioning within its industry grouping, and its risk tolerance [380, 390]:

- A "*market leader*" strategy (also termed a "*first-to-market strategy*") has a goal of being the first to introduce new products, processes, or services into the marketplace. This can provide an early, temporary, monopoly on the new offering but is expensive, in terms of R&D, and is risky, in that market acceptance is not assured. Market leaders also tend to focus more on breakthrough innovations as opposed to incremental innovations, although in practice, most end up pursuing a blend of both. Research has shown that both approaches can pay off, with incremental innovation usually bringing in most of the revenue but breakthrough innovations usually bringing in most of the profits [380]. This suggests a possible strategy of seeking a balance of incremental and breakthrough innovations. Examples of or-

79 In this connection, one also encounters the term "sustainable innovation," meaning innovation that improves an organization's sustainability performance (whether measured in terms of economic, environmental, or social aspects, but increasingly the latter two). This is also termed sustainability-oriented innovation (SOI), or responsible innovation. For a case-study example, see [389].

ganizations that have become known for their "market leader" strategy include Intel, Apple, and Disney.

- A *"fast follower"* strategy (also termed a *"second-to-market strategy"*) has a goal of being an early participant in the marketplace as a new product, process, or service evolves "up" its lifecycle trajectory. Rather than being the first into the market, this strategy involves a form of imitation of the market leader's first movement (see e.g. [391]). This eliminates the need for front-end applied research and can be less risky, but it still requires substantial development engineering work and a marketing and sales approach that can attract customers away from the market leader. Examples of organizations that have become known for this strategy include Zenith, Google, and Facebook. Some examples are provided in Chapter 1 of reference [204].
- A *"cost minimization"* strategy (also termed a *"late-to-market strategy"*) has a goal of producing a low-cost product, process, or service later in the lifecycle trajectory. Somewhat similar to the fast follower approach, this strategy involves a form of imitation of the market leader's first movement, eliminates the need for front-end applied research, can be less risky, and still requires substantial development engineering work and a marketing and sales approach that can attract customers away from the market leader. Examples of organizations that have become known for this strategy include General Motors and "no-frills" airlines like Southwest and Ryanair.
- A *"specialist"* strategy (also termed a *"market segmentation strategy"*) has a goal of serving niche markets with customized versions of the basic product, process, or service, usually later in the core technology's lifecycle trajectory. Here, again, this strategy requires substantial development engineering work and a marketing and sales approach that can target the niche markets and attract customers away from the market leader. Examples of organizations that have become known for this strategy include Starbucks, Volvo, and Silicon Valley Specialists.

Depending on an organization's risk tolerance and intended positioning within its industry, such as illustrated above, its allocation of technological innovation-generating resources can be assigned among, for example, incremental, adjacent, and radical innovation initiatives (see e.g. [392, 393]).

An example of innovation strategy mapping is provided by that of Mohan Swahney, who has divided companies' innovation approaches into four categories, based on whether innovation is developed internally or externally and how structured or informal their innovation processes tend to be (see Davis [373]). This is illustrated in Figure 7.2. In this framework are the following:

- *"Explorers"* tend to be in emerging, rapidly expanding markets or redefining their markets, but in either case are in markets demanding speed and agility. They also tend to be focussed on customers, partners, and markets generally for new insights and opportunities, users of rapid prototyping and testing of large

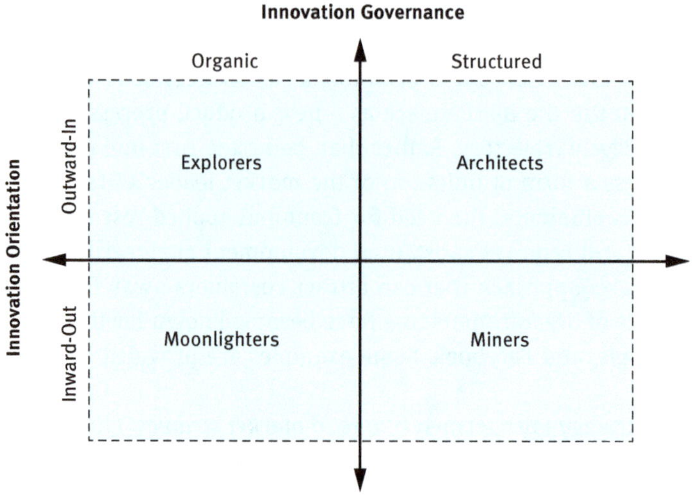

Figure 7.2: Illustration of innovation strategy mapping (adapted from Swahney as cited in Davis [373]).

numbers of relatively small ideas, and innovating through a diffuse, customer-facing process. Jaruzelski and Dehoff have referred to such organizations as "*need seekers*" [381].

- "*Architects*" tend to be in mature markets, with intensive capital and resource requirements and with a centralized, structured approach to innovation. They also tend to be focussed on customers and competitors for new insights and opportunities, outsourcers of development and prototyping, and innovating through a "*top-down*,"[80] formal process. Jaruzelski and Dehoff have referred to such organizations (and also the moonlighters, described next) as "*market readers*" [381].
- "*Moonlighters*" tend to be large companies with lots of technology and process expertise, that look inward to develop innovations based on internal ideas and know-how. They also tend to be focussed on a strong innovation culture that supports internal experimental and development project initiatives, including internal "moonlighting," being willing to fund internally generated innovation projects, and innovating through "*bottom-up*"[81] processes. Jaruzelski and Dehoff have referred to such organizations (and also the Architects, described above) as "*market readers*" [381].

80 Top-down innovation is the process of seeking out, identifying, and evaluating potential market opportunities and then challenging the organization to come up with and develop concepts for innovative new products, processes, or services that could align with the selected market opportunities.

81 Bottom-up innovation is the process of originating and developing concepts for new business products, processes, or services and then evaluating them for market potential.

- *"Miners"* tend to be very large companies with lots of technology and process expertise, very large internal business units, and strong silos. They also tend to be weak on internal innovation, necessitating the use of external innovation resources and/or organizations, focussed on finding innovation opportunities beyond the organization, and innovating through formal, centralized processes working through the external innovation organization. Jaruzelski and Dehoff have referred to such organizations as *"Technology Drivers"* [381].

Another organizational approach to innovation strategy is based on an organization's intended positioning with respect to other organizations in its industry grouping. For example, an RTO might have a strategy that sets out its key areas of focus in a way that also shows how it intends to position with regard to other organizations in the innovation continuum. Figure 7.3 shows an example that uses the innovation continuum (*"technology S-curve"*) concept and the linear model of innovation[82] (see [290]). In Figure 7.3, the lightest-shaded region labelled "concepts/fundamental" refers to discovery research (*"basic research"*), and the blue-shaded ellipse is drawn to only slightly overlap this area to indicate that this is not an area of RTO focus but rather an area from which the RTO would draw new knowledge and understandings by collaborating and/or partnering with academic institutions. The next-shaded region labelled "feasibility/development" refers to applied research, development engineering, and proof-of-concept at the laboratory bench-scale (*"R&D"*), and the blue-shaded ellipse is drawn to fully overlap this area to indicate that this is a key area of RTO focus. Similarly, the next-shaded region labelled "piloting/demonstrations" refers to pilot testing, scale-up engineering, and full-scale field/plant demonstration (*"piloting and demonstration"*), and again, the blue-shaded ellipse is drawn to fully overlap this area to indicate that this is a key area of RTO focus. The right-hand region labelled "commercial innovation" refers to commercial-sector activities (*"products and services deployed in the marketplace"*). The blue-shaded ellipse is drawn to only slightly overlap this area to indicate partly that only some RTO activities would be focussed in this area (i.e. commercial testing and analyses) and partly that this is an area in which the RTO would collaborate and/or partner with primary industry and service companies to assist with technology transfer and commercial deployment.

In connection with the management of innovation processes in an organization, the term Holistic Innovation Management has been advanced to capture the combined processes of strategic, total, open, and collaborative innovation processes [394–396]. In some of the literature, holistic innovation management is also taken to include aspects of sustainability [396]. This is a kind of enterprise innovation ecosystem approach, and as such holistic innovation management represents an organizational, strategic version of the open innovation and ecosystem approaches discussed in Section 5.1 [395–397].

82 In this case, the linear model is used because of its simplicity and the fact that the developmental stages referred to are still legitimate and distinct.

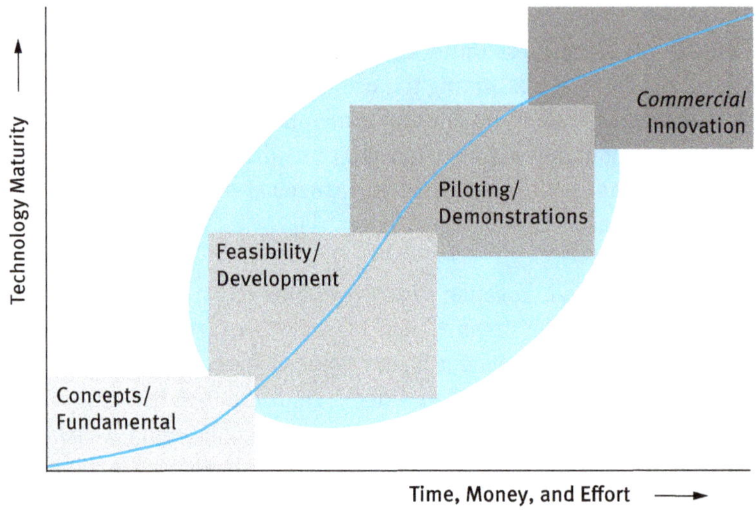

Figure 7.3: Illustration of an RTO strategy based on the innovation continuum and a linear model of innovation. The blue ellipse is drawn to show key areas of focus and key areas of overlap with non-focus areas. Courtesy of Saskatchewan Research Council, 2016, [290].

Strategies for maximizing innovation in organizations. The literature is replete with articles and books on innovation strategies. Only a few perspectives and examples will be provided here.

A classic illustration and categorization of strategic choices was published by Igor Ansoff in 1957 [398]. Ansoff defined four basic product/market strategies, as illustrated in Figure 7.4. In a "*market penetration*" strategy, the idea is to sell more current products into the currently served market or markets (such as by increasing or improving quality, productivity, or marketing). In a "*market development*" strategy, the idea is to sell more current products into a new market or markets (such as with new marketing and sales efforts). In a "*product development*" strategy, the idea is to sell new products into the currently served market or markets, while in a "*product diversification*" strategy, the idea is to sell new products into a new market or markets. The product development and diversification strategies usually involve R&D, adopt and adapt, or even purchase or licensing-in of technologies, and then commercialization. The so-called Ansoff Matrix, shown in Figure 7.4, also illustrates the need to choose a degree of risk appetite and tolerance. Some examples are discussed in [399].

Mapping organizational innovation strategy performance. The nature of an organization's innovation strategy can determine not only the probability of its being able to achieve innovation performance but also the probability of being able to sustain it. Reflecting on such strategies as the ones just described, Jaruzelski and Dehoff concluded [381] that:

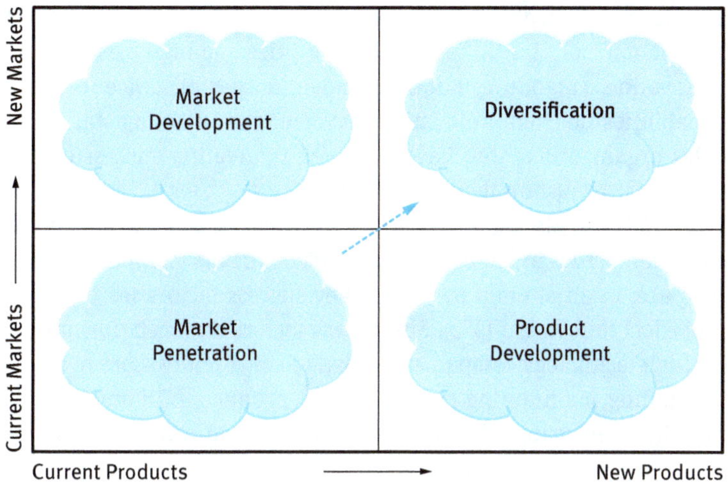

Figure 7.4: Illustration of Ansoff's Market/Product Matrix. The broken arrow indicates the direction of increasing risk.

> *"Companies that can craft a tightly focused set of innovation capabilities in line with their particular innovation strategy – and then align them with other enterprise-wide capabilities and their overall business strategy – will get a better return on the resources they invest in innovation."*

An example of innovation performance mapping, that is, comparing the relative innovation performance and capabilities of organizations, is shown in Figure 7.5.

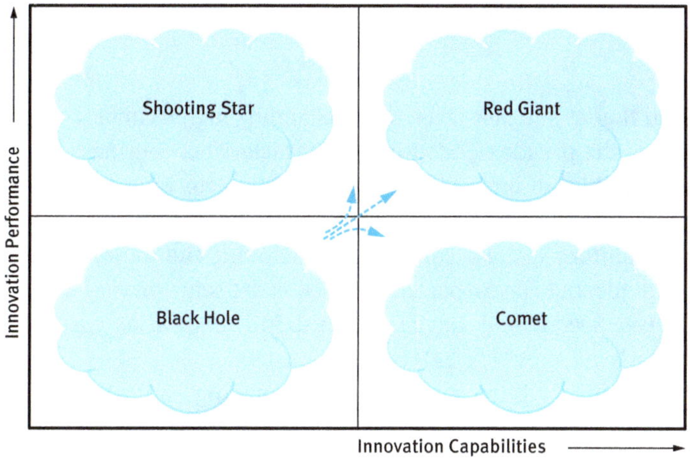

Figure 7.5: Illustration of innovation performance mapping.

In this model, organizations are rated as having both high innovation capabilities and high performance (red giants), both low innovation capabilities and high performance (black holes), high innovation capabilities but low innovation performance (comets), or low innovation capabilities but high innovation performance (shooting stars). The latter category includes organizations that have achieved innovation success(es) but which do not have the ability to sustain them (see [400].

Between the 1960s and 1980s, Maidique and Hayes interviewed over 250 executives from US high-technology companies in biotechnology, semiconductors, computers, pharmaceuticals, and aerospace, in an attempt to discern key success factors for organizations requiring technological innovation [382]. Six success factors emerged, for most of which the outstanding high-technology companies seemed to consistently score highly:

1. Business focus, including maintaining closely related products in dominant and secondary product lines, focusing R&D in only a few areas at a time, and maintaining consistent organizational priorities
2. Adaptability, meaning the willingness (risk tolerance) and ability to make rapid, major changes when necessary
3. Organizational cohesion so that the business focus can be realized, including good internal communications and cross-organizational cooperation
4. Entrepreneurial culture
5. Sense of integrity, including a commitment to long-term relationships with key stakeholders including employees, shareholders, customers, suppliers, and local communities
6. Engaged top management, not "hands-on" to the point of micro-managing, but rather engaged, in the sense that top management is deeply interested and engaged in the organization's entrepreneurial and innovation development culture and processes

Maidique and Hayes found that at a higher level, the outstanding high-technology companies were able to manage the paradox of continuity characteristics (business focus, organizational cohesion, and sense of integrity) with change characteristics (adaptability, entrepreneurial culture, and engaged management). Subsequent studies up to and including the 2000s have reinforced these conclusions, including the importance of being able to manage Maidique and Hayes' paradox, which is the same or very similar to the "*Success Paradox*" (see Section 4.9), Davila and Epstein's "*innovation paradox,*" and Christensen's "*innovator's dilemma*" [79, 241] (*see* Section 4.9).

Finally, in addition to being willing and capable of making radical strategic changes, an organization also needs to be able to determine when they are necessary. Some examples of senior management initiating and implementing radical strategic change are provided by Tushman et al. [401], Quinn [383], and Smith and Wright [402].

Possibly the most important things that an organization's leader can do to enhance technological innovation is to set bold, stretch organizational goals, foster creativity, and demonstrate a willingness to accept (and manage) the commensurate risks to the organization.

Possibly the most important thing the senior management team can do is to manage the paradox of continuity characteristics (business focus, organizational cohesion, and sense of integrity) with change characteristics (adaptability, entrepreneurial culture, and "engaged" management).

7.4 Innovation culture

Culture refers to people's values, aspirations, expectations, assumptions, their sense of "the way things are done around here," and even their sense of "the way things should be done around here." The importance of culture has already come up in the discussion of innovation strategy. Of the major groupings in Table 7.1, the one related to culture is probably the most important because an organization's culture will tend to either inhibit or enable the implementation of strategies and operational plans. Kerzner [122] provides some case studies of the innovation culture at Disney.

The best starting points for building and maintaining an innovation culture involve setting the "tone from the top" in terms of organizational strategy built on mission, values, and vision and clearly identifying in, or for, these whether or not (and why) innovation is seen as being important to achieving organizational success. With a clear strategy in place, attention can then be turned to operational and tactical aspects.

Some tactical features associated with building and maintaining an innovation culture include [380]

- the ability to focus on customers;
- teamwork, communication, and collaboration with others;
- availability of resources;
- ability to select good ideas;
- creative people;
- freedom to innovate;
- ability to measure innovation;
- encouraging small and big ideas;
- innovation goals;
- culture of risk tolerance;
- organizational structures;
- diversity; and
- balancing incremental improvements and breakthrough discoveries.

Teamwork and collaboration can be contributors to, or even critical components of, successful technological innovation, particularly if the teams have good leadership and diversity [380]. In this context, diversity refers to having people with different knowledge, skills, experiences, styles, and perspectives. It can also be an advantage, or even critical, for the team(s) to have people from all parts of the innovation process, such as R&D, development engineering, production, marketing, and distribution.

Similarly, external collaboration can be a contributor, or even critical, to successful technological innovation. According to surveys conducted by AMA, nearly half of modern technological innovation is in the open innovation model (*see* Section 4.2) and involves external collaborations [380]. Teamwork and collaboration enable the kind of iterative and feedback loops described in non-linear models of innovation, such as the *chain-linked model* described in Section 3.1.

Many organizations have experimented with creating innovation teams in dedicated ideation and collaboration spaces. The term "*skunkworks*" has been used to describe the workings of a small team of people, often within an organization's R&D department, that are fairly unconstrained, unstructured, and aimed at testing and developing radical ideas. Some skunkworks are officially supported and may be well funded, while others are completely unofficial and may not have any explicit funding. In either case, they may be highly secret. For example, Lockheed Aircraft Corporation's World War II "*Skunk Works*" project developed the American XP-80 Shooting Star jet fighter in 1943.

In addition to teams and teamwork, champions[83] tend to be needed as well. In Chapter 1, it was noted that the initial response of people and societies to radical new ideas is usually a vigorous resistance. In a 1963 study of successful technological innovations, Donald Schon found that a critical success factor was having a "product champion" that had the drive, skills, and persistence to find ways to overcome such resistance [403]. Schon noted that "the new idea either finds a champion *or dies*" (the italics are Schon's emphasis), and the champion has to work through whatever formal or informal organizational processes are necessary to develop the new technology [403]. There are a number of other articles on the importance of innovation champions/promoters, and their roles in the new product development (NPD) process, including [404–407].

In the 1960s and 1970s, an investigation named "Project SAPPHO" attempted to discover the differences between successful and unsuccessful innovations by comparing pairs of successful and unsuccessful technological innovation attempts for which both members of each pair were aimed at the same market [408–410]. Altogether, 72 such pairs were studied (SAPPHO Phase I and Phase II). From the Project SAPPHO studies, four kinds of key individuals within an innovating organization were identified and considered:

1. The "*technical innovator*," who is associated with making the most important technical contributions to the development and/or design.

83 Another term for innovation or product champion is "*innovation promoter*."

2. The *"business innovator,"* who has management responsibility for the overall progress of the innovation initiative.
3. The *"product champion,"* who – through active and enthusiastic internal promotion – makes decisive contribution(s) to the progress of the innovation through critical development stages, and
4. The *"chief executive,"* who is the head of the executive structure of the organization.

Although the role of the product champion, as defined above, was found to be important, the most important role was found to be that of the business innovator and their "power, respectability, status, and experience" within the organization [406].

Since the time of project SAPPHO, other works, such as that of Trias de Bes and Kotler [411], have been published that give their own recommendations for the composition of innovation development teams, but they share a common theme of recommending a rich diversity of skills, experience, and thinking styles.

Also from the Project SAPPHO studies, five success factors emerged that seemed to differentiate successful from unsuccessful technological innovation champions. The successful innovation champions [408–410]:
1. had a better understanding of user needs;
2. paid more attention to marketing;
3. performed their development work more efficiently (not necessarily more quickly);
4. made more use of outside technology and advice; and
5. were usually more senior and had greater authority.

Organizational culture is critical because it will either inhibit or enable the implementation of strategies and operational plans.

Rothwell's "10 Cs" for successful implementation of technological innovations are [98] the following:

"*. . . effective Communications to gain Consensus for Change, Champions to sustain Continuous Commitment to Change,* [and] *a Culture that is Customer Centred.*"

7.5 Innovation and managing operations

This book is not about management *per se*, so this section will simply introduce some features and tools that might be considered to be important elements in a "toolbox" for the management of innovation in operations:
- Engaging employees
- The technical ladder
- Business intelligence
- Technology foresight
- Technology roadmapping

- Legal factors

Employee engagement and skill development. Engaging and developing employees are organizationally important, so that they will be able to apply their knowledge and skills, adapt to changes, generate new ways of thinking, and develop and implement solutions [412]. This is in addition to the features described in the previous section of this chapter because, as Richard Florida writes, in "*The Rise of the Creative Class – Revisited*" [169]:

> *". . . creative people and knowledge workers respond well to organizations that provide solid values, clear rules, open communication, good working conditions, and fair treatment . . . they don't want to take orders, but they do want direction . . . The commitments of creative people are also highly contingent, and their motivation comes largely from within."*

Several companies offer programmes through which employees can submit innovation ideas that may be selected for funding assistance and further development. IBM has called theirs "Genesis Grants" [413], Ericsson Canada's is called "Idea Box" [414], and the Saskatchewan Research Council's (SRC's) is called the "Innovation Fund" [290]. Such programmes can have the triple benefit of engaging and developing employees while ensuring lots of new ideas are constantly being generated and introduced into the front end (funnel) of the NPD process (see Section 4.8).

Some key skills associated with building and maintaining an innovation culture are listed in Table 7.2. Of course, most people don't excel at all of these competencies, so teams need to be created, led, and managed, and some of these can be acquired

Table 7.2: Some key skills associated with building and maintaining an innovation culture [412, 415].

Skills	Needed to Support
Problem/opportunity identification, creativity, and problem-solving	New idea generation
Communication, relationship-building, collaboration, and partnership	Relationships that support any or all aspects of the innovation process
Project implementation and management	Translating ideas into strategies, capabilities, products, processes, and/or services
Applied research, engineering development, prototyping, piloting and demonstration, manufacturing design and operation, etc.	Activities throughout the new product development (NPD) process
People leadership and management	The ability to lead and manage teams
Business management	The economic viability of new products, processes, and/or services

Table 7.2 (continued)

Skills	Needed to Support
Intellectual property management	Enabling legal protections and freedom to market and sell new products, processes, and/or services
Risk assessment and risk management	Calculated, managed risks, and entrepreneurism
Capital raising	Generation of internal and/or external financial resources
Marketing and sales	Accessing, developing, and/or maintaining markets and sales

and/or enhanced through training, external collaborations, partnerships, and/or subcontracting.

The technical ladder.[84] Some of the key skills listed above require dedicated trained, skilled, and experienced specialist-professionals. Many organizations have experimented with a "*dual-ladder*" approach to career development, by which professionals can choose whether to pursue the traditional management path or to remain on a professional/technical path. The idea is to motivate and retain experienced professional/technical staff, who do not want to go into management, by offering them additional career stages that either offer theoretically equivalent status and rewards or at least increasing status and rewards. Of course, the technical ladder rungs are not really equivalent to the management ladder rungs because the former do not have the power and authority of the latter. Numerous other shortcomings of the dual-ladder approach have been discussed elsewhere [416]. Perhaps, the greatest potential shortcoming, one that can completely destroy the credibility of the entire system, is if senior management yields to the temptation to use the technical ladder as a place to "shelve" failed or surplus managers. Nevertheless, having a technical ladder can be a progressive strategy if managed openly and with integrity.

Business intelligence (or "competitive intelligence"). Business intelligence is essentially knowledge about developments and trends relevant to an organization's competitive position. For example, such knowledge could relate to discoveries, inventions, and/or technologies; products, processes, and/or services; partners, competitors, and/or suppliers; and customers and markets. Kenneth Sawka has referred to the business intelligence function of an organization as the "*Department of Surprise Avoidance*" [417].

Business intelligence thus encompasses data, information, and knowledge. Business intelligence as a process involves such activities as data mining and collecting, data processing and analytics, querying, and reporting. Although such processes are

84 Also known as the "*professional ladder*" or "*individual contributor ladder.*"

greatly aided by computerized and online tools, they require human analysis and judgement. Although the terminology varies, the goal is to get from data to actionable knowledge. Examples of the data translation and interpretation process include the following:

Data → Analysis → Insights

Data → Information → Knowledge → Wisdom

The data sources are, in principle, very numerous and diverse, spanning, for example, news media, social media, advertising materials, government reports, regulatory and legal filings, conference and trade-show materials, technical and industry publications, patent filings, and even interviews with industry players and analysts. For more information, see [418–420].

Business intelligence generally deals with the present and can sometimes provide insights into possible near-future scenarios. Looking further than a year or two ahead is the domain of technology foresight and technology forecasting.

Technology foresight. Technology foresight and technology forecasting have evolved into useful tools for organizations interested in technological innovation. Technology forecasting is about making informed judgments about the probable nature and pace of technological developments and is intended to be an aid to decision-making.

Foresight is a systematic process involving environmental and horizon (futures) scanning aimed at anticipating future events in order to be able to develop strategies to encourage, prevent, change, or simply manage the impacts of such events. In this sense, foresight is much more than simply the opposite of hindsight. The term was originally introduced by Irvine and Martin in the 1980s [421] to describe foresight activities intended for use as a public policy tool for setting priorities in science and technology, but its use has expanded beyond the government sector [422]. Compared with technology foresight, innovation foresight (also termed *"innovation system foresight"*) has to look further than the development of new technologies and into their commercialization, which also involves social and political dimensions.

For a brief historical overview of the development of technology foresight, see reference [423]. A set of descriptions of major advances in the evolution of technology foresight methodologies since the beginning of the 1990s, called the five generations of technology foresight, has emerged in the literature [422, 424]. These are based in part on Rothwell's [98, 126] five generations of models of technological innovation (see Section 3.1):

- *1st-generation foresight* focusses on the prediction and diffusion of technologies and often involves the "technology-push" model of innovation.
- *2nd-generation foresight* adds the marketplace and the activities of the participants in technology development, be they industry, intermediary, government,

and/or academic. This level of sophistication often involves the "market-pull" model of innovation.
- *3rd-generation foresight* adds interactions with society and often involves a "coupling" model of innovation (such as the "chain-linked" model).
- *4th-generation foresight* adds the distributed roles of the involved parties in innovation systems, and often involves the "integrated" model of innovation.
- *5th-generation foresight* adds structural and policy issues and often involves the "*system integration and networking model*" model of innovation.

The principal kinds of future events anticipated by foresight activities usually involve technological, economic, environmental, political, social, and/or ethical futures. Two terms that occur frequently in foresight processes and scenarios are "*Wild Cards*" and "*Weak Signals*" [425]. In foresight scenarios, Wild Cards[85] are events that are perceived to be of low probability of occurrence but potentially high impact should they actually occur. Weak Signals are unclear observables that provide a warning of the probability of possible future events, including Wild Card events.

Some examples of mapping the innovation foresight process are given by Andersen and Andersen [422].

Anticipated future events can be analysed using a number of tools, including the Delphi, SWOT, and social, technological, economic, environmental, and political (STEEP) methods, among others [422]:
- The *Delphi Method* involves surveying a group of experts in a field for their opinions about a possible future situation or scenario. A summary of the entire group's responses and rationales is provided as feedback, and then the group is surveyed again one or more times until something approaching a consensus (the "*expert consensus*") is achieved.
- *SWOT* analysis involves assessing a real or imagined, current or future situation and its potential impacts from each of the four perspectives as an aid to planning. For example, when faced with a new possible market situation, an organization might assess its own strengths and weaknesses in the context of that situation and consider the relative merits of avoiding or pursuing the potential opportunity or defending or fleeing from the potential threat.
- *STEEP* analysis involves assessing the named environments and factors as an aid to making business decisions, including those regarding technology commercialization and product, process, or service opportunities. Other factors and combinations are sometimes considered, such as legal, ethical, demographic, ecological, regulatory, and/or values.[86]

85 Also termed "*outliers*," "*black swan events*," or simply "*black swans*."

86 Leading to other acronyms, including LEPEST, PEST, PESTEL, PESTLE, SLEPT, STEEPLE, STEEPLED, STEER, STEP, and STEEPV; explanations for these are provided in Section 11 and reference [2].

The results of such foresight activities have been applied in a range of business areas, including strategy development, priority-setting, business planning, change management, and team and culture building [426]. Some case studies of national foresight programmes (UK, Germany, Hungary, Columbia, Brazil, and Venezuela) of the 1990s and 2000s are described in references [427–430].[87]

Following a review of the foresight literature up to 2011, Rohrbeck and Gemünden identified three broad roles that corporate foresight can play that can improve the innovation capacity of an organization [431]:

- A "*strategist role*," helping to create a vision, provide strategic guidance, consolidate opinions, assess and reposition innovation portfolios, and identifying the new business logic and business models of competitors.
- An "*initiator role*," helping to trigger innovation initiatives by identifying new customer needs, new technologies, and the product concepts of competitors.
- An "*opponent role*," challenging the organization's innovators to create better, more successful innovations by challenging basic assumptions, challenging current R&D projects (including challenging whether they are truly "state of the art"), and scanning for, and identifying, market disruptions that could threaten current and future innovations.

A key application of the combined results of business intelligence and technology foresight is in what is called "*technology roadmapping*" (TRM).

Technology roadmapping (TRM). TRM is a process by which an organization, or even an entire industrial sector, sets out predictions for its future technology needs (or opportunities) and maps out one or more pathways to achieving them (hence the metaphor of a road map). When implemented, the organization or sector then pursues one of the pathways, several in parallel, or a hybrid. TRM seems to have originated in the electronics industry, beginning with Motorola in the 1970s [61], and then with others in the same industry in the 1980s [432, 433], following which it spread to other sectors.

TRM is often industry sector and/or market driven, involves a team of technical and industry specialists and/or experts, and generally adopts a time horizon of 10–20 years. Garcia and Bray provide a detailed description of the TRM process [434]. The US Department of Energy website contains links to numerous R&D and technology roadmaps for specific programme areas [435]. The Alberta Chamber of Resources provides an example of an industrial sector's technology roadmap, in this case for the oil sands industry [436]. Figure 7.6 provides an illustration of challenge identification from an oil sands industry technology roadmap exercise [436]. Some other examples are discussed in references [437–439].

87 Some more broadly focussed, innovation-related case studies are provided in Section 10.2.

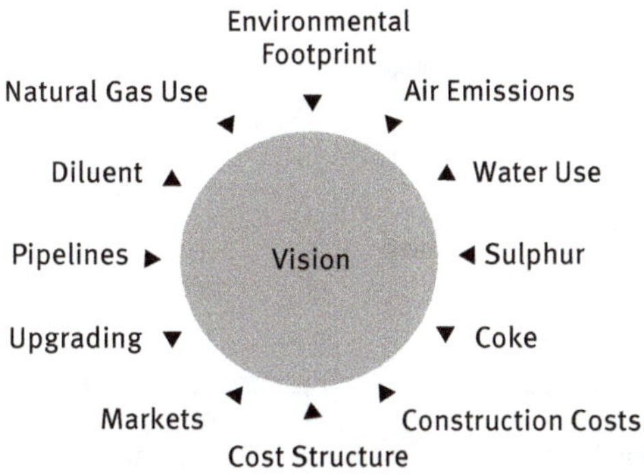

Figure 7.6: Illustration of challenge identification from an oil sands industry technology roadmap. Courtesy of Alberta Chamber of Resources, 2004, [436].

TRM can also be used at the organizational level, as an aid to strategic planning. It can help develop consensus on future technology needs, identify risks, provide a high-level plan of approach for developing and/or adapting and adopting technologies, and serve as a guide to setting priorities and investment decision-making. Although there is no single "right way" to structure a TRM, an organizational TRM might contain the following main elements (see Figure 7.7):

- Markets, including competition, trends, and opportunities
- Products, processes, and or services, including current status, trends, and forecasts
- Technologies, including current status, trends, and forecasts
- Future state, summarizing the desired future state, including the key opportunities, barriers and risks, and a strategic approach

Willyard and McClees provide a "corporation" example of applying TRM at Motorola [440]. Pataki et al. provide some additional examples, including examples illustrating the use of the STEEP and SWOT foresight tools in TRM [432]. Moehrle et al. provide other examples, including examples illustrating the use of the DELPHI foresight tool and the TRIZ thinking and problem-solving tool in TRM [439].

Legal factors in technological innovation. The pursuit of technological innovation raises some legal issues, particularly with regard to IP. IP includes almost any definable intellectual creation, including products, processes, services, literary and artistic works, symbols, names, images, designs (including logos), and other forms of know-how. The three main ways of managing IP are through legal protections, maintaining secrecy, or by publishing to the public domain.

Legal protections. Some forms of IP are legally protectable under IP law,[88] such as by *contract, copyright, patent, trade dress*, or *trademark*. A copyright is a legal protection for ownership of the original author(s) of published or unpublished artistic or

88 Depending on the country concerned, the intellectual property rights that may be protected by legislation include patents, trademarks, copyrights, industrial designs, integrated circuit topographies, and plant breeders' rights.

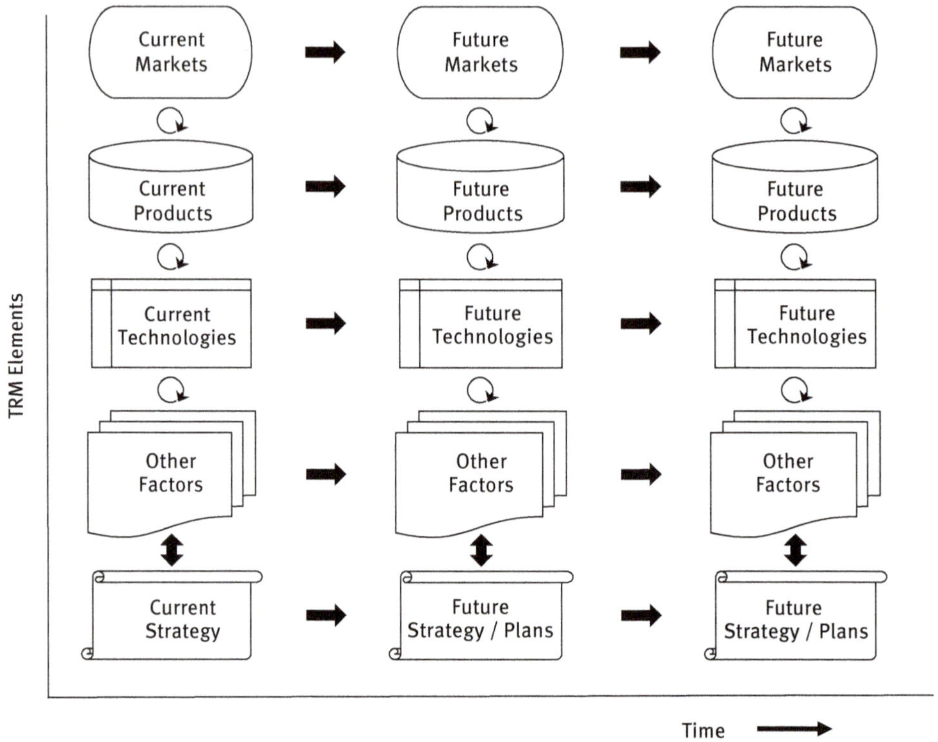

Figure 7.7: Some key structural elements in an organizational technology roadmap. In this illustration, "*products*" could be any combination of products, processes, and/or services. Adapted from information in [432].

literary creations, including computer software. The specific rights and terms are determined by a given country's copyright law. A patent is a grant of property rights on new, useful, and non-obvious inventions by a government. The benefit to the patent owner is that the grant of a patent excludes, for a certain number of years, others from making, using, selling, offering to sell, or importing the invention in the specific country granting the patent. For this reason, many inventions are patented in multiple countries. The downside is that this protection is made in return for government making the details of the patented invention public. Patents are awarded for processes, machines, manufactured items, compositions of matter, designs related to manufactured items, and for certain kinds of new invented or discovered plants. Trade dress refers to the physical appearance of a product and/or its packaging. If the physical appearance of a product is unique, unusual, and/or widely recognized by the public, then its trade dress may be legally protectable in some countries. A trademark is a word, phrase, name, design, symbol, logo, or device associated with a business brand, product, or process and which is used for differentiation from other brands or products in the same general line of business. The term "*service mark*" is sometimes used when a business service is meant. Registered trademarks and service marks are legally protected in the countries within which they are registered.

Maintaining secrecy. Some forms of IP can be protected by holding them in secrecy, such as concepts, discoveries, and *trade secrets*, which can include non-patentable inventions. A trade secret can be any kind of technological or business know-how that is protected from competitors by keeping it confidential. Compared with patents, the benefit to the invention owner is that the knowledge is not made public, and the downside is that the onus is on the invention owner to maintain that secrecy (otherwise someone else could potentially patent the invention, thereby restricting or preventing the original invention owner from using it).

Publishing. Finally, in the case of an invention, design, or know-how, another strategy is to openly publish (disclose) the invention in order to prevent anyone else from patenting it and potentially restricting its use by the original invention owner.

Depending on how, or if, any specific IP is protected, there are a number of options available in terms of how to deal with it. Examples include the following:

- *Non-disclosure agreements* (NDAs) are legal contracts aimed at protecting information considered to be proprietary, or otherwise business confidential. NDAs can be useful because they enable parties to share confidential information with each other but otherwise maintain confidentiality because the parties to the contract agree not to disclose to others the information covered by the agreement.
- *Material testing agreements* are somewhat similar to NDAs in that they are legal contracts aimed at protecting information considered to be proprietary (or otherwise business confidential) when materials are transferred from one party to another for testing, evaluation, or further development. Here again, the parties to the contract agree not to disclose to others the information covered by the agreement.
- *Licensing*. A licence agreement entitles a person, group, or organization to make, use, or sell a piece of IP, usually in return for some form of compensation (royalties).[89] Such licences can be exclusive or non-exclusive, and in either case, there may be limitations, such as geographic territory, technical field of use, product application, or time.
- *Cross-licensing* is a mechanism by which one party grants royalty-free licence(s) to another party in exchange for reciprocal rights to the latter's technologies. These can be useful because they provide a simple and practical mechanism for exchanging technologies while maintaining some protections.
- *IP portfolio mining* is a business intelligence process that involves scanning and evaluating IP, including, but not limited to, patents, held within an organization and/or external to an organization. The purpose is to identify IP that is dormant,

89 Sometimes, such licences are granted royalty-free, as part of a bundle with another product, process, or service or following an up-front, one-time payment.

underutilized, or that could be utilized in new ways, as part of an organization's innovation management programme. Classic examples are the identification of patented technologies from another country that are either not protected in the organization's home country or could be licensed for use in its home country.

A business plan is built around value creation and/or value capture, and IP factors can be important for both. In this connection, it can be helpful to construct a value chain for a potential new product, process, or service and map areas of strength, weakness, opportunity, and threat from an IP perspective. An example is given by Chesbrough [156].

Summary. Some examples of companies that have used management innovation to grow their businesses include [230, 380]:
- *Dell*, in the computer industry;
- *Southwest*, in the airline industry;
- *Apple*, in the music industry;
- *Motorola*, which has an "*invention leadership programme*" (launched in 1995) that aims to enhance creativity and develop disruptive innovations; and
- *3M*, which has a "Pacing Plus" programme (launched in the mid-1990s) that requires each business unit to develop and deliver at least one new product that changes their basis of competition.

Some success indicators related to the leadership and management of innovation are discussed in Section 8.2.

7.6 Some innovation barriers

Beyond the technical challenges discussed in earlier chapters, a common occurrence is for organizations to hit a so-called "*innovation barrier*,"[90] which can be any challenge or obstacle to the successful process of innovation, up to and including successful technology commercialization [29]. This usually means that a critical operational component is lacking or insufficiently developed. The nature of this component could be, for example, financial, legal, technological, manufacturing, cultural, or regulatory.

An inability to access capital is the most common innovation barrier for SMEs, whereas larger organizations can typically deal with the financial issues but have difficulties in other areas. For example, there may be an issue related to leadership and management, or with strategy. An organization's strategic and operating plans may conflict with the difficulties, costs, risks, and time scales associated with the pursuit of technological innovation. A 2012 Conference Board of Canada survey [441] found that

90 Also termed "*bottleneck*" or "*reverse salient*."

the most prevalent challenges to innovations in large firms are "corporate culture" and "fear of risk."

Most, if not all, such innovation barriers have to be successfully identified and managed before complete technological innovation is regularly and reliably possible. When this is accomplished in specific projects, one sees references to having achieved a breakthrough, or a so-called *"New-to-the-World Solution."* When this is accomplished at the organizational level, the entire organization should advance.

Some of the most important factors thought to act as barriers to innovation in organizations are shown in Table 7.3. Most of the barriers shown in this table relate to strategy, to operations, or to mismatches between the two. Once identified, these can be relatively easy to fix. More pernicious, and much more difficult to change, are cultural barriers to both technological and non-commercial innovation, some of which are discussed further in the next few paragraphs.

Table 7.3: Some key barriers to successful innovation in organizations [122, 230, 370, 380, 442–445].

Strategic conflict	The strategic plan is incoherent and/or conflicts with the difficulties, costs, risks, and time scales associated with the pursuit of innovation.
Lack of ambition	The strategic plan is focussed on maintaining the *status quo* and lacks boldness.
Inconsistent leadership	Frequent changes in executive leadership (people and/or direction) driving short-term planning, strategies, and implementation.
Risk aversion	Fear of risk, risk aversion, low risk appetite, and/or low risk tolerance conflicting with the risks and time scales associated with the pursuit of innovation (creating a "caretaker" approach).
Culture mismatch	Organizational culture conflicts with the difficulties, costs, risks, and time scales associated with the pursuit of innovation.
Creativity challenge	Being non-creative internally ("idea poor") and/or unable to take advantage of externally sourced creativity.
Cultural implementation challenge	Inability to undertake innovation processes and/or to collaborate with partners (i.e., being "implementation poor") due to internal bureaucratic resistance and/or lack of knowledge constraints.
Directional viewing challenge	Having a predominating cultural rearward-looking focus on things that worked well in the past, rather than a forward-looking focus on things that might work well in the future.

Table 7.3 (continued)

Operations conflict	The annual operations plan (or even the organization itself) conflicts with the difficulties, costs, risks, and time scales associated with the pursuit of innovation.
Absence of goals and measures	A version of operations conflict involving the lack of innovation-related goals, priorities, and/or measures.
Targets versus reality conflict	A version of operations conflict involving the counter-productive implementation of unrealistically "stretch" and/or "impossible" targets in an attempt to drive innovation.
Short-term mindset trap	A kind of "Performance Trap," which can occur when a company *is not* performing well and as a result focusses on short-term tactics at the expense of long-term strategy (*see also* the "Business Model Trap").
Business Model Trap	Another kind of "Performance Trap," which can occur when a company *is* performing well and focusses on its current successes at the expense of other opportunities (*see also* the "Short-Term Mindset Trap").
Commitment Trap	There are two versions of this: (1) when a company fails to commit to a new opportunity and extensive research, analysis, and/or testing inhibit or prevent progress or (2) when a company over-commits to an opportunity that no longer appears prospective, inhibiting or preventing retreat.
Micro-management challenge	Micro-management of R&D, product development, production, and/or marketing and sales teams.
Quality challenge	Insufficient quality practices/controls leading to rework and additional costs.
Structural implementation challenge	Inability to undertake innovation internally and/or to collaborate with partners (i.e. being "implementation poor") due to policy and/or resource constraints.
Commercialization challenge	Inability to monetize innovations, whether directly or indirectly (i.e. being "commercialization poor").
Corporate myopia – the primary culture trap	A corporate culture that focusses on current "*day-to-day*" *status quo* business practices to the exclusion of potential or even prospective future business practices (hence the analogy with nearsightedness). A variation of this is the "*not invented here (NIH)*" syndrome.
Corporate immune system	A corporate cultural trait in which the internal "*system*" reacts against anything that is perceived to be a threat to the continuance of current norms and practices and/or a threat to the overall system.
Technological myopia	When management overestimates the improvement potential of existing technology. It can easily take 5–7 years to discover this error, creating a barrier to the development of alternatives.
Misreading market signals	The tendency to discount the early signs of market penetration by new competitors, such as when a company defines its market too narrowly.

Corporate myopia – the primary culture trap. A corporate cultural trait that focusses on current *"day-to-day" status quo* business practices to the exclusion of potential or even prospective future business practices (hence the analogy with nearsightedness). Although such a culture can help with maintaining focus on the things that currently provide business value, it can also repel new ideas and practices that could benefit or even be critical to the organization in the future. According to Foster, *"several major corporations still give top R&D priority to the defense of their existing product lines"* [132]. A variation of this is the *"not invented here (NIH)"* syndrome, in which individuals, groups, or even entire organizations come to believe that they have a monopoly on knowledge in their field, leading them to resist ideas, inventions, and/or innovations that have been developed by others [442]. This frequently leads to either missing opportunities or to rediscovering, reinventing, or redeveloping things that were already available elsewhere (often referred to as *"reinventing the wheel"*).

The corporate immune system. A corporate cultural trait in which the internal *"system"* reacts against anything that is perceived to be a threat to the continuance of current norms and practices and/or a threat to the overall system. This is somewhat similar to the Corporate Myopia but involves being more generally resistant to changes of any kind, simply due to an aversion to change, as opposed to resulting from an inordinately short-term focus. Although such a culture can help protect against things that could harm the organization, it can also repel new ideas and practices that could benefit the organization. A 3M example is provided in reference [373]. Another example is provided by IP models that inhibit collaboration and, therefore, new approaches such as Innovation 2.0 (see Sections 3.1 and 4.2).

Technological myopia. When management overestimates the improvement potential of existing technology. For example, when a new technology has been successfully introduced into the marketplace and it is progressing rapidly "up" the technology "S-curve" (see Section 3.2), it can be *"tempting to assume that the rapid progress that occurs midway through the S-curve will continue"* [132]. According to Foster, the difference between reality and expectations is usually recognized too late, often something like 5–7 years too late on a 10- to 20-year technology S-curve [132]. The time lag alone can create a barrier to the development of alternatives. A variation on this theme is that technology development that is meant to address a particular challenge or opportunity may be no longer relevant by the time the development is completed because it simply took too long ("future technology myopia" [446]).

Misreading market signals. This is the tendency to discount the early signs of market penetration by new competitors. According to Foster, this sometimes occurs when a company defines its market too narrowly [132]. For example, in the United States, in the mid-1960s, all automotive tires were bias-ply tires, and many companies assessed their market performance in terms of the bias-ply tire market. Had they instead defined it as tires of any kind, they might have noticed the entry in to the market of the radial-ply tires, which went from a 20% to an 80% share of the market within 4 years [132]. Such a tendency is made all the more dangerous by the seemingly ever-increasing pace of change in the business sectors, in large and complex organizations, and in global events and affairs in general[91] [445].

In summary, the greatest barrier to technological innovation in an organization may be risk aversion, and the greatest antidote may be culture. Table 7.4 provides some questions that can be asked to test for the presence of innovation barriers.

The alternative to extreme risk aversion is not wild risk taking, but rather a balanced approach to risk management (not avoidance). Risk management practiced well involves everyone in the organization, not just the leaders and managers.

According to the 2005 AMA survey referenced earlier [380], possibly one of the trickiest balances to strive for in an organization is holding employees accountable without making them risk-averse. This suggests a focus on culture management. As has been noted several times in this chapter, culture can make or break organizational attempts to achieve innovation because culture so strongly affects the innovation processes themselves. Not surprisingly, several studies have found that organizations having a high degree of alignment between their culture and their innovation strategy experienced significantly better enterprise-value growth over time than have those with a low degree of alignment [373, 441]. This is well illustrated in a quote from Louis V. Gerstner Jr., who was the CEO of IBM from 1993 to 2002 [447]:

> *"The thing I have learned at IBM is that culture is everything."*

Table 7.4: Fifteen questions stakeholders should ask about an organization's innovation strategy. Adapted, in part, from [415, 448].

1	*Mission:* What are we here to do? How does our mission drive our focus?
2	*Vision:* What is our vision? Does it drive what to be best at, and of what "best" means?
3	*Benchmarking:* How will we know when we're the best?
4	*Awareness:* Does our organization need innovation? Why?
5	*Strategy:* Is innovation part of our strategy? Is it a core competency?
6	*Positioning:* Is our strategy to just stay alive or is it to be an innovation leader?
7	*Smarter-Not-Harder:* Do we consider smart new solutions over costly old ones?
8	*Groundwork:* Do we have "the right stuff": people, process, structure, tools, & safety?
9	*Approach:* Do we create new innovations or "adapt and adopt" others?
10	*Degree:* Do we aim for radical innovations or incremental innovations?
11	*Foresight:* Do we anticipate future trends, and are we adapting accordingly?
12	*Sustainability:* Do we invest to build/maintain our long-term innovation capacity?
13	*Perverse Policies:* Do we weed out policies/practices that inhibit innovation?
14	*Voice of the Customer:* Do we work with customers/clients to identify new innovations?
15	*People:* Do we remember that "*it's all about the people*" and engage our employees?

91 There is even an acronym for this, VUCA – volatility, uncertainty, complexity, and ambiguity.

7.7 Some icons of technological innovation

Having introduced some of the approaches that can be used in terms of the processes of technological innovation, it seems appropriate to pause and consider some examples of organizations that seem to consistently accomplish technological innovation well. Chapter 8 describes some approaches to measuring technological innovation, but for the purposes of this section, it is interesting to look at some of the organizations that are widely perceived by their peers to be leaders in this field.

Several organizations have polled senior executives for their views on the most innovative companies. The Boston Consulting Group, for example, regularly surveys more than 1,000 senior 'innovation' executives from about 70 countries. The top results from their 2023 survey [449] are shown in Table 7.5. Comparison with the corresponding table in the first edition of this book shows that of the top 20 companies of 2005, a few have moved up or down in the ranking, but most of the rest (15 out of 20) have been replaced entirely. Apple, on the other hand, continues to dominate the rankings, and has done so with amazing consistency since 2005 [450]. To cite another example, Table 7.5 also shows the "top 10" results from a 2018 Strategy[92] and "*Global Innovation 1000*" study of the world's 1,000 largest corporate R&D spenders [451]. Here again, there have been many changes since the 2010 survey results [153, 381]. highlighted in the first edition of this book.

These kinds of survey rankings do not tend to correlate well with such measures as R&D spending, sales, or R&D intensity (i.e. R&D spending as a percentage of sales); however, the top-ranked companies do tend to outperform their nearly 1,000 industry peers in terms of such measures as 5-year revenue growth, 5-year average earnings before interest, taxes, depreciation, and amortization as a percentage of revenue, and 5-year average market cap growth [377, 381].

Table 7.5: Examples of "most innovative companies" rankings.

Ranking	Boston Consulting Group, 2023 [449]	Strategy&, 2018 [451]
1	Apple	Amazon
2	Tesla	Alphabet
3	Amazon	Volkswagen
4	Alphabet	Samsung
5	Microsoft	Intel
6	Moderna	Microsoft
7	Samsung	Apple
8	Huawei	Roche Holding

92 Formerly Booz & Company.

Table 7.5 (continued)

Ranking	Boston Consulting Group, 2023 [449]	Strategy&, 2018 [451]
9	BYD Co.	J & J
10	Siemens	Merck
11	Pfizer	
12	J&J	
13	SpaceX	
14	Nvidia	
15	ExxonMobil	
16	Meta	
17	Nike	
18	IBM	
19	3M	
20	Tata Group	

Some innovation case studies for some famous innovation icons (present and past) are provided in references [145, 377, 452] and in the following list.[93] Note that the technological innovations for which these companies are known span each category of products, processes, and services.

3M Inc. An American company founded in 1902 as the Minnesota Mining and Manufacturing Company. 3M started out making abrasive sand for grinding wheels. When that didn't work out as well as planned, they shifted into sandpapers and developed waterproof sandpaper, and when that didn't work out as well as planned, they shifted into adhesives and developed masking and Scotch™ tapes. These early experiences caused 3M to develop a strong technological innovation culture and to develop strong systems and process to support innovation, leading to a number of famous products, including Post-it® notes, Scotch-Brite™ products, and Magic™ tape. For several decades, 3M has been able to generate at least 25% of its revenues from new products developed and commercialized within the previous 5 years (see [122, 453–455]).

Apple. From the original Macintosh personal computer and the firm establishment in the marketplace of the graphical user interface[94] (GUI), Apple has produced a series of evolutionary and/or disruptive innovations, including hardware innovations like the iPod, iPhone, iPad, and the Apple Watch and electronic innovations like icons (where would the GUI be without icons?), iTunes, and the App Store; often ranked as one of the most innovative companies (see [3, 61, 65, 122, 450, 456]).

Amazon. From its initial successes in online books sales, Amazon added a plethora of other platforms, including those that let almost anyone sell almost anything while (if they want) using Amazon's own capabilities in warehousing and shipping. Their business model enabled Amazon to

93 Some more broadly focussed, innovation-related case studies are provided in Section 10.2.

94 The GUI seems to have been invented by Ivan Sutherland in the United States, in 1963, then enhanced with the "mouse" by researchers at Stanford, then adopted and adapted by Xerox in the 1970s.

become a dominant online retailer: by 2013, its online annual sales volume of nearly US$70 billion nearly doubled that of the next three contenders combined [61, 457, 458].

Bell. It no longer tops the innovation lists of recent years, but Bell was one of the pioneers of US technological innovation. Bell was one of the first two (along with GE) large companies to establish dedicated industrial research laboratories in the United States. This was originally intended to help defend their existing markets but shifted soon thereafter to improve their manufacturing processes and to create new products and services. In the case of Bell, their industrial research lab (founded in 1911) was initially focussed on developing an electronic repeater for its telephone networks and on advancing the company into the radio market [145].

eBay. Possibly, eBay is the most famous innovator of global e-commerce products and services. eBay has gone from its founding in 1995 to becoming the giant of online auctions to a much broader online marketplace (that bears some similarity to Amazon's). They made their platform easier to use with their acquisition of PayPal, which enables convenient digital payments, and then developed eBay Enterprise, their multichannel retailing product for other enterprises (see [65, 459, 460]).

GE. This is one of the first two large companies (along with Bell) to establish dedicated industrial research laboratories in the United States. This was originally intended to help defend their existing markets but soon thereafter to improve their manufacturing processes and to create new products and services. In the case of GE, their industrial research lab (founded in 1900) was initially focussed on helping the company dominate the electric lighting market [145]. More recently, GE has been able to manage and sustain a culture of innovation through two generations of executive leadership. GE's theme of the new millennium has been "Imagination at Work" (see [61, 145, 461–463]).

Google. This was launched in 1998, Google has become the most popular web-based search engine, and Google has since diversified into other online products, such as Gmail, Google Maps, and YouTube. It is frequently ranked as one of the world's most innovative companies (see [3, 61, 65, 464–466]).

Microsoft. This was launched in 1975, Microsoft initially developed software products, the best known of which is their original operating system, *MS-DOS* (rebranded by IBM as *PC-DOS*). Possibly even more important was their licensing innovation by which they sold MS-DOS to IBM but retained the rights to further develop it themselves and to also sell it to others. As Microsoft evolved, they also brought to market such well-known software products as *Microsoft Windows*, *Microsoft Office*, and *Internet Explorer*. Microsoft later diversified into hardware products such as the *Xbox* and the *Microsoft Surface* (see [65, 467–469]).

Procter & Gamble (P&G). Another venerable icon, P&G was founded in 1837 to sell household and personal care products, beginning with candles and soaps. In 1879 J.N. Gamble (a son of the founder) invented a new and inexpensive white soap that became the company's most famous invention. In 1882, P&G began to nationally promote[95] this new "*soap that floats,*" called *Ivory*, which eventually became established as P&G's most famous technological innovation. As the company expanded, its internal R&D function kept producing new products (in 1890, e.g. P&G was selling more than 30 different types of soap). Among these were the first soap designed for washing machines (*Chipso*), the first synthetic detergent (*Dreft*), the first all-vegetable shortening for cooking (*Crisco*), and the first safety razor (*Gillette*). In the 1940s and 1950s, P&G introduced two more of its most famous technological innovations: *Tide* washing detergent and *Crest* toothpaste. P&G has been a leader in successfully applying the principles of open innovation, through its dedicated portal called "*Connect + Develop.*" In 2006, for example, about 45% of the initiatives in their NPD process contained substantive components that were sourced externally, and more than 35% of their new products had significant components that originated externally (see [65, 453, 470–472]).

Toyota. Toyota is generally known for quality, reliability, and dependability, but they have also been highly regarded for their technological innovation. Particularly in the 1980s, Toyota implemented innovations in automotive design, manufacturing processes, and in its market positioning. Another reason Toyota isn't necessarily famous for technological innovation is that its innovations are typically in the nature of an almost continuous stream of incremental and evolutionary innovations (such as the *Prius* hybrid), as opposed to disruptive innovations. One of their strengths is reputed to be Toyota's employee engagement, including its employee suggestion programmes (see [473, 474]).

Whirlpool Corp. Another long-standing corporation, Whirlpool attributes its modern-day success to taking a strategic view of technological innovation and to fully embedding innovation into their culture. In 1999, Whirlpool began a strategic transformation from being manufacturing driven to being consumer driven. This change required a leap of faith that consumers would be willing to pay a premium price for new and innovative products and services, and it required that Whirlpool take resources away from its ongoing business in order to devote them to developing new product and service innovations (see the *innovator's dilemma*, Section 4.9). Whirlpool has a formal NPD process, and for each stage, Whirlpool assumes that there will be a 1 in 10 survival rate. For ideation, Whirlpool uses structured ideation sessions, called "*idea labs,*" and an "*innovation pipeline*" (I-pipe) to log all ideas generated and to track their progress through the NPD process. In addition to assisting in the NPD process, the I-pipe enables employees to see what ideas have been explored and their current status, while management can use aggregate data to take a more strategic view (see [475, 476]).

95 One of P&G's marketing plans involved sponsoring radio programmes in the 1920s and 1930s. The associated P&G commercials for Ivory soap led to these radio programmes becoming known as "soap operas."

8 Can innovation be measured?

> *". . . anything unmeasured is invisible to policymakers."*
>
> Diane Coyle, "When capitalisms collide,"
> *Nature*, **2019**, *574 (14 October)*, 322–324.

> *". . . not everything that can be counted counts, and not everything that counts can be counted."*
>
> American Sociologist William Bruce Cameron
> In *Informal Sociology*, Random House, 1963, p. 13

Can innovation be measured? Yes. This applies to innovation in its broadest sense, and particularly to technological innovation.

Is it easy to measure innovation? No. One might think of financial measures (Figure 8.1) and that is part of the answer, but it is more complicated than that. There are at least four main challenges in this area:

1. There is no single indicator that can be used to measure innovation. A 2006 review by the Australian Parliament [304] observed that "*Due to the complexity of innovation systems, there is no single indicator that is capable of assessing all elements of innovation. Instead an array of measures is needed.*"
2. There are a vast number of potential indicators that can be measured. It is a challenge to identify a modest-sized set that can actually provide an overall measure.
3. Of the many things in innovation that can be measured, some are very easy to measure, but most of the easiest things to measure are actually the wrong things to measure.
4. Last but not least, another complication is the absence of standardization in this area. A 2013 Conference Board of Canada survey and report on business innovation [444] found that:

> *"Inconsistent definitions and models of innovation, and a lack of adequate measurements or 'metrics' to track performance, hamper managing innovation effectively. Various proxies for innovation (e.g., R&D, patents, publications) are seen as barometers for innovation. But they can also be seen as impediments that can 'muddle' one's understanding of true innovation performance. Innovation is complex and multi-dimensional."*

The development of indicators for tracking the progress of S&T (and later, innovation) began with OECD and the U.S. National Science Foundation in the 1960s. A historical overview of this development is given by Godin [477]. In this chapter, some approaches to measuring technological innovation will be described, and reference will be made to four categories of indicators as shown in Table 8.1.

https://doi.org/10.1515/9783112206720-008

Figure 8.1: Measuring innovation is much more complicated than just valuing a new idea. (Courtesy pxhere, image #1434871, CC0 public domain.).

Table 8.1: Categories of innovation indicators.

Leading or lagging indicator categories	Sub-categories	Explanation	Examples
Leading indicators	Capacity indicators	Creating the "base capacity" that will enable technological innovation to occur	Investment in knowledge
	Input indicators	Providing fuel to stages of the process in the hope that this will increase the rate of technological innovation	R&D intensity
Lagging indicators	Output indicators	Outputs from one or more stages of the innovation process	Triadic patents issued.
	Impact indicators	Indicators of successful technological innovation happening	Technology balance of payments

In general, the most readily measured and most commonly reported indicators are *"leading indicators*1,"[96] meaning that they are used to represent possible future performance:

96 Advocates and practitioners of workplace health and safety will recognize that this usage is by analogy with the leading and lagging indicators of best-practice occupational health and safety (OH&S) programmes.

- *"Capacity indicators"* relate to creating the *"base capacity"* that will enable technological innovation to occur. Some examples of base capacity are having a pool of highly qualified people, an array of suitable R&D facilities, and having a high degree of connectedness.
- *"Input indicators"* are relevant to base capacity but just as importantly also provide fuel to stages of the technological innovation process. Such fuel is important not only as an enabler but potentially as an accelerator of the rate of innovation. Examples include business- and government-led R&D expenditures, the number of technological breakthroughs, and the number of new technologically based companies formed. These latter two are only indicators of potential because the breakthroughs, for example, are only relevant if they lead to technological innovation, and the "tech" start-up companies, for example, are only relevant if they both survive as businesses and produce genuine and significant technological innovations.

Leading indicators have to be used with great caution since they may represent elements that are a necessary part of a healthy innovation system, but increases to any one, or even many, of the leading indicators will not necessarily create increases in the desired outcomes of more technological innovation.

Other indicators are *"lagging indicators,"* meaning that they represent the past. As such, they provide historical context and trending information but may or may not serve as reliable indicators of future performance:

- *"Output indicators"* relate to the outputs from one or more stages of the innovation process but do not, by themselves, necessarily provide indicators of actual technological innovation occurring. Examples include published discovery- or applied-research findings, patent applications, and numbers of patents and/or trademarks granted.
- *"Impact indicators"* are actual indicators of successful technological innovation happening. Examples include numbers of new products, processes, and services introduced into one or more marketplaces.

In the following sections, some specific indicators are discussed for regions and/or countries, commercial enterprises, and intermediary organizations (including RTOs). Innovation indicators for government labs and academic institutions are not explicitly discussed because there are not usually any direct connections between discovery research and technological innovation. Of course, discovery research is important, but it creates a base from which innovations can be developed, rather than a direct link in the chain. According to a US Congress evaluation, the factors that would need to be taken into account "are too complex and subjective; the payoffs too diverse and incommensurable; and the institutional barriers too formidable to allow quantitative models" [478].

8.1 Regional and national innovation indicators

The Innovation Theory of Economic Growth as conceived by Schumpeter and advanced by Solow-Swan and others (see Section 1.3) asserts that technological innovation is essential to long-term economic health and economic growth, meaning economic competitiveness with other economies. Regions and countries that wish to have world-class or even world-leading standards of living have therefore tended to focus on building strong economies with strong employment performance, hence a focus on technological innovation (although governments usually just use the broad term "*innovation*").

In the 1990s, it was said that "*Nations do not compete against one another – companies do*" [479]. This may have been true at the time; however, countries have increasingly been competing against one another for economic health and growth, using a range of tools – policies, subsidies, trade agreements/barriers, and even espionage (i.e. national security agencies conducting espionage for economic benefit). Strategies and tactics aside, all this interest in "*innovation*" has led, in turn, to interest in measurements, targets, and benchmarking of "*innovation*" at the regional, national, and international levels. For international benchmarking, many countries make comparisons to OECD countries, partly because so much of the available information in this area is collected and reported by OECD.

Stories frequently appear in the media on the topic of whether the amount and/or rate of development of technological innovation in a region or country is increasing or decreasing, and concerned with how such rates compare with those of other regions or countries (see, e.g. reference [480]). The so-called "*Innovation Gap*" is the real or perceived gap between desired and actual innovation performance, where the desired performance is normally set by comparison with other regions or countries. This topic is complicated by the fact that there are no generally accepted measures for the amount of technological innovation or for the amount of non-commercial innovation.

As a result, regions and countries, in their efforts to measure and benchmark innovation performance, have inevitably had to use a combination of direct and indirect indicators and even leading indicators[97] (such as measures of resources and activities directed towards innovation). Ideally, the principal indicators would be appropriate "*lagging indicators,*" meaning that they represent the past (such as measures of new products, processes, and services already introduced into marketplaces), hence reality rather than prediction. But it is much more common to use "*leading indicators,*" meaning that they are used to represent possible future innovation perfor-

97 Also termed "*proxy indicators.*" These are not measures of innovation or the innovation performance of an economy, but they are used as indicators that, taken together, may be indicative of current and/or future innovation performance.

mance (such as measures of resources and activities directed towards technological innovation) [479, 481–483].

A number of indirect indicators have been proposed and/or are being used for regions and nations, as illustrated in Table 8.2 [479, 481–485]. Some of the most common are as follows:[98]

- *R&D Intensity*, the relative amount of R&D spending from all sectors in an economy. An example of this is *"gross domestic expenditure on R&D"* (GERD), which is the total expenditure on R&D performed within a region or country during a given period of time. GERD is calculated by adding the expenditures in each of the four performing sectors: business, not-for-profit, government, and academic. GERD includes R&D performed within a region or country and funded from abroad, but it does not include payments for R&D conducted outside the region or country.
- *Business R&D Intensity*, the relative amount of R&D spending by just the business sector. An example is the ratio of business-enterprise-funded R&D to GDP.
- *Government R&D Intensity*, the relative amount of R&D spending by just government. Examples include *"government-financed GERD"* and *"government budget appropriations or outlays for R&D"* (GBAORD). Government-financed GERD, as the name implies, is simply the portion of GERD that is financed by the government. GBAORD is obtained from government budgets by identifying all the budget items involving R&D and estimating their R&D content in terms of funding. GBAORD therefore represents government R&D spending intentions rather than actuals.
- *Changes in R&D Spending*, such as indicated by changes to any of the above indicators over time.

The above list illustrates a second weakness of such indicators. Not only are they indirect, by focussing solely on R&D spending, they are also based on only a few segments of a linear model of innovation, which, as discussed in Chapter 2, is rarely how technological innovations are actually developed. As leading indicators, these still have some value because R&D spending by businesses and governments at some level is a necessary, but not sufficient, condition for technological innovation, as discussed in Chapters 3, 4, and 5. The problem is that it is rarely clear when enough is enough, and beyond a certain base level of spending, other factors (the other elements in the nonlinear innovation models), become more important.

98 Regional and national indicators are frequently reported as a ratio in which the indicator is divided by the GDP of the region or country. GDP is the total economic productivity of a region or country during a specified period of time and is different from business sales in that it takes into account the sales value of the goods and services produced less the value of the goods and services consumed during the production. Thus, GDP can potentially increase with increased production, increased sale prices, and/or reduced costs of production.

Table 8.2: Examples of innovation indicators for regions or countries [481–483, 486].

Indicator	Explanation	Background
Multifactor productivity (MFP)	MFP is an indicator of the impact of changes due to technological innovation and sometimes as an indicator of the amount of technological innovation	An industrial measure of the change in output per unit of combined inputs (labour, materials, and capital), due to organizational and/or technological innovation
Total factor productivity (TFP)	TFP is an indicator of the influence of technological innovation on labour productivity	
Innovation "market" share	The market share of commercialized products/processes/services	Could be in terms of revenues, profits, or numbers of organizations involved
R&D intensity	The ratio of gross domestic expenditure on R&D to gross domestic product (GERD/GDP)	A broad indicator of R&D investments from all sectors in an economy
Business R&D intensity	The ratio of business-enterprise-funded R&D to gross domestic product (BERD/GDP)	An indicator of R&D investments by and within the business sector
Government R&D intensity	The ratio of government budget appropriations or outlays for R&D to gross domestic product (GBAORD/GDP)	An indicator of R&D investments by governments
Investment in knowledge	The ratio of total knowledge investments to gross domestic product	"Knowledge investments" include higher education, R&D, and software
ICT investment intensity	The ratio of ICT expenditure to gross domestic product	Expenditures in the information and communication technology (ICT) sector
R&D personnel	The number of researchers involved in R&D per 10,000 personnel in the labour force	An indicator of the total number of people directly involved in R&D in an economy
Patent applications	The number of patent applications* per 10,000 personnel in the labour force	An indicator of the total number of practical inventions being protected in an economy
External patent applications	The number of external patent applications per 10,000 personnel in the labour force	An indicator of the total number of significant**, practical inventions being protected
Triadic patents issued	Patents in triadic patent families issued per 1 million people in the population	Triadic patent families are groups of patents that have been granted in multiple (three or more) countries on the same invention

Table 8.2 (continued)

Indicator	Explanation	Background
Percentage of patents with foreign co-inventors	100 times the ratio of the number of patents having foreign co-inventors to the total number of patents issued in a country	An indicator of the degree of connectedness and collaboration among a country's inventors and those in other countries
Trademarks	The number of trademarks registered per 10,000 personnel in the labour force	An indicator of the total number of new products and services being protected in an economy
Connectedness Index (ICT)	A blended indicator calculated based on such product and service factors as the market-ready supply of infrastructure, networks, and systems, demand, price, and usage	Example: The Conference Board of Canada Connectedness Index
Ease of Entrepreneurship Index	A blended indicator, calculated based on such factors as barriers to competition, regulatory and administrative opacity, and administrative burdens	Example: The Conference Board of Canada Ease of Entrepreneurship Index
Venture capital intensity	The ratio of venture capital investments to gross domestic product	
Technology balance of payments	The net technology transactions (purchasing power parity US$) per 10,000 personnel in the labour force	The balance of sales versus purchases of technology (such as patents, licences, designs, trade-marks, and trade secrets) in an economy

Each of these would be with regard to a specified period of time.
* This can be taken to be either the national patent applications (meaning by both residents and non-residents) or the resident (only) patent applications. The former includes an indication of potential technology diffusion to come from other countries).
** Here, significant means inventions of such impact that they are worth protecting in other countries beyond that in which the inventions were made.

Similarly, some of the indicators relate to elements that form essential infrastructure for technological innovation, such as investments in "knowledge,"[99] R&D personnel, and "connectedness,"[100] but, again, these are necessary, but not sufficient, aspects. So, beyond certain base levels, the other elements in the non-linear innovation models become more important. To the extent that the degree and nature of S&T, for exam-

99 "Knowledge" investments are usually taken to include higher education, R&D, and software.
100 The availability and use of information and communications technology (ICT) to enable communications, information flows, and trade.

ple, are components of the commercialization and sales of technological innovations, there exist dedicated indicators. An example of this is the set of "main science and technology indicators" (MSTIs) developed and used by OECD. The OECD MSTIs cover selected inputs, such as R&D spending and numbers of researchers; activities, such as R&D activities by sector; outputs, such as patents; outcomes, such as exports or export market share; and impacts, such as balances of trade or payments, and international trade, jobs, and GDP growth [486].

Inspection of the other indirect indicators listed in Table 8.2 shows that most of the remainder focus on patents and trademarks and are therefore based on an assumption that there are correlations among inventions, patentable inventions, and technological innovation. As discussed in Section 2.1, this is rarely the case because, in practice, most inventions are only minor improvements on the prior art and therefore not patentable, and most patented inventions do not have significant commercial value and are therefore not connected with technological innovation. To the extent that patenting is a component of the commercialization of some technological innovations, there exists a host of related indicators ranging from numbers of patent applications (as an indicator of inventive activity), to numbers of patents granted in multiple countries (as an indicator of practical invention production having international potential), and even to indicators of the relative strength of a country's IP laws (e.g. the Ginarte-Park Index [487, 488]).

Productivity. A key, direct, lagging indicator for regional or national innovation is productivity. For example, "*multifactor productivity*" (MFP) is an industrial measure of the change in output per unit of a number of combined inputs (including labour, materials, and capital). Any change in MFP is due to factors other than the combined inputs, such as technology changes, change of scale efficiencies, production changes, management changes, and changes due to technological innovation. Although it is only one of these other factors, MFP is sometimes used as an indicator of the impact of changes due to technological innovation and sometimes as an indicator of the amount of technological innovation [489]. This could be for an industry or for an individual enterprise.

A related measure that underlies labour productivity is termed "*total factor productivity*."[101] This is obtained by dividing labour productivity into three principal components representing contributions from educational attainment, capital input, and TFP, where TFP represents the influence of technological innovation [322].

Two indicators of overall technological innovation in a region or nation are *multifactor productivity* (MFP) and *total factor productivity* (TFP).

101 TFP is also termed "*Solow's Residual.*"

Competitiveness. Another key, direct, lagging indicator for regional or national innovation is competitiveness, that is, the ability of a region or country to comprise competitive environments that offer and supply products, processes, services, or jobs that effectively compete with others in the same market(s). This follows from the Schumpeterian model of technological innovation discussed in Section 1.3. Accordingly, the so-called "*innovation gap theory*" proposes that a region or country that achieves new technological innovations will have a competitive advantage over their competitors until the gap is closed. This theory is sometimes extended by assuming that other regions or countries will have to import the new innovations until they can be replicated or substituted within their own regions. In this sense, an innovation gap can stimulate international trade.

Competitiveness can be a difficult comparison to make due to the varied nature and developmental history of different regions or countries (see, e.g. Chapter 8 in reference [490]). The approach taken by the World Economic Forum (WEF) is to consider each country's stage of economic development maturity, using three broad categories of competitiveness drivers and different weightings for calculating a global competitiveness index (GCI) within each category [491]. In the WEF categorization, *stage 1*, or *factor-driven economy*, refers to countries having GDP per capita of less than US$2,000; *stage 2*, or *efficiency-driven economy*, refers to GDP per capita within the range US$3,000–8,999; and *stage 3*, or *Innovation-Driven Economy*, refers to GDP per capita of more than US$17,000. *Economies in transition* lie between stages 1 and 2 or between stages 2 and 3.

Depending on the category, the WEF weightings are adjusted in an attempt to make the GCI appropriate to a given country's degree of economic development and, therefore, the manner in which it competes globally. For example, a factor-driven economy competes mostly based on the strength of its "basic requirements" such as natural resources, infrastructure, and/or labour, whereas an efficiency-driven economy competes mostly on the strength of its "efficiency enhancers," such as higher education and training, technological readiness, and market size, and an innovation-driven economy competes mostly based on the strength of its "innovation and sophistication factors," such as capacity to innovate, value-added manufacturing, and value chains [491].

The GCI itself is a composite index whose principal components are "basic requirements," covering institutions, infrastructure, and the economic environment; "efficiency enhancers," covering higher education and training, the labour market, technological readiness, and market size; and "innovation and sophistication factors," covering business sophistication and capacity to innovate [491]. As an illustration, Canada's GCI ranking places substantially more weight on the value of its innovation-driven performance: basic requirements (20%), efficiency enhancers (50%), and innovation and sophistication factors (30%), whereas for India, the GCI ranking shifts its emphasis to factor-driven performance: basic requirements (60%), efficiency enhancers (35%), and innovation and sophistication factors (5%) [491]. A GCI ranking illustration is provided in Chapter 9.

Competitiveness indicators can also be taken to the level of factors affecting entities within a region or a country's innovation ecosystem. For example, the "*ease of entrepreneurship index*" is a blended indicator calculated based on such factors as barriers to competition, regulatory and administrative opacity, and administrative burdens. A specific example is the Conference Board of Canada's "*ease of entrepreneurship index*" [492].

An indicator of overall technological innovation in a nation is the *Global Competitiveness Index* (GCI).

Finally, as the measurement and indicator tools for technological and non-commercial innovation continue to evolve, there is also a broadening underway by which indicators beyond technological innovation are being developed, such as for social innovation (see, e.g. the discussion in reference [493]).

An indicator of the net benefits to society of investment in applied R&D is the "*Social ROI on R&D*"[102] (SROI on R&D). The SROI on R&D is usually expressed as the increased value produced for society associated with an investment, divided by the amount invested. The value component needs to be specified. It can comprise such things as social value, economic value, environmental value, or a combination of these. In the case of social or environmental value, some kind of financial proxy is usually developed in order to be able to complete the calculation. For example, one form of SROI on R&D is net incremental industrial productivity achieved within a specific time period divided by investment in applied R&D over the same time period.

According to a U.S. Congress evaluation, economists have found a strong positive correlation between private sector R&D spending and economic growth, finding "*private returns in excess of 20% per year and social returns in excess of 40%*"; however, they had "*not been able to show comparable returns, and at times been unable to show any returns*" on federal government department and academic R&D expenditures [478].

8.2 Innovation indicators for commercial enterprises

Should companies attempt to measure and track their innovation performance?

According to a 2013 survey conducted by the Conference Board of Canada, almost 40% of surveyed companies do not use any innovation metrics [443]. Except for the small fraction of companies that genuinely do not need to pursue technological innovation, this is almost certainly a mistake. Some studies (not many seem to have been

102 Also termed "*Social Rate of Return on R&D*" or "*Social Internal Rate of Return on R&D*."

conducted) have reported a correlation between the use of innovation metrics and overall corporate performance [414, 443]. Some companies measure only a few key things, while others measure a broad mix of indicators.

More difficult questions are which and how many indicators to follow. Some first principles for selecting a balanced portfolio of innovation indicators might be the following (adapted from Cohn, 2013 [443]):

1. Align with corporate strategy
2. Cover all key aspects of innovation in the organization
3. Address historic, present, and forward-looking innovation capabilities
4. Focus on the most important (highest business impact) aspects
5. Measure consistently over a significant period of time
6. Enable benchmarking with comparable organizations

The simplest lagging indicators for commercial enterprises are measures of the extent to which they have introduced new or significantly improved products, processes, or services into the marketplace within a certain period of time. The reported use of this kind of indicator dates back to at least the early 1950s, when Maclaurin, for example, developed an indicator he called "*technological progressiveness,*" by which he classified the performance of different industries as high/medium/low in terms of having introduced "important" new or improved products or processes [62, 105].

3M, for example, uses two key company-wide innovation metrics: the "*New Product Vitality Index,*"[103] covering the last five years of activities, and the ratio of R&D investments to sales. In addition, there are secondary innovation metrics for each business unit to measure its specific innovation targets [414].

Return on innovation investment (also termed "*ROII,*" or "*ROI2,*" or "*ROInn*") is an overall measure of the financial ROI in innovation for one or more (usually specified) innovations within a specified time period. This evolved from companies' need to compare their performance with "benchmarks."[104] A simple ROII is the revenues derived from sales of a new product, process, or service, less the cost of sales associated with manufacturing, marketing, selling, and shipping or providing them, with the difference being divided by the costs associated with ideating, developing, and commercializing the innovation. A slightly different example of ROII for intermediary organizations is given in Section 8.3:

103 3M's NPVI is used to help quantify the amount of its business and/or business growth originating from products that were introduced to the marketplace within the previous 5 years. The formula is NPVI = (sales from products introduced within the past 5 years)/(total sales).

104 Benchmarking refers to the process of comparing an organization's processes and performance measures to the best practices of a specified peer group.

$$\text{ROII} = \frac{(\text{revenue from new product sales}) - (\text{cost of sales})}{(\text{product development expense})}$$

A related measure is the return on product development expense (RoPDE), in which the product/process/service development expense is subtracted from the numerator in the ROII equation:

$$\text{RoPDE} = \frac{(\text{gross margin}) - (\text{product development expense})}{(\text{product development expense})}$$

Here, the gross margin is the revenue from new product sales minus the cost of sales.

Several approaches have been developed for assessing some combination of the capacity, effort, and/or potential for technological innovation at the individual business level [479]. These tools can involve large numbers of indicators. For example, one survey tool used by the Conference Board of Canada, the "*Index of Corporate Innovation*," has 25 capabilities or aptitudes [494], whereas another survey tool focussed more on innovation management practises has 75 [495]. A very short such index tool, using only three indicators, could be the following:

1. Over the past five years, how many new or significantly improved products, processes, or services did your company introduce to market?
2. For the answer to the above question, what proportion did this represent of the total number of company products, processes, or services that were available for sale?
3. Over the past five years, what proportion of company revenues represented the sales of new or significantly improved products, processes, or services?

There are, of course, many more potential indicators. Technological innovation measures can be divided into four broad categories: measures specific to technological innovations per se, financial and/or market performance indicators, business process metrics, and measurements of employee performance. Tables 8.3 and 8.4 show some examples of lagging and leading innovation indicators[105] that have been proposed for organizations [83, 111, 230, 370, 380, 443, 496, 497]. Some indicators for the implementation of open innovation in SMEs, specifically, are suggested in reference [498].

One can also look at indicators of innovation leadership management. The lagging indicators, in this case, would probably be the lagging indicators of innovation itself (see, e.g. Table 7.3 in reference [499]). There is some work on leading indicators of innovation management, framed in terms of leading and managing an innovation

105 In some work, a distinction is drawn among inputs, outputs, and outcomes. For comparison, the leading indicators discussed here would be approximately the same as (inputs + outputs) and the lagging indicators would be approximately equal to the outcomes.

Table 8.3: Examples of Lagging Innovation Indicators for Organizations. Each of these would be with regard to a specified period of time. Some of these indicators could also be calculated per organization unit or product line [92, 111, 230, 370, 380, 443, 497].

Lagging indicators	Explanation
Multifactor productivity (MFP)	MFP is an indicator of the impact of changes due to technological innovation and sometimes as an indicator of the amount of technological innovation. *See* Table 8.2.
Total factor productivity (TFP)	TFP is an indicator of the influence of technological innovation on labour productivity. *See* Table 8.2.
Customer/client satisfaction	Probably the most important indicator of all
New products, processes, or services launched	Could include the numbers of new products, processes, or services or their longevity, values, or product line growth
Sales due to innovation	Could include the numbers and/or proportions of total sales of new products, processes, or services
Financial returns from innovation	Revenues and/or profits from the sale of new products, processes, or services
Productivity gains from innovation	Productivity gains from the internal implementation of new products, processes, services, or technologies
Innovation resource effectiveness	Ratio of revenues from commercialization of new products/processes/services to the number of full-time equivalent employees dedicated solely to technological innovation
Incremental revenues per "innovation" employee	Technological innovation revenues per employee devoted to innovation initiatives
Market size/share gains from innovation	Market size and/or market share gains from the sale of new products, processes, or services

Table 8.4: Examples of leading innovation indicators for organizations. Most of these would be with regard to a specified period of time [111, 230, 370, 380, 443, 497].

Leading indicators	Explanation
Customer satisfaction gains	Improvements in customer satisfaction due to technological innovation-driven improvements
Executive involvement	An indicator of an organization's interest and/or commitment to innovation
R&D spending	An indicator of an organization's investment in R&D
Technology investments	Investments in machinery, equipment, and advanced technology, forming an indicator of an organization's investment in R&D

Table 8.4 (continued)

Leading indicators	Explanation
Researchers	An indicator of the total number of people directly involved in R&D in the organization
New idea generation rate	A measure of the rate of employee generated, or led, ideas for improvement and potential technological innovation
Estimated market value of innovation in the "pipeline"	An indicator of the expected future return on technological innovation spending
Innovation process pipeline flow	The number of new product concepts in each stage of the new product development process
Speed to market	Average time from idea to market launch per new product/process/service launched
Value to market	Average ratio of new product/process/service revenue to R&D and production costs
Product innovation intensity	Ratio of R&D costs for new products/processes/services launched to the overall corporate R&D expenditures
Patent applications, patents issued	An indicator of the total number of practical inventions being developed and protected by an organization
External patent applications, patents issued	An indicator of the total number of "significant" practical inventions being developed and protected (i.e. significant enough that they are worth protecting in other countries beyond that in which the inventions were made)
Triadic patents issued	Triadic patent families are groups of patents that have been granted in multiple (three or more) countries on the same invention
Trademarks	An indicator of the total number of new products and services being protected by an organization
External awards for innovation	May provide comparison with innovation performance of peer organizations
Existence of internal Innovation champion(s)	An individual who brings internal and external players and processes together to drive business processes needed to identify and evaluate opportunities
Customer, supplier, and partner engagement for new ideas	Leading innovators bring these stakeholders "in" to help identify and evaluate opportunities. Helps ensure understanding of the markets
External partnerships and collaborations	External partnerships and collaborations with customers and suppliers, as an indicator of an organization's level of collaboration with outside parties

Table 8.4 (continued)

Leading indicators	Explanation
Formalized business Development processes	Leading innovators use formalized business processes and practices to help identify and evaluate opportunities
Global reach	Extent to which global activities are used to source new ideas and opportunities, not just markets for exported goods and/or services
Product, process, or service feature/ performance gains	Improvements in features and/or performance due to technological innovation
Spinoffs or new operations from technological innovation	Spinoffs and/or new operations from new products, processes, or services

strategy, an appropriate organizational culture, a supportive environment for the needed creative processes, and an NPD investment practice (see reference [500]).

8.3 Innovation indicators for intermediaries (including RTOs)

To the extent that intermediary organizations are engaged in enabling technological innovation, it should be, and is, possible to track appropriate indicators for this and to use them as mission-performance indicators. Doing so is, however, a relatively recent phenomenon.

In the 20th century, RTOs became interested in the economic impacts of their work, but they tended to believe that there was no way to measure them at an organizational level. In the 1940s, for example, the NRC found that it was "*difficult to estimate the costs of research and extremely difficult to estimate the returns from this investment*"; however, sometimes, "*a single application of scientific knowledge does lend itself, for a little while, to estimations of dollar value that are clear, simple, and convincing*" [501]. Similarly, the SRC reported in 1950 that "*It is difficult to assess the return on monies invested in research; certainly no attempt will be made to arrive at an estimate in this report . . . A successful conclusion to almost any of the projects . . . will offer . . . a potential return far beyond the amounts invested*, and in 1956 that "*It is not possible to assess precisely the value of the applications, but it is probable that they result in savings . . . that are greater than the total expense of the Council up to the present and currently foreseen*" [290].

Also, in the 20th century, other intermediary organizations, such as government agencies and universities, started to measure and report on leading indicators comprising.

– Inputs, such as the numbers of students entering, faculty/researchers being hired, "Chairs" and "Centres of Excellence" being established, new buildings built, new

instruments purchased, and the numbers and sizes of research grants and contracts obtained.
- Activities, such as programmes and projects underway, interactions with others, quality of the scientific and engineering research, effectiveness and/or efficiency of programme management, and the effectiveness of user facilities (in terms of uptake, etc.).
- Outputs, such as numbers of students graduated, research papers presented and/or published, projects and/or contracts completed, new technologies, patents applied for and/or granted, technology/market evaluations, technologies licensed or sold, and companies started-up and/or spun-off.

Inspection of the above examples shows that they are almost all leading indicators at best. Even the few that come close to technological innovation results, such as technologies licensed or sold and companies started up and/or spun off, are still only at the output, not the impact stage (Figure 8.2), although successfully commercialized technologies and successful spin-off companies certainly contribute economic impacts over the time they are active in the marketplace.

Example: INO and Spin-offs. One of Canada's leading RTOs is L'Institut national d'optique (INO, National Optics Institute), which was created in 1985. As the name implies, most of its work relates to research, development, demonstration, and commercialization of optics and photonics technologies (the latter referring to the creation, manipulation, and detection of light in practical applications). Although INO regularly develops and transfers technologies to its industry clients, it also has a strong record of transferring its technologies into spin-off companies, the first being Nortech Fibronic in 1989. In many, if not all, cases the pattern has been that one or more INO employees turn private entrepreneur(s) and establish the spin-off company with INO's assistance[106] [502]. Such assistance has included transfer of the appropriate technologies under a licencing agreement plus a further agreement under which INO provides additional applied R&D support to the new company, which helps ensure their success while providing additional revenue for INO. In many cases, these new companies have even continued to work with, and even for, INO on technology development projects. By 2025, INO had nucleated 36 spin-off companies (see references [503–505]), most of which have been successful.[107] This is a remarkable success rate given that the average success rate for start-ups globally is only about 10%.

As more and more spin-off successes of its own technologies were achieved, INO added an *Entrepreneur-in-Residence* programme to facilitate coaching for the fledgling new external companies, which ultimately led to the creation of Quantino™ – a dedicated, in-house entity, to serve as a business incubator for the optics/photonics sector by providing incubatees with such services as: business coaching, technical development support, laboratories and specialized facilities (optical lab, electronics lab, mechanical lab, workspaces), access to the business ecosystem, and

106 In fact, several INO employee/entrepreneurs have left to found the spinoff companies and then later left the successful spinoffs to return to INO to either reprise the cycle or move up into senior management.

107 Successful meaning that the companies are still in business or have been acquired by other firms with the business line remaining active.

even access to federal and provincial funding support [504–508] With Quantino in place, INO has been able to assist external, as well as internal, entrepreneur/incubatees [506, 509]. In 2023/24, for example, they supported 27 external companies in this manner [504], with numerous success stories published [506].

Even in the 21st century, government and academic institutions do not normally seem to measure or report the numbers of such technologies or companies that fully enter the competitive marketplace, nor statistics on whether they survive for reasonable periods of time.

Some intermediary organizations, such as government agencies and universities, rely on indirect rather than direct economic measures. For example, the activities conducted by an intermediary organization will, to some extent, be multiplied in the economy through increased spending on the jobs, goods, and services needed to conduct those activities. Indirect economic measures are estimates because they attempt to account for indirect effects (sometimes termed *spillover effects*) that are usually not directly or easily measurable. Indirect economic measures are also of limited value because they rarely amount to more economic impact than would have accrued if the money had simply been spent directly in the economy, without any R&D having been performed at all.

Research and technology organizations. In the late 1900s and early 2000s, RTOs in particular came under increasing pressure to measure and demonstrate the outcomes and the impacts of their R&D work and, therefore, the ROI that their work delivers regardless of whether the original investments came from internally or externally generated funds, from the public or private sectors, or both. As a result, many RTOs began to track the project-specific impacts that their industrial clients were achieving, in such areas as product development and/or production cost savings, sales and export increases, job creation and layoff/bankruptcy avoidance, and NPDs and launches. This helped the RTOs to benchmark internally; however, project-specific often meant client-specific, and the RTOs were usually not permitted to publicly report the results in any way that might reveal the identities of the clients themselves. As a result, the best practices for impact reporting for RTOs were initially based on "Indicator-Based Frameworks" and "Case Studies" [510].

For economic impacts, some RTOs have advanced to a higher level by attempting to conservatively estimate actual incremental (direct) economic impacts.

An example of the development of an economic impact audit process for RTOs that would ensure that solid data would be collected that could provide a credible indication of direct economic impacts was begun at the Alberta Research Council in the late 1990s [511, 512] and further developed by the SRC in the early 2000s [513, 514]. The result was an R&D impact assessment tool that relies on "*the voice of the customer*" to provide data that can be aggregated and scaled up to provide conservative estimates of R&D ROI attributable to the RTO. Such audit methodologies tend to have high credi-

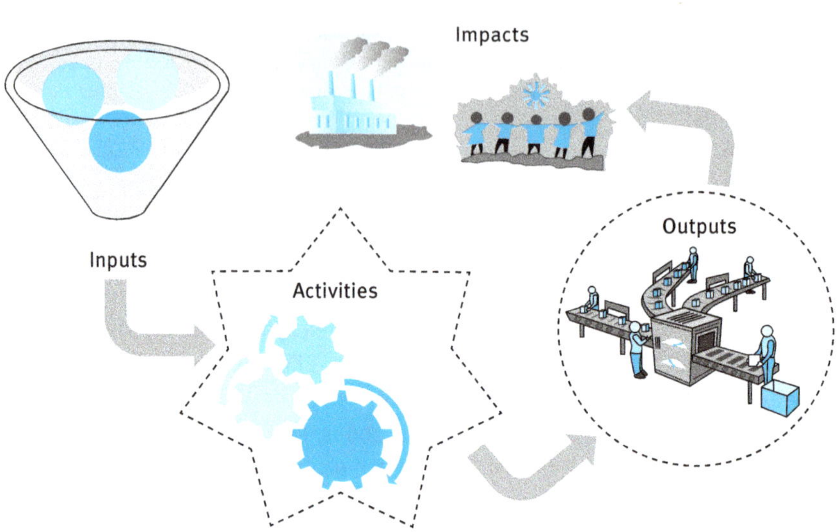

Figure 8.2: Illustration of the pathway to technological innovations and economic activities, from inputs and activities to outputs and impacts.

bility because the data come directly from the clients, who are the direct users of the RTO's services and probably the best judges of the value of such services [513, 514].

In this example methodology, since it is not practical to interview every client every year, an annual audit-basket selection is made of past projects that have been completed for private-sector clients. The kinds of data collected include sales and other revenues, cost savings, new revenue flowing into the economy, job creation and maintenance, and increased productivity – all resulting from the technological innovation assistance provided to the clients by the RTO. The bulk of the data is collected directly from clients through face-to-face interviews conducted for programmes that are either mature or completed (since there is generally a lag between the time an R&D project is conducted and the time at which the results have been implemented into commercial practice and business returns on investment are realized; see Figure 8.3). The result of these interviews is a repertoire of "*Impact Audit Reports*" for each client, outlining details on the work, its benefit to the client, as well as economic data. In addition to hard data, some clients are able to provide qualitative information on how well the RTO's work benefited their business. Although such qualitative information cannot be easily aggregated, it does provide additional feedback on difficult-to-measure yet valued contributions to the clients' successes. Finally, a key feature of these case studies is to have a senior executive or manager from the client firm authenticate and sign-off on the reports (see reference [514]), making them auditable and helping to ensure accuracy of reporting on the impacts and also on the attribution of the RTO's contribution(s).

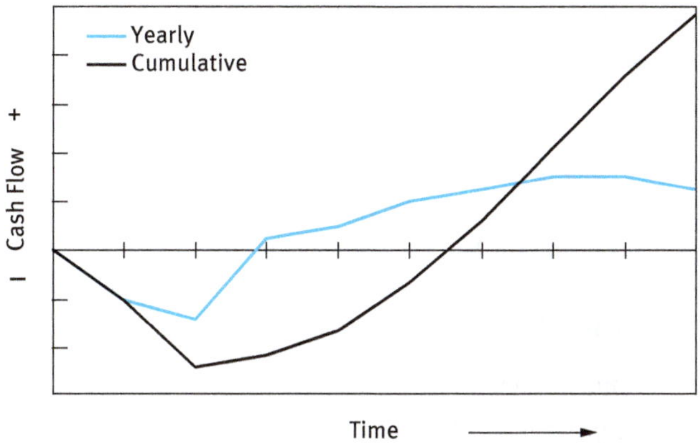

Figure 8.3: The lag time in realizing impacts from R&D. Courtesy Saskatchewan Research Council [413].

Once the impact data have been collected, several kinds of calculations can be performed to obtain conservative estimates of overall incremental, direct economic impacts for the audit year. Figure 8.4 shows an example of such results as reported by the SRC. These results show that between 2003 and 2017, SRC achieved over $7.6 billion in cumulative direct economic plus employment impacts in Saskatchewan.

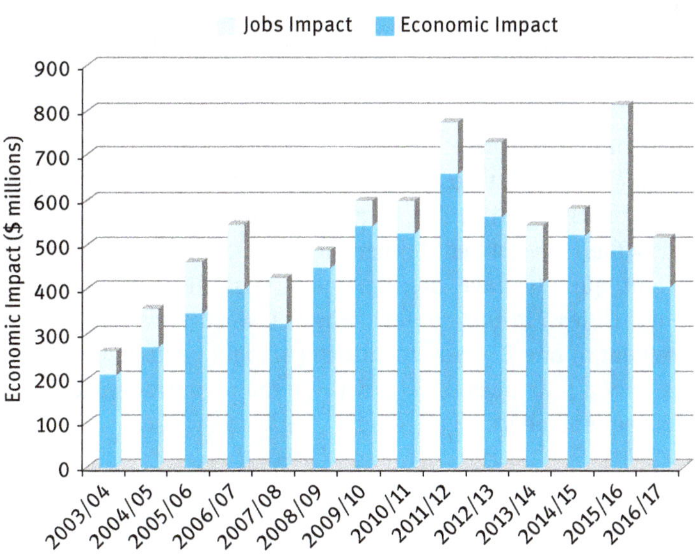

Figure 8.4: Annual incremental direct economic and jobs impacts from innovation-enabling on the part of the Saskatchewan Research Council.

The above approach to estimating economic impacts of RTOs is just one of several. Examples of different econometric approaches, as applied to RTOs, are provided for the European RTOs in reference [515] and for the SENAI ISI institutes of Brazil in reference [516].

Impact Ratios. For RTOs, or any other intermediary organizations that receive government funding,[108] measures of direct economic and jobs impacts can be translated into returns on innovation investment (ROII, or ROI2, or ROInn; *see* Section 8.2). RTOs themselves tend to use slightly different language to refer to the ROInn, such as "*mandate effectiveness ratio*" (MER) or "*Mission Effectiveness*":

$$\text{MER} = \frac{(\text{RTO's annual incremental economic impact})}{(\text{Annual 'base finding' investment in the RTO by government})}$$

A related measure is the "*financial effectiveness ratio*" (FER):

$$\text{FER} = \frac{(\text{RTO's annual incremental economic impact})}{(\text{Total annual revenue of the RTO})}$$

Using the FER normalizes the impact by the size of the organization, and is useful when making comparisons with organizations that do not receive annual base finding from government.

Worldwide, an increasing number of RTOs are measuring and reporting their impact effectiveness, although the methodologies used can vary widely. The mandate effectiveness ratios being reported by RTOs range from lows of around 1 to highs of about 35.

Given that RTOs, regardless of incorporation status, are critically dependent on a substantial level of patient, base government funding, achieving an MER of not much more than 1 could be economically disastrous. The WAITRO international benchmarking study of best practices for RTOs [291], referred to in Section 5.2, found that RTOs that were not able to satisfy their government funders' needs tended to lose their government funding and subsequently have their primary missions fundamentally changed and/or go out of business. In this context, it is noteworthy that in the five years following an assessment of one RTO's economic impact and value to their government owner as being only a few times their cost [287], the organization suffered huge funding cuts and consequent job losses [517–519].

108 This refers to the direct, non-targeted, "base funding" provided by government, whether granted or contracted, as distinct from government revenues arising from specific R&D contract projects.

8.4 A cautionary note

Before concluding this chapter, it is probably worth pointing out that there is an inherent risk involved in the over-emphasis and/or over-use of innovation indicators. The risk arises because organizations and their people will be asked to adapt to goals and targets in ways that seek to maximize the measured results. If this is done by doing the kinds of things that the measures are intended to be reflective of, then this is fine. The danger is that there are often ways to make "the numbers look good," without actually having done what was intended in the setting of the measures and targets in the first place.

Cohn and Good have referred to this as "*value degradation*" and report that "*People and organizations tend to adapt behaviour to maximize measured results, even if they are not the most important results. Using the same metrics over long periods sometimes means that they no longer reflect the critical competitive imperatives of the company even as people become ever more skilled at showing 'progress' through the metrics they use*" [443].

British economist Charles Goodhart put this even more bluntly, writing that "*when a feature of the economy is picked as an indicator of the economy then it inexorably ceases to function as that indicator because people start to game it*" [520]. This is "*Goodhart's law,*" which is more popularly stated as "*When a measure becomes a target, it ceases to be a good measure.*"

Cohn and Good's antidote for the "*value degradation*" effect is for an organization, or region, or country to change their key indicators from time to time in order to avoid "*metric mastering for show rather than for value.*"

Goodhart's Law: "When a measure becomes a target, it ceases to be a good measure."

9 Looking forward

It is probably a rare region or country that does not feel that it is in some way underperforming when it comes to technological innovation and productivity growth compared with other competing regions or nations. This could possibly apply to Singapore, which is frequently ranked no. 1 by the WEF in terms of global competitiveness [521, 522]. For the rest of the world, comparative global competitiveness is a perennial issue of concern, if not preoccupation.

The WEF's top 15 most globally competitive economies for 2019 are shown here [522]:

- Singapore
- United States
- Hong Kong
- Netherlands
- Switzerland
- Japan
- Germany
- Sweden
- United Kingdom
- Denmark
- Finland
- Taiwan, China
- Korea, Rep.
- Canada
- France

While the issue for the leading nations is how to at least sustain their position, for other countries, the challenge is to try to improve their positioning.

Part of the difficulty is complexity, in that there are a number of necessary components, most if not all of them need to be present, and these components need to be effectively linked into larger systems – regional and national innovation ecosystems. Having only one or a few of the necessary components is not sufficient for reliable, sustainable, economic success. For example, although there is a strong link between technological innovation and economic growth, there is no correlation between strength of a country's discovery-science base and economic performance. Worse yet, a 2005 OECD study found that countries that perform well in terms of their science base *do not* perform well in terms of innovation [120]. The issue is not that a strong discovery science base is not necessary – it is. The issue is that other components are also needed, and they all need to be effectively inter-connected and dynamic.

https://doi.org/10.1515/9783112206720-009

9.1 Evolving technological innovation models and systems

There is only a strong link between R&D and technological innovation if the right kind of R&D is performed. For example, the United Kingdom and Canada have strong academic research systems and are leaders among the G7 countries in spending on R&D in academia. However, their records of translating new knowledge into technological innovation and economic competitiveness have been relatively weak by international standards. This is because discovery research and even applied R&D alone are not sufficient. Nor is it sufficient when applied R&D is used to develop new ideas or even when new ideas are developed into new products or services. It is only when a new product/service is commercially successful that innovation has been produced.

Mission-oriented, applied R&D must be connected with businesses' ability to take products and services to the marketplace. Since technological innovation can only be accomplished in concert with businesses' ability to take products and services to the marketplace, the key S&T question is how to use S&T to help businesses, and the other key players in the ecosystem, to accomplish technological innovation.

Technological innovation models continue to evolve. As complex, non-linear models like the quad-helix model continue to gain favour and influence the thinking of all players in the innovation ecosystem, ever more advanced models are being developed and tested, such as n-tuple helix models. Similarly, ecosystem models for regional and national innovation systems, and cluster systems, are being tested and expanded virtually everywhere. Linking government funding and policies to more advanced technological innovation models and to more advanced innovation ecosystem models, and then effectively applying those models in practice, should enable significant improvements to ROII results and to regional and national productivity and competitiveness results.

An opportunity for smaller economies. An indicator of the amount of discovery research in a country is its GERD. GERD for the OECD countries plus Argentina, China (People's Republic), Romania, Russia, Singapore, South Africa, and Taiwan was about US$2.6 trillion in 2021 [523]). Meanwhile, the proportion of GERD by government, for discovery research, averaged across OECD countries is about 31% (2015 data [524]). An estimate for global GERD specific to discovery research is therefore approximately US$800 billion per year.

This global effort is producing a rapidly growing pool of discoveries, knowledge, and understanding that has great breadth and depth. Furthermore, it is widely available, either free or at low cost. This is because almost 100% of discovery research gets published, almost immediately, in the open literature. This information is now easily searched, courtesy of the internet and potentially aided by artificial intelligence tools.

Armed with information searching and access capabilities, some countries may, in the future, begin to divert some of their government-GERD funding away from discovery research[109] and into data/information mining (*"technology mining"*; see Section 9.2) and applied R&D. This approach is sometimes referred to as *"indirect commercialization."* It could be a mechanism through which small- and medium-sized countries can better compete with larger countries, at least in the areas of technological advances, inventions, and technological innovations.

9.2 Some emerging frontiers in technological innovation

"The future ain't what it used to be."

L.P. "Yogi" Berra,
American Professional Baseball Catcher, Manager, and Coach

As new technologies and individual technological innovations arise and advance, several current trends illustrate the potential for new clusters, and for waves of disruptive technological innovation.

Gene editing. Practical gene editing became possible with the development by Jennifer Doudna and Emmanuelle Charpentier of the CRISPR[110] technology in 2012, for which they won the Nobel Prize in Chemistry in 2020. CRISPR enables the genomes of living organisms to be modified, and the technology is regarded as having with huge potential in such fields as drug development, genetic-disease treatment and crop engineering [525, 526].

Key enabling technologies (KETs). KETs are a small set of technologies that are expected to be the most important "building blocks" for future technological innovation across all industrial sectors, including but not limited to manufacturing [527]. As such, these technologies are expected to play critical roles in the evolution and sustainability of leading-edge economies. Examples of currently identified KET areas include advanced materials; photonics and optoelectronics; quantum, digital, and information technologies; artificial intelligence (AI) and robotics; nanotechnology, life science, and biotechnologies; and advanced engineering and fabrication technologies [528]. Some examples of industries for which KETs could lie at the heart of disruptive, or "game changing" technological innovations include aerospace, agriculture, automotive, building construction, food, healthcare, mining and minerals, oil and gas, specialty chemicals, and textiles.

Technology mining. As mentioned in the previous section, the current rate of global GERD that is specific to discovery research, about US$800 billion per year, may produce new knowledge and discoveries faster than people can figure out what to do

109 Knowing that their reduced funding of discovery research funding will barely make a "dent" in the global total and therefore not significantly reduce the development of new knowledge and discoveries.
110 Clustered Regularly Interspaced Short Palindromic Repeats.

with all this information. This suggests a possible opportunity, namely, to dedicate teams or even entire organizations to mine this readily available treasure trove for pieces of technology that could be used, adapted, and/or bundled (possibly by others) into new inventions, solutions, and technological innovations.

Nanotechnology. Nanotechnology has found its way into a myriad of products in virtually every sector including, for example, agriculture and the environment; mineral processing and petroleum; manufacturing; biotechnology, pharmaceuticals, and medicine; foods; and personal care [529, 530]. In the not-too-distant future, there are likely to be newer, smaller, and more efficient membranes and catalysts, computer memory and circuits, solar and fuel cells, display and lighting systems, therapeutic agents, sensors (including biomedical sensors), and "smart" materials, to name just a few [529, 530].

Automation and AI. Graham Brown-Martin has speculated that anything that can be measured and based on rules will eventually be automated and that the only jobs available in the future will be the ones that machines can't do [531]. Among the most rapid changes along this line are the robotics and automation being incorporated in the manufacturing sector under Industry 4.0 [349, 350, 532–534]. Many expert systems, neural nets, and the like are referred to as "quasi-intelligent," emphasizing their lack of ability to perform at the level of human intelligence. However, automated factories and self-driving vehicles are already emerging. Can robo-accounting, robo-law, and robo-teaching be far behind?

AI continues to advance even further than just implied in the above paragraph. At some point in time, the sophistication of AI could reach that of the so-called "*technological singularity*," which refers to the point at which machines with greater than human intelligence come into being (this term was coined by Vernor Vinge in 1993 [535]). Although whether or not the singularity has already been reached (or can ever be reached) is being debated the concept, at least remains alive [351]. For example, a form of AI already exists with the advent of "*deep learning*," in which computers use massive data sets and neural networks to learn through experience [536]. Brown-Martin wrote in early 2017 that "*Learning how to code isn't going to help that much when machines are coding themselves.*" Within 3 months of the date of that statement, an article appeared in the technology literature describing the advent of software that can design machine-learning software and, thus, computers that can teach other computers [537]. Some of these computers are already doing things *that are not understood* by the engineers who built them [538]. In a 2017 article titled "The Dark Secret at the Heart of AI," Will Knight writes: "*No one really knows how the most advanced algorithms do what they do*" [538].

Quantum computing. As of the time of writing, quantum computing is still in its infancy, but many countries are investing heavily in it, hoping to gain an early-mover advantage, or at worst a strong defensive position. This is because it carries the promise of both useful new capabilities and inherent dangers. In the defence industry, for example, there is a hope that quantum technology could make conventional radar systems virtually obsolete by proving enhanced object detection while remaining invisible to the objects themselves (like "stealth" jets, and drones, ships). In the communications industry, quantum computing is expected to be able to crack conventional encryption methods, but provide new and less easily hacked encryption.

Commercialization and militarization of space. In the early 1960s, commercial satellites began to be used for the global transmission of television signals, while the early 1980s brought civilian GPS navigation, the mid-1990s brought satellite internet, and the early 2000s brought satellite radio services. Throughout this period, and indeed beginning in the late 1950s, have been a variety of military satellite programs. In principle, both commercial and military satellites may be orbiting in any or all of the low earth orbit *(LEO; altitudes 160–2,000 km),* medium earth orbit *(MEO; altitudes 2,000–35,786 km),* or geostationary orbit *(GEO; altitude 35,786 km),* or high earth orbit (HEO; altitudes above 35,786 km). Although estimates vary, in 2024, there were probably 8,000–10,000 active satellites in earth orbit, not counting some thousands of inactive satellites. As the number of satellites grows, particularly in LEO, the probability of unintended collisions is no longer insignificant. Beyond that, is an emerging myriad of weapon-bearing satellites with such capabilities as launchable projectiles, robotic grappling arms, and directed energy (such as electromagnetic pulse, laser, or microwaves) [539, 540].

Think globally, Act locally. This phrase[111] refers to an admonition to consider the health of the entire planet before embarking on local actions. As concerns for the planet's health and the sustainability of human ecosystems rise, some organizations may find a successful niche in technological innovations that balance corporate social and environmental responsibility with commercial success. Such sustainability-oriented innovations could be quite disruptive (in a positive, Schumpeterian sense).

9.3 Technological innovation in the future

> *"It's tough to make predictions, especially about the future."*
>
> Attributed to L.P. "Yogi" Berra,
> American Professional Baseball Player, Manager, and Coach

Notwithstanding the above judgement, if the past is any indicator, some future technological innovations will arise out of things that have already been imagined by science-fiction and fantasy-fiction writers (see, e.g. references [541–543]). In his book *Future Shock,* Alvin Toffler wrote "*. . . if we view it as a kind of sociology of the future, . . . science fiction has immense value as a mind-stretching force for the creation of the habit of anticipation*" [544]. Here are some examples:

111 The phrase "think global, act local" is attributed to Patrick Geddes, who introduced it in the context of town planning in 1915. The origin of the phrase "think globally, act locally" in an environmental context is still disputed.

- *Submarines*, as we now know them, were envisioned by Jules Verne in his book *20,000 Leagues Under the Sea* (1870), although the general concept was recorded much earlier, in Leonardo da Vinci's drawings (*circa.* 1508).
- *Interactive video games* that respond to a user's position and gestures were envisioned in the movie *Minority Report* (2002).
- *Handheld tablet computers* were used in episodes of *Star Trek* (the original series) and in the movie *2001: A Space Odyssey* (both in the 1960s).
- *Flip-style, handheld mobile phone/radios* were used in episodes of *Star Trek* (the original series, in 1966).
- Simple versions of the *tricorder* envisioned in *Star Trek* episodes from the 1960s have been appearing for several years, in the form of handheld diagnosis devices and personal health and activity trackers. More recently, such devices have become increasingly sophisticated. An example is DxtER, which uses AI, is handheld, and enables the user to monitor five real-time health vital signs and diagnose 34 diseases without a physician [545]. Another example, KOSMOS, provides ultrasound, electronic stethoscope, and electrocardiogram (ECG) results in real time with AI-assisted interpretation enabling the interpretation of ultrasound images with heart sounds and simultaneous ECG tracings [546]. Many other such technologies are currently under development [350].
- *Interactive, natural-sounding computers* with which users can have conversations and ask questions, such as have been featured in many science fiction stories, including the original *Star Trek* series, are already available. Examples from the 2010s include IBM's "Watson" [547], Apple's "Siri" [548], and (briefly) Amazon's "Alexa" (the voice of Amazon Echo, [549]), which brought the term "chatbot" into everyday use (Wooldridge [351]). By the 2020s, AI engines were merging chatbot intelligence with live web-searching capabilities, enabling search capabilities beyond the limitations of their underlying big-data resources.

Example. In November, 2022 ChatGPT, an AI chatbot was released. Although it suffered from some serious problems (like its tendency to "hallucinate"), within three months it had more than 100 million users. By 2024, an improved version was claimed to score in the 94th percentile on the Standard Aptitude Test (SAT), 88th percentile on the legal version (LSAT), and 90th percentile on the Uniform [Legal] Bar Examination (in the U.S., UBE) [550].

- *Miniaturized devices.* In the 1966 movie *Fantastic Voyage*, a miniaturized submarine was injected into a patient to perform *in situ* brain surgery. By 1999, a first wireless-camera pill had been developed, and by the 2000s improved models were already being used for "capsule endoscopy" for gastrointestinal tumour recognition and therapy) [207].
- Simple versions of the food replicators featured in numerous science fiction stories, including the original *Star Trek* series, are now appearing in the form of 3D food

printers. Some of the early adopters were bakeries [551]. Of course, 3D printing[112] technology has advanced in many other respects as well, in such areas as the kinds of materials that can be printed (including biocompatible, chemically resistant, and mechanically flexible and/or strong) and also in the scale of the printing (with the advent of "large-scale" 3D printing machines or assemblies that can print large components for the automotive and marine industries, and even home-building).

- *Drones*. The concept of drones (unoccupied aerial vehicles, UAVs) has been illustrated for us in many movies, including the original *Star Wars* movies (in that case, flying droids). One of the early applications for such drones involved carrying cameras, followed by package transport and delivery and, more recently, weapons [207].
- Reminiscent of the fictional characteristics of automatic translators in *Star Trek* episodes, movies, and their spin-offs and also in the TARDIS of the *Doctor Who* series, have come about via *AI-powered language translators* [351]. In fact, functional, real-time AI-powered language translators have become a practical reality in the last two decades, and their capabilities have been steadily improving.
- *Androids* (humaniform robots, Figure 9.1) have been imagined in science fiction stories since the late 1800s. Realistic, single-purpose androids have already been demonstrated, such as one that gives directions to shoppers in a Tokyo shopping mall [552]. They have become multi-tasking, and they continue to become more sophisticated.

Figure 9.1: First-generation Ameca robot (Engineered Arts Ltd., 2022). A 2nd-generation version was released in 2025 (image used courtesy of Willy Jackson under Creative Commons Attribution-Share Alike 4.0 International license at https://commons.wikimedia.org/wiki/File:Ameca_Generation_1.jpg).

112 Also termed "*Additive Manufacturing*."

Example. In August, 2025, the first World Humanoid Robot Games was held in Beijing, with the participation of hundreds of robots from 16 countries. The opening ceremonies included a choreography led by 15 robots, and a parade in which 25 robots marched autonomously in formation, without remote control [553].

Looking further forward, some technological innovations that have been imagined and/or are just beginning to emerge, include the following:

- Expanding functions for robots, especially as computerized AI advances, in which robots become increasingly integrated into daily lives, including humaniform and nano-sized robots [53, 535, 536, 541].
- Accelerated biotechnology developments assisted (or catalysed) by AI, such as "AI-assisted drug discovery" and "AI-based medical diagnostics" [350, 554].
- 3D metal printing is becoming a reality with the demonstration of several approaches by which metal powders, or metal particles imbedded in polymer binders, are used to build up a part which is then sintered to fuse the metal [555]. This enables alloys, like steel, to be fabricated layer by layer in a 3D printer. If the technology becomes practical for large-scale, high-volume manufacturing, it could create a disruptive change to the entire manufacturing industry.
- The first "*cloaking devices*," which can shield objects from optical view by bending light around them, have already been built on a very small scale [556, 557].
- The first "*tractor beams*," which can move physical objects, have been built. For example, a laser beam has been used to move gold-coated, hollow-glass spheres by tens of centimetres, and in two directions [558].
- The first demonstration of "*teleportation*" was published in 2015, in which quantum entanglement was used to teleport the quantum states of a photon over 100 km of optical fibre [559].
- Not yet demonstrated is the "*space elevator*," a system for lifting people or materials from earth to a space station in geostationary orbit without the need for rockets. Versions of this have been imagined by Russian author Konstantin Tsiolkovsky (*Speculations about Earth and Sky and on Vesta*, 1895) and British author Arthur C. Clarke (*The Fountains of Paradise*, 1979). A tether-based system could be used for this if a sufficiently light and strong tether material could be developed, and it has been proposed that carbon nanotubes might be suitable [560].
- Other space-related initiatives include plans to further explore and potentially inhabit the moon and possibly Mars.

Example. China's *Chang'e* lunar exploration program[113] has been active since 2004. It has progressed from lunar orbiters in 2007 and 2010, to lunar landers in 2013, 2019, with lunar sample-return missions in 2020 and 2024 (with several of the lander missions deploying rovers) [561, 562]. The next few missions are expected to involve orbiters, landers, rovers, flying probes, and the testing of lunar-base construction technologies with robots and/or human crews. China, in partnership with Russia and potentially other countries, plans to construct a research station at the moon's south pole by about 2030 [563].

Figure 9.2: There is plenty of scope for many more technological innovations in the future.

Projecting even further ahead, one can imagine that practical applications may one day be derived from what are currently only theories (Figure 9.2). For example, we can already break matter, whether living or inanimate, down into its constituent particles, or molecules, or atoms (via classical chemistry and physics), and we now have a fairly good understanding of the means by which such entities can be caused to self-assemble in various ways (via colloid and interface science).[114] Current theories suggest that we should be able to move particles from one physical location to another without them having to physically travel there (quantum mechanics), to change their behaviours from those of particles to those of waves (wave-particle duality), and/or to cause them to travel in a way that bends time (special theory of rela-

113 *Chang'e* (pronounced "chung-ur" in English) was named after the Chinese goddess of the moon (originally known as *Heng'e*).

114 Now more commonly referred to as nanoscience and nanotechnology.

tivity). Who knows what practical problems or opportunities may be addressed with the aid of with such knowledge?

Some of the trends noted in these last two sections are exciting, frightening, or both, but things will continue to change. In a competitive world, with competitive markets, people, organizations, and countries alike will need to forge ahead or at least "keep up," lest they be left behind.

> *"Panta Rhei" (All Things Change)*
> *Heraclitus of Ephesus,*
> *Greek philosopher, circa. 500 BCE*

10 Appendices

10.1 Histories and case studies of some RTOs

The websites of many, if not most, RTOs provide a link to a section on their organization's history. In some cases, there are separately published histories. Here are some examples of separately published RTO histories and case (and/or benchmarking) studies[115].

Country	RTO	History	Case or benchmarking study
Australia	Commonwealth Scientific and Industrial Research Organization (CSIRO)	[564–566]	[567]
Belgium	Inter-University Micro Electronics Centre (IMEC)		[568]
Brazil	*Serviço Nacional de Aprendizagem Industrial – Institutos SENAI de Inovação* (SENAI-ISI ; National Service for Industrial Training-Innovation Institutes)		[516, 569, 570]
Canada	Alberta Research Council	[571]	
	FPInnovations	[572]	
	National Research Council of Canada (NRC)	[501, 573]	[574]
	Research and Productivity Council (RPC)	[575]	
	Saskatchewan Research Council (SRC)	[290]	
	Vineland Research and Innovation Centre	[576, 577]	
Denmark	*Godkendt Teknologisk Serviceinstitut* (GTS; Approved Technological Service Institute)		[292, 578, 579]
Finland	*Valtion teknillinen tutkimuskeskus* (VTT; National Technical Research Centre)		[574, 578, 580, 581]
France	Centre national de la recherche scientifique (CNRS; National Centre for Scientific Research)	[582–584]	[585]
Germany	Fraunhofer-Gesellschaft (Fraunhofer Institutes)		[292, 568, 574, 586, 587]
	Institut für Produktionstechnik und Automartisierung (IPA)		[588]

115 A case study involving four large, anonymized European RTOs is given in Giannopoulou, E., "The Role of Research and Technology Organizations (RTOs) in Open Service Innovation: A Dual Perspective," Thesis, Economics and Finance, Université de Strasbourg, 2017.

https://doi.org/10.1515/9783112206720-010

(continued)

Country	RTO	History	Case or benchmarking study
Hong Kong	Hong Kong Productivity Centre (HKPC)		[587, 588]
India	Council of Scientific and Industrial Research (CSIR)	[589]	[590]
Italy	*Centro Infornaazione Tessile di Emilia Romagna* (CITER; Textile Information Centre of Emilia Romagna)	[591]	[587, 588]
Japan	Advanced Industrial Science and Technology (AIST)		[292, 586]
Korea	Korea Institute of Science and Technology (KIST)		[587, 588]
Netherlands	Holst Centre		[568]
	Nederlandse Organisatie voor Toegepast Natuurwetenschappelijk Onderzoek (TNO; Dutch Organisation for Applied Scientific Research)	[592]	[292, 574, 578]
Norway	*Stiftelsen for industriell og teknisk forskning* (SINTEF; Foundation for Scientific and Industrial Research)	[593]	[578]
Russia	62 Russian RTOs		[594]
Singapore	Singapore Institute of Standards and Industrial Research (SISIR)		[587, 588]
South Korea	Electronics and Telecommunications Research Institute (ETRI)		[292,568]
Sweden	*Institutet für Verkstadsteknisk Forskning* (IVF)		[587, 588]
	Research Institutes of Sweden (RISE)	[595]	[578, 596]
Taiwan	Industrial Technology Research Institute (ITRI)	[597]	[568, 586–588, 598, 599]
Tanzania	CAMARTEC, COSTECH, SIDO, TEMDO, and TIRDO[116]		[600]
Thailand	National Science and Technology Development Agency (NSTDA)		[599]
United Kingdom	Pera International		[587, 588]

116 These are: Tanzania Commission for Science & Technology (COSTECH), Tanzania Industrial Research and Development Organization (TIRDO), Small Industries Development Organization (SIDO), Centre for Agricultural Mechanization and Rural Technology (CAMARTEC), and Tanzania Engineering and Manufacturing Design Organization (TEMDO).

(continued)

Country	RTO	History	Case or benchmarking study
United States	Argonne National Laboratory	[601]	
	Battelle Memorial Institute (BMI)	[602]	[588]
	Los Alamos National Laboratory	[603, 604]	
	Oak Ridge National Laboratory	[605]	
	SRI International (formerly Stanford Research Institute)	[606, 607]	

10.2 Some examples of innovation case studies

This section provides over 190 literature citations to references, covering nearly 900 case studies concerned with innovation. Some very early case studies of specific innovations have already been referenced in preceding sections and can be found in Usher [96, 124] and Schmookler [18] (see also Sections 6.1 and 7.70; these are also included in the references below).

This section does not provide a complete collection, as the number of innovation case studies is huge and the number published each year seems to be steadily growing (about half of the case studies referenced below were published since 2015). Those that are referenced here illustrate a wide range of case studies related to commercial- and non-commercial innovation, as well as to issues within different aspects of innovation ecosystems. The reader is cautioned however that the nature and style of these case studies vary widely. Unfortunately, there seems to be no generally accepted standard for case studies. In a review of more than 800 articles involving case-study research in innovation management published in leading journals over a 20-year period (1997–2016), Goffin *et al.* [608] found that:

> *"the quality of case study research has often been low, although it has improved over time . . . Many articles: did not justify why case study research was appropriate; did not apply theoretical sampling criteria; were not transparent on how conclusions were drawn from the data; did not consider validity and reliability adequately; and did not go beyond description in their interpretation."*

Therefore, although these case studies provide a wealth of real-life, experience-based information, one should bear in mind the caution *caveat lector*[117].

117 Latin for "reader beware."

Geographic focus area

Africa	Algeria [609], Burkina Faso [610], Burundi [610], Ghana [611], Kenya [610, 612], Madagascar [612], Morocco [610], Nigeria [612], Rwanda [610, 612], Sierra Leone [612], Senegal [610], Somalia [610], South Africa [337, 610, 612, 613, 614], Tanzania [385, 610, 612], Uganda [610, 612], Zambia [612]
Asia	Bangladesh [615], China [24, 257, 335, 397, 610, 611, 616–626], Hong Kong [616], India [206, 335, 337, 590, 610, 627–632], Indonesia [385, 610], Israel [610], Japan [586, 610, 611, 633–637], Jordan [610], Malaysia [614, 638], Myanmar [610], Pakistan [610], Singapore [610, 611, 629], South Korea [61, 490, 568, 611, 628, 632–634, 639–641], Syria [610], Taiwan [24, 568, 586, 598, 599, 610, 642–646], Thailand [599, 610], Turkey [610], United Arab Emirates [611], Vietnam [611, 647]
Europe	Austria [263, 610, 633, 634, 648], Belgium [568], Czech Republic [649, 650], Denmark [24, 263, 280, 335, 407, 578, 610], Estonia [635], Europe and/or European Union [137,151], Finland [24, 152, 282, 578, 581, 616, 633, 634, 635, 651–653], France [24, 263, 610, 633, 634, 648], Kosovo [24, 610], Germany [24, 233, 279, 329, 335, 388, 568, 586, 610, 648, 654–656], Greece [614, 633, 634], Italy [24, 279, 587, 610, 616, 648, 657, 658, 659, 660], Hungary [661], Netherlands [335, 389, 568, 633, 634, 662], Norway [578, 631, 633, 634, 663–665], Poland [349, 616, 635], Portugal [616, 666–669], Russia [150, 670], Spain [616, 632, 648, 671], Sweden [150, 263, 596, 632- 635, 672, 673, 674, 675], Switzerland [24, 116, 117], the United Kingdom [15, 22, 24, 97, 147, 157, 205, 263, 271, 279, 330, 610, 611, 633–635, 663, 674, 676–679]
North America	Canada [263, 317–319, 490, 610, 626, 635, 636, 674, 680–709], Guatemala [610], Mexico [388, 610, 710], Nicaragua [710], Panama [610], the United States [15, 18, 24, 48, 49, 61, 89, 90, 92, 93, 122, 145, 164, 214, 239, 263, 279, 281, 323–333, 335, 388, 389, 610, 611, 626, 628, 635, 696, 711–718]
Oceania	Australia [279, 516, 567, 610, 633, 634, 719], New Zealand [263, 633, 634, 720]
South America	Argentina [710], Bolivia [710], Brazil [127, 334, 387, 388, 610, 616, 628, 636, 710, 721–724], Chile [388, 610, 611, 710], Columbia [337, 388, 610, 710, 725, 726], Ecuador [610, 616], Venezuela [710]

Scope

City/village/ smart city	[206, 610–612, 619, 640, 646, 652, 670, 695]
Community/ region	[24, 205, 263, 317–319, 385, 407, 609, 610, 612, 616, 629, 636, 645, 657, 658, 661, 662, 665, 673, 674, 675, 680, 689, 692, 695, 697–700, 703–706, 709, 710, 718]
National	[15, 18, 22, 24, 48, 49, 61, 89, 90, 92, 93, 97, 127, 145, 147, 150, 152, 214, 257, 271, 279, 280, 281, 282, 323–332, 334, 335, 337, 349, 385, 397, 490, 516, 567, 568, 578, 581, 586, 587, 590, 596, 598, 599, 613–615, 618, 623, 624, 626, 627, 635, 637, 644, 647–649, 651, 653, 654, 664, 669, 671, 675, 676, 678, 679, 684, 687, 690, 691, 693, 696, 701, 702, 707, 708, 710, 711, 714, 716, 720, 722, 723, 724, 726]
MNEs	[24, 61, 116, 117, 122, 157, 164, 233, 239, 389, 615, 622, 625, 631, 632, 639, 650, 654–656, 659, 660, 663, 672, 683, 684, 688, 690, 694, 711, 715, 717, 719]
SMEs	[205, 263, 638, 642, 648, 658, 667, 669, 677, 681, 682, 684, 685, 686, 688, 693, 721, 723]
International	[24, 137, 151, 333, 336, 610, 611, 616, 628, 632–634, 668, 723, 727, 728, 729, 730, 731]

Sector

Academia/education	[206, 271, 397, 616, 620, 630, 635, 663, 664, 678, 679]
Agriculture/aquaculture/ biotechnology	[18, 22, 61, 137, 147, 257, 263, 282, 318, 385, 612, 613, 642, 667, 684, 702, 704, 720, 723, 731]
Aviation/transportation	[15, 18, 24, 61, 97, 122, 214, 324, 325, 702, 708, 713, 717]
Biomedical/pharmaceuticals	[257, 263, 318, 326, 617, 632, 683, 713]
Broadcasting/entertainment	[24, 122, 318, 626]
Energy/utilities	[15, 18, 24, 145, 157, 205, 319, 335, 336, 337, 407, 490, 645, 663–665, 681, 683, 689, 690, 704, 713, 714, 722, 730]
Financial services	[24, 239, 622, 644, 687, 696, 697, 711, 724, 726]
Government/policy	[24, 89, 90, 92, 93, 263, 279–281, 319, 328, 616, 633–635, 652, 673, 707, 710, 714, 728]
Healthcare	[24, 151, 205, 318, 609, 612, 632, 636, 643, 651, 674, 690, 691, 700–702, 706, 712, 713, 725]
Information and communication technol. (ICT)	[24, 145, 257, 263, 282, 318, 326, 333, 334, 385, 407, 598, 616, 619, 622, 624, 626, 631, 632, 641, 646, 653, 665, 681, 682, 683, 690, 694, 713, 724]

(continued)

Manufacturing	[15, 18, 22, 48, 49, 61, 97, 116, 117, 122, 127, 137, 150, 152, 164, 214, 233, 257, 263, 318, 323, 327, 329–332, 349, 387–389, 407, 587, 624– 626, 632, 637– 639, 647, 648, 650, 654–656, 659, 660, 662, 668, 669, 670, 671, 672, 673, 675, 677, 681–686, 688, 690, 691–693, 699, 701–703, 708, 709, 713–716, 721]
Social	[24, 205, 610–612, 615, 621, 623, 627, 628, 629, 640, 674, 695, 697, 698, 727, 729]
Sports	[680, 691]
Tourism	[612, 614, 616, 622, 626, 642]

Specific aspects

Entrepreneurship and enterprise hubs	[206, 397, 616, 630, 661, 697, 698, 718]
Collaboration, clusters, innovation networks, and intermediaries	[151, 205, 263, 271, 279, 280, 281, 317, 318, 319, 407, 490, 516, 567, 578, 581, 586, 587, 590, 596, 599, 616, 619, 624, 627, 635, 645, 654, 657, 658, 664, 667, 670, 675, 677, 696, 699, 703–705, 719, 727]
Business/innovation models and planning	[24, 116, 117, 127, 150, 152, 157, 164, 205, 239, 349, 387, 397, 622, 625, 626, 635–638, 648, 656, 662, 665, 697, 698, 716, 717, 721, 722, 724, 726]
Diffusion of innovation	[137, 333–337, 620, 676, 692, 712]
Peripheral innovation	[639]
Service innovation, KIBS	[24, 407, 614, 621, 631, 643, 693, 695, 700, 726]

10.3 Some famous, revolutionary technological innovations

There are a number of references on the topic of world-changing technological innovations (see, e.g. references [732–736]); the table below contains only an illustrative selection. As of the time of writing, it appears likely that recent innovations such as artificial intelligence will find their way onto such lists.

Era		Innovation and principal innovator(s)	Significance
Pre-industrial	Mid-15th century	Printing press (Johannes Gutenberg)	The mechanized process for transferring ink from movable type to paper made rapid printing possible and enabled mass production of books, which accelerated the diffusion of knowledge and ideas.
	1735	Marine chronometer (John Harrison)	Enabled navigators to establish their longitude at sea, revolutionizing global merchant-seafaring.
First Industrial Revolution	1764	Spinning Jenny (James Hargreaves)	Began the shift from cottage- to factory-based manufacturing.
	1775	Steam engine (James Watt)	Revolutionized transportation and machinery.
	1835	Telegraph (Samuel Morse)	Enabled instantaneous, long-distance communication.
	1859	Electrical Battery (Gaston Planté)	Although the voltaic pile was invented by Alessandro Volta in 1799, the first practical, commercializable battery was developed by Planté. It enabled portable electronics.
Second Industrial Revolution	1860	Internal-combustion engine (Étienne Lenoir)	The first practical, commercialized such engine. It revolutionized transportation.
	1876	Telephone (Alexander Graham Bell)	Revolutionized communication by allowing instant speech transmission over long distances.
	1879	Electric light bulb (Thomas Edison)	The first practical, commercialized light bulb. Enabled factories to operate continuously, day and night.
	1903	Powered aeroplane (Wright brothers)	Enabled rapid, long-distance commercial aviation.
	1904, 1906	Diode electron tube (John Fleming) and triode electron tube (Lee de Forest)	Enabled the first age of electronics, sound amplifiers, and radio.
	1913	Efficient mass production (Henry Ford)	Ford's moving-assembly-line mass-production, with sub-divided labour and interchangeable parts, revolutionized manufacturing.

(continued)

Era		Innovation and principal innovator(s)	Significance
	1913	Electric refrigeration	There were many inventors and innovators involved in such key aspects as evaporative cooling (1748), vapour-compression (1834), and gas-liquefaction systems (1876). These enabled commercial refrigeration (circa. 1850), and Fred Wolf's domestic refrigerator (1913). It transformed food transportation, storage, and lifestyles.
	1943	Ball-point pen (Laszlo Biro)	Biro's first ballpoint pens enabled RAF air-navigators to write at high altitudes on damp paper, at awkward angles, and for 200,000 words without refilling. They could also be used underwater. By 1949, they were outselling fountain pens internationally.
Third Industrial Revolution	1947	Transistor (John Bardeen, Walter H. Brattain, and William B. Shockley)	Enabled the second age of electronics, computers and the digital revolution.
	1954	Industrial robot (George Devol and Joseph Engelberger)	Revolutionized manufacturing.
	1956	Oral contraceptives (Gregory Pinkus and John Rock)	Although condemned in some quarters, "the pill" gave a revolutionary boost to the liberation of women.
	1958/59	Semiconductor integrated circuit (Jack Kilby and Robert Noyce)	Enabled the third age of electronics with miniaturization, improved performance, and reduced cost in computers, smartphones, and televisions, among others.
	1959	Automatic photocopier (Chester Carlson)	The Xerox 914 required only the push of a button to make seven copies per minute, eliminating the need for time-consuming copying of text by retyping or handwriting, while also being able to reproduce images.
	1969–1990	Internet	There were many inventors and innovators involved in such key aspects as packet switching (1960s); TCP/IP communication protocols (1970s); ARPANET (1969); and the World Wide Web (1990, Tim Berners-Lee). It revolutionized communication, information-transfer, and global business.

(continued)

Era		Innovation and principal innovator(s)	Significance
	1970	Personal computer (John Blankenbaker)	Brought computing to small- and medium-sized businesses and catalysed the development of new generations of software applications.
	1981	3D printing (Hideo Kodama)	Transforms manufacturing with layer-by-layer production of objects from digital designs with speed and flexibility. It has transformed rapid prototyping and enables remote manufacturing and build-on-demand parts.
Fourth Industrial Revolution	2010s	Cloud computing (many innovators)	By the 2010s, cloud computing had advanced beyond distributed computing to massive storage, multi-clouds, and the underpinnings for machine learning, IoT, and AI.
		Practical gene editing (Jennifer Doudna, Emmanuelle Charpentier and Feng Zhang[118])	The CRISPR technology provides a means to edit the genes of animals, plants, fungi, and microbes, with huge potential in such fields as genetic-disease treatment and crop engineering.
	2020s	Internet of Things: IoT (many innovators)	By the 2020s, IoT had advanced from connected devices and connected homes to smart factories, autonomous vehicles, and connected healthcare.

10.4 Selected books on technological innovation by topic

There is a massive literature on innovation now available, even for the narrower field of technological innovation. Here are some examples – restricted to books only – in some of the sub-fields of technological innovation. This collection is not comprehensive and is intended only to provide starting points based on reference citations that have already been made elsewhere in the text. Some of the citations refer to chapters in books but, in this context, it is to the books themselves that reference is made.

118 Doudna and Charpentier received a Nobel Prize in Chemistry (2020) for developing CRISPR in 2012, while Feng Zhang received the first patent in 2014.

Topic (within technological innovation)	Reference (first author, date, and citation)
Historical origins	Holland, 1928 [58]; Schumpeter, 1939 [15]; Schumpeter, 1950 [16]; Usher, 1954 [96]; Schmookler, 1966 [18]; Reich, 1985 [145]; and Pithan, 2022 [56]
Creative thinking	Osborn, 1953 [171]; Gordon, 1961 [178]; de Bono, 1967 [174]; de Bono, 1972 [175] ; de Bono, 1985 [176]; Altshuller, 2004 [72]; Johansson, 2004 [180]; Rabe, 2006 [744]; Florida, 2012 [347]; Brown, 2009 [51]; and Franklin, 2023 [184]
Clusters	OECD, 1999 [311]; Asheim *et al.*, 2006 [256]; Potter and Miranda, 2009 [263]; Klimova, 2016 [315];
Concepts and models	Copp, 1993 [713]; Garud, 1997 [454]; Maxwell, 2000 [235]; Farson, 2003 [240]; Kaye, 2003 [679]; Fitzgerald, 2010 [121]; Thota, 2011 [202]; Carayannis, 2013 [268]; Davila, 2014 [241]; Malerba, 2016 [326]; Phani, 2017 [630]; Smil, 2023 [361]
Diffusion of innovations	Rogers, 1971 [208]; Rogers, 2003 [77]; Mahajan and Peterson, 2011 [210]
Economic cycles/waves	Kondratieff, 1984 [338]; Grinin, 2016 [340]
Ecosystems	Edquist, 1997 [246]; Håkonsson, 2007 [150]; Mercier-Laurent, 2011 [8]; MacGregor, 2012 [269]
Foresight and futures	Standage, 2005 [53]; Georghiou, 2008 [430]; Johnston, 2008 [423]
Government policy	Campbell, 2004 [710]; OECD, 2011 [255]; Vonortas, 2015 [306]
Handbooks and encyclopaedias	Fagerberg, 2009 [34]; Shavinina, 2003 [128]; Tibbals, 1994 [736]
Industry 4.0	Bali, 2022 [534]; Duda, 2022 [349]; Sharma, 2023 [533]; Elngar, 2025 [350]; Ratnesh, 2026 [530]; Sindhwanni, 2024 [532]
Innovation systems (regional)	Braczyk, 1998 [253]; Asheim *et al.*, 2006 [256]; OECD, 2011 [255]; Mitra, 2020 [207]; Breznitz, 2021 [598]; Kim, 2021 [611]
Innovation systems (national)	Schmoch, 2006 [303]; Jonsson, 2015 [384]; Reynolds, 2019 [570]
Management	Tushman, 1988 [132]; Rabe, 2006 [744]; Howells, 2005 [331]; AMA, 2006 [380]; Maital, 2007 [377]; Govindarajan, 2010 [364]; Betz, 2011 [61]; Sattler, 2011 [499]; Sehested, 2011 [745]; Trias de Bes, 2011 [411]; Deschamps, 2014 [399]; Chen, 2023 [395]; Damberg, 2025 [396]
Marketing and sales	Moore, 1991 [143]
New product/process development	Belliveau, 2002 [229]; Cooper, 2011 [223]; Roberts, 2002 [181]; McGrath, 2011 [750]
Open innovation	Chesbrough, 2003 [160]; Chesbrough, 2006 [162]

(continued)

Topic (within technological innovation)	**Reference (first author, date, and citation)**
Service innovation	Macaulay, 2012 [24]; Mustafa, 2019 [25]
Strategy and business models (innovation in)	Foster, 1986 [131]; Christensen, 1997 [79]; O'Sullivan, 2009 [183]; Verganti, 2009 [104]; Govindarajan, 2012 [154]; Miller, 2012 [453]; Moehrle, 2013 [439]; Verloop, 2013 [81]; Kerzner, 2019 [122] ; Pandiarajan, 2022 [632]; Cantone, 2023 [717]; Martin, 2024 [666]
Strategy (in innovation)	Johnston, 2003 [182]; Libecap, 2008 [204]
Sustainable innovation	Koops, 2015 [385]; Mousavi 2019 [389]
Technology transfer and commercialization	Touhill, 2008 [186]; Schaufeld, 2015 [193]
Topics within non-technological innovation	
Social	Nicholls, 2015 [628]; Cheah, 2020 [629]; Kim, 2021 [611]; Amatullo, 2022 [610]
Tourism	Correia, 2024 [616]; Valeri, 2024 [614]

11 Dictionary of innovation technology

Now, more than 75 years since Schumpeter's time, a growing lexicon is associated with innovation because it applies to so many fields of commercial endeavour. The field has also broadened beyond Schumpeter's original scope because various parties, particularly governments, have focussed on ways to help make organizations of all kinds more efficient and/or effective regardless of competitive concerns. This has even extended to the notion of helping to make non-commercial entities more efficient and/or effective, such as not-for-profit organizations (NFPs) including charities, non-government organizations (NGOs), and even government departments and agencies. In this book, the unqualified term "*innovation*" refers to Schumpeter's original definition of innovation, which was exclusively about developing and introducing new commercial products, process, and services into the marketplace [15, 16]. However, in some usage, this will be referred to as "*technological innovation*" in order to make a distinction from "*non-commercial innovation*," which refers to any aspects of improving an organization's efficiency or effectiveness that manifest themselves in ways other than the introduction into the marketplace of new commercial products, process, or services. In this way, the modern, broader uses of the term *innovation*, such as the definition given in the OECD *Oslo Manual*, can be taken to include technological and/or non-commercial innovation. Table 1.1 illustrates some of the breadth of specifically named "innovation" terms.

This section provides brief explanations for nearly 1,500 terms and acronyms that may be encountered in a study of the fundamental principles, application approaches and strategies, and commercial aspects of technological innovation. Even this coverage inevitably represents only a personal selection of the terms that could have been included were there no constraints on the size of the book.[119]

I have tried to include as many important terms as possible, and cross-references for the more important synonyms and abbreviations are also included. The difficulty of keeping abreast of the innovation vocabulary has been worsened by the tendency for the language itself to change and broaden since the term "*innovation*" was first coined in the 1930s. Some older terms that are no longer in common use, or worse, that now have completely new meanings, are included as an aid to the reader of the older innovation and economics literature and as a guide to the several meanings that many terms can have. The meaning of the term "*innovation*" itself is still in flux, although some standardization is beginning to occur. I have also included a few terms from the current and possible future waves of economic and technological advances, such as nanotechnology, smart technologies, and robotics, since these tend to recur in writings and conversations about the future of innovation. Finally, a selection of brief biographical introductions to some of the pioneers in the field is also included (see Table 1.3).

119 This section represents an update and modest expansion of reference [2].

https://doi.org/10.1515/9783112206720-011

Specific literature citations are given when the sources for further information are particularly useful, unique, or difficult to find. Some particularly helpful sources include references [26, 76, 186, 485]. There are a few other published dictionaries, glossaries, and typologies of innovation, such as references [2, 9, 10, 27, 202, 737]. Any or all of the above can be good starting points for further information.

11.1 Numeric

10/5 rule	A venture capital "rule of thumb" that a commercialized product, process, or service should yield a 10-times return on investment within a period of 5 years. *See also* venture capital.
1st-generation innovation model	The technology-push model. *See* first-generation innovation model, generations of innovation, linear innovation models.
1st-generation nanotechnology	*See* generations of nanotechnology.
1st-generation of technology foresight	*See* generations of technology foresight.
1st Industrial Revolution	*See* technological ages.
1st wave	The first Kondratieff wave, spanning the Industrial Revolution era and lasting from about 1780 to about 1830. Key developments during this wave included the steam engine and general industrialization [343]. *See also* Kondratieff waves.
20th- to 21st-century innovation	*See* innovation (20th–21st centuries).
20/30 rule	A market potential "rule of thumb" that an invasionary technology should perform its function 20% better and 30% cheaper than the pre-existing, competing technologies. *See also* best available technology [76].
2nd-generation innovation model	The market-pull model, *see* second-generation innovation model, generations of innovation, and linear innovation models.
2nd-generation nanotechnology	*See* generations of nanotechnology.
2nd-generation of technology foresight	*See* generations of technology foresight.
2nd Industrial Revolution	*See* technological ages.
2nd wave	The second Kondratieff wave, spanning the industrial production era and lasting from about 1830 to about 1880. Key developments during this wave included railways, steel, and heavy engineering [343]. *See also* Kondratieff waves.
3M Inc. (1902–present)	An American company founded in 1902 as the Minnesota Mining and Manufacturing Company. The company's original foci were on mining for grinding-wheel abrasives and later sandpaper products. Over time, 3M developed innovation as a core competency, leading to a number of famous products, including Post-it® notes, Scotch-Brite™ products, and Magic™ tape. 3M has continued to be a champion and a leader in making innovation central to its business strategy. At one point, 3M had a goal of

	having 30% of total sales be from new products developed and commercialized within the previous 5 years [455]. *See* New Product Vitality Index.
3rd-generation innovation model	The coupling model, *see* generations of innovation, third-generation innovation model, non-linear innovation models.
3rd-generation nanotechnology	*See* generations of nanotechnology.
3rd-generation of technology foresight	*See* generations of technology foresight.
3rd Industrial Revolution	*See* technological ages.
3rd wave	The third Kondratieff wave, spanning the scientific revolution era and lasting from about 1880 to about 1930. Key developments during this wave included electricity, chemistry, and the chemical industry [343]. *See also* Kondratieff waves.
4th-generation innovation model	The Integrated Model. *See* fourth-generation innovation model, generations of innovation, and non-linear innovation models.
4th-generation nanotechnology	*See* generations of nanotechnology.
4th-generation of technology foresight	*See* generations of technology foresight.
4th Industrial Revolution	*See* technological ages.
4th pillar organization	*See* fourth-pillar organization.
4th wave	The fourth Kondratieff wave, spanning the technical revolution era and lasting from about 1930 to about 1970. Key developments during this wave included automobiles, mass production, and the petrochemical industry [343]. *See also* Kondratieff waves.
5th-generation innovation model	The systems integration and networking model. *See* fifth-generation innovation model, generations of innovation, and non-linear innovation models.
Fifth generation of technology foresight	*See* generations of technology foresight.
5th wave	The fifth Kondratieff wave, spanning the information and telecom revolution era and lasting from about 1970 to about 2010. Key developments during this wave included microcomputers, information technology, and telecommunications technology [343]. *See also* Kondratieff waves.
6th-generation innovation model	The open innovation model. *See* sixth-generation innovation model, generations of innovation, and non-linear innovation models.
6th wave	The sixth Kondratieff wave, spanning the current era and predicted to last from about 2010 to 2050. Key developments during this wave might include environmental technology, genetic engineering, nanotechnology, robotics, and health technologies [343]. *See also* Kondratieff waves.
7th wave	The predicted seventh Kondratieff wave, spanning a future era and predicted to last from about 2050 to 2090. It has been speculated that the seventh wave may be driven by a merging of technology and intelligence to

create autonomous robots, whether at nanoscale, microscale, or macroscales, that are capable of independent action, self-repair, and replication [343]. *See also* Kondratieff waves.

11.2 A

Abernathy-Clark model — An innovation model that defines forms of evolutionary innovation and distinguishes between an innovation's effect on organization's technological knowledge and resources and its effect on the scale of the technological advance and whether the competing products, processes, or services remain somewhat competitive or are made obsolete. In the Abernathy-Clark Model, organizations that simply enhance their technological knowledge and resources are most likely to achieve modest incremental innovation at best (they termed it "regular innovation"), whereas those that develop completely new technological knowledge and/or resources and use them to make huge, game-changing (i.e. market-changing) technological advances, are most likely to achieve disruptive innovation (they termed it "architectural innovation"). In between these extremes fall two categories of organizations. One category represents organizations that develop completely new technological knowledge and/or resources but only use them to make more competitive products that do not displace or render obsolete their competing products. Such companies achieve evolutionary innovations (they termed it "revolutionary innovation"). The other category represents organizations that develop advances in their technological knowledge and/or resources and are able to introduce products that displace or render obsolete their competing products in a small market niche. Such companies achieve niche innovation. The Abernathy-Clark Model has been used to explain how incumbent companies tend to be in good position to implement incremental innovations, as they can leverage their existing technological knowledge and resources, whereas new companies entering a marketplace tend to be in a good position to implement disruptive innovations, as they do not need to change their existing technological knowledge and/or resources in order to do so [86]. *See* Figure 2.5. *See also* incremental innovation, evolutionary innovation, disruptive innovation, and Henderson-Clark Model.

ABG Model — *See* Triple Helix Model.

Ab initio — In science and engineering, the Latin term *ab initio* generally refers to developing models and/or making calculations based on first principles; without experimental data (other than fundamental physical constants).

Absorptive capacity — *See* technology acceptor capacity.

Academic revolution — *See* second mission, entrepreneurial university.

Accelerator — *See* business accelerator.

Acceptance threshold — The transition realm of customer perceptions, between considering a technology to be unacceptable (or unattractive, or not useful) and acceptable (or attractive, or useful). Unacceptable technologies can

	sometimes be turned into acceptable technologies through an "*itemized response*" process of analyzing each negative or unacceptable feature (whether real or perceived) of a technology and finding ways to change or address each such drawback. Also termed *"threshold of acceptance." See also* technology readiness index and technology acceptance model.
Acquisition of technology	The purchase of, or barter for, externally derived knowledge of how to effectively use a product, process, or service (knowhow), a way of conducting or controlling a manufacturing activity (a practice or process), or a thing to be manufactured, used, or consumed (product). Two broad categories are sometimes distinguished in this context: embodied and disembodied technology. Embodied technology refers to technology that is acquired indirectly because it is embedded in machines, instruments, new employees, or contractors that incorporate or contain the knowledge. Disembodied technology refers to technology that resides in trade secrets, books and other publications, patents, licenses, trademarks, software, and the like. *See also* knowledge and embedded knowledge.
Active materials	*See* smart materials.
Activities	(Innovation) Any or all scientific, engineering, technological, organizational, financial, or commercial activities that lead, or are intended to lead, to development and deployment of innovation(s). Innovation activities include R&D plus subsequent activities such as experimental development, prototyping, pilot testing, demonstration, and preproduction. *See also* inputs, outcomes, outputs, reach, research and development, metrics, and impacts [76].
Adapt and adopt	*See* adopt and adapt.
Adjacent innovation	The commercialization of a product, process, or service that is already being sold in one market into a new market, particularly where the new market is in some sense "near" the established market. This term is sometimes alternatively used as a synonym for *incremental innovation*, in the sense of a minor improvement to a product, process, or service in an existing market. *See also* incremental innovation, disruptive innovation, and evolutionary innovation.
Administrative innovation	A kind of *organizational innovation*, administrative innovation refers to new knowledge applied to developing and implementing new and improved organizational structures and administrative processes, including business strategy. Also termed "*business process innovation*" or "*business structure innovation*," respectively. Such new organizational processes should have some kind of efficiency or productivity benefit, even if they are internal and have little or no connection to technological innovation. Example: the Total Quality Management processes pioneered by W. Edwards Deming in the United States in the 1980s. A third form is *Business Culture Innovation*, in which an organization sets about to change its internal culture in order to improve its organizational performance. *See also* organizational innovation and innovation.
Adopt and adapt	A technology development and/or innovation strategy in which one adopts an existing technology from another business, industry, or market and adapts it to the innovator's needs. Also termed *"adapt and adopt." See also* Ansoff Matrix and fast follower.
Adoption	*See* technology adoption life cycle.

Advanced technology	*See* high technology.
Advanced technology development	*See* research and development.
Agency theory	*See* theory of the firm.
Aghion, Philippe (1956–present)	A French economist who was co-awarded the 2025 Nobel Prize in Economic Sciences for, in his case, having co-identified the prerequisites for sustained growth through technological progress (his co-awardees were Joel Mokyr and Peter Howitt).
Agile development	A product development approach that involves collaboration among self-organizing, cross-functional teams. This approach facilitates adaptive planning, evolutionary development, responsiveness, and continuous improvement. The "*agile development*" concept seems to have originated in the computer software industry.
Agnostic marketing	A term coined by Clayton Christensen, referring to the inability to know in advance whether, how, or in what quantities customers will use a disruptive innovation product until some first customers try using it. This suggests maintaining a very broad (i.e. agnostic) view of what the ultimately successful market could be [79].
Algorithm of inventive problems solving	(ARIZ is the Russian acronym) One of the tools used to try to solve invention problems in Altshuller's Theory of Inventive Problem Solving. *See* Altshuller, creative thinking models, and theory of inventive problem solving.
Alliance capitalism	The trend for some groups of companies to form cooperative relationships and/or strategic alliances in order to accelerate their technological innovation processes and to penetrate international markets. The alliance companies can include partners, suppliers, distributors, and even competitors.
Alpha release	*See alpha* test.
Alpha test	One of several levels of pre-release testing conducted as part of the development of a new product, process, or service [190, 191]. "*Alpha testing*" is usually conducted internally, once all of the intended features have been built in, and usually includes testing for all aspects of integration and performance. If this version passes the testing, it is referred to as an "*alpha release*" of the product. "*Beta testing*" is conducted by a selected number of external end-users. This is considered to be pre-release testing and provides a mechanism for obtaining end-user feedback while providing the marketplace with a preview of the intended product. This version is referred to as a "*beta release*" of the product. "*Gamma testing*" is sometimes conducted by customers in a limited market with a product that is almost but not quite ready for full market release. This is somewhat like a next level of *beta testing* and may be aimed at testing a specific feature of the product, such as an aspect of product safety, before the full release. This version is referred to as a "*gamma release*" of the product.
Altov, Genrikh	*See* Altshuller.
Altov, Henry	*See* Altshuller.

Altshuller, Genrich (Saulovich) (1926–1998)	A Soviet mechanical engineer and inventor best known in the innovation world for his *Theory of Inventive Problem Solving*, a systematic approach to invention based on a series of approaches that, taken together, aid in thinking about problems in unconventional ways that he identified as being key to most patented inventions (prior to the 1970s). This approach is also known by its acronyms in Russian (TRIZ) and English (TIPS). Altshuller is also known as the "father of TRIZ." Pseudonyms include Genrikh Altov and Henry Altov [72].
Angel investor	*See* seed capital.
Annual return	*See* return on investment.
Ansoff Matrix	A tool for business growth planning in which Igor Ansoff [398] defined four basic product/market strategies. See Figure 7.4. In a *Market Penetration* strategy, the idea is to sell more current products into the currently served market or markets (such as by increasing or improving quality, productivity, or marketing). In a *Market Development* strategy, the idea is to sell more current products into a new market or markets (such as with new marketing and sales efforts). In a *Product Development* strategy, the idea is to sell new products into the currently served market or markets, while in a *Product Diversification* strategy, the idea is to sell new products into a new market or markets. The Product Development and Product Diversification strategies usually involve R&D, adopt and adapt, or even purchase or licensing-in of technologies, and then commercialization.
A posteriori	*See a priori*.
Applied research	*See* research and development.
Applied research organization	*See* research and technology organization.
A priori	A Latin phrase meaning "from the earlier." This phrase is usually used as an adjective to describe a type of knowledge. For example, *a priori* knowledge is a knowledge that can be deduced from reason and logic. The converse of *a priori* is *a posteriori*, meaning "from the latter." For example, *a posteriori* knowledge is knowledge that comes only from experience and observation. Much of the knowledge in mathematics is *a priori*, whereas much of the knowledge in the other natural sciences and engineering is *a posteriori*.
Architects	A characterization of one of four kinds of organizational approaches to innovation strategy. "*Architects*" tend to be in mature markets, with intensive capital and resource requirements and with a centralized, structured approach to innovation. They also tend to be focussed on customers and competitors for new insights and opportunities, outsourcers of development and prototyping, and innovating through a top-down, formal process. Jaruzelski and Dehoff have referred to such organizations (and also to "Moonlighters") as "*Market Readers*" [381]. *See* innovation strategy mapping. *See also* top-down innovation.
Architectural innovation	A form of *Evolutionary Innovation* in which improvements are made to the linkage(s) between the components in a product, process, or service, but not to the components themselves. This is in contrast to another form of evolutionary innovation, *Modular Innovation* (also termed "*component innovation*"), in which improvements are made to one of more of the

	components in a product, process, or service, but not among the linkages between those components [87]. *See* Figure 2.6. In some literature, the term architectural innovation is used to mean disruptive innovation, but this is not recommended. *See* evolutionary innovation. *See also* innovation, disruptive innovation, incremental innovation, Abernathy-Clark Model, and Henderson-Clark Model.
ARIZ	Algorithm of Inventive Problems Solving (ARIZ is the Russian acronym). One of the tools used to try to solve invention problems in Altshuller's Theory of Inventive Problem Solving. *See* Altshuller, creative thinking models, and theory of inventive problem solving.
A-round capital	*See* start-up capital.
Art technology	A technology that is developed or deployed using know-how, craft skill, or experience rather than formal scientific or engineering knowledge, as would be the case with scientific or engineering technologies [76].
Assets	Owned things having significant value owned by a person or organization, such as "fixed assets" such as machinery, large equipment, buildings, and land; "current assets" such as cash, short-term securities, accounts receivable, and inventory; and "intangible assets" such as patents, copyrights, trademarks, and licenses.
ATAR model	*See* Awareness, Trial, Availability, Repeat model.
Augmented product	*See* product platform.
Avoiders	A category of potential customer or technology adopter in a psychological model of technology adoption. *See also* technology acceptance, technology acceptance model, Technology readiness index, and technology adoption life cycle.
Awareness, Trial, Availability, Repeat model	(ATAR model) A sales- or profits-forecasting model based on advertising and brand-awareness data. It was developed for consumer product sales, but could be used for process or service sales as well. A version of the model is: Sales = (purchasing unit #) × (% aware) × (% available) × (% trialing) × (% repeating), and profit = number of units sold × (revenue per unit − cost per unit), where *purchasing unit #* is the number of purchase participants (people, households, companies, or departments) in a sales region; *percent aware* refers to the fraction of purchasing units that have heard about the new product; *percent available* refers to the probability that the product is available to the purchasing unit; *percent trialing* refers to the fraction of purchasing units that have purchased at least once; and *percent repeating* refers to the fraction of purchasing units that have purchased more than once. Reference [202].

11.3 B

Back of the envelope calculation	*See* "Does it pencil?"
Barrier to innovation	*See* innovation barrier.
BASIC	Brazil, South Africa, India, and China. *See* BRIC nations.
Basic requirements	*See* Global Competitiveness Index.
Basic research	*See* research and development.
Bayh-Dole Effect	A reference to the tendency for university patenting activity to increase following the enactment of legislation enabling universities and/or their researchers to patent inventions arising out of federally funded research programmes. The name refers specifically to the US Bayh-Dole Act (1980), but the term has been applied to similar situations in other countries as well. Reference [738].
Bell, Alexander Graham (1847–1922)	A Scottish-born Canadian inventor and teacher of the deaf who is best known to the innovation world for his invention of the practical telephone and improvements to the commercial phonograph (the graphophone). The Bell Telephone Company was established in 1877 to commercialize his telephone invention.
Benchmarking	The process of comparing an organization's processes and performance measures to the best practices of a specified peer group. Originally a business management tool, benchmarking has also been applied to non-commercial organizations such as not-for-profit organizations (NFPs) including charities, non-government organizations (NGOs), research and technology organizations (RTOs), and even government departments and agencies. In any of these cases, the purposes of benchmarking are usually to assess organizational performance, efficiency, or effectiveness against best practices and/or to improve organizational performance. Also termed *"best-in-class benchmarking," "best practice benchmarking,"* or *"process benchmarking."*
BERD	Business Enterprise Expenditure on R&D.
Best available technology	The best state-of-the-art technology available at a given point in time. In some industries, regulations require purchases to meet a *Best Available Technology* standard. This can be a way to differentiate a product, process, or service in the marketplace. A synonym is Best Practicable Means. *See also* 20/30 rule. Reference [76].
Best-in-class benchmarking	*See* benchmarking.
Best practicable means	*See* best available technology.
Best practice benchmarking	*See* benchmarking.
Beta release	*See beta test and alpha test.*
Beta test	A test of a prospective design model product, process, or service conducted prior to commercial production and release [186]. *Beta* testing generally also involves having the users provide feedback on their experiences with the new product, process, or service. *See* alpha test. *See also* prototype.
Beta version	*See beta test.*
BI	Business intelligence. *See* competitive intelligence.
Bibliometrics	*See* scientometrics.

Big four nations — Brazil, Russia, India, and China. *See* BRIC nations.

Black box — In the innovation world, the "*black box*" refers to the technological innovation process(es) as a system with unknown components and unknown processes. In some economic models, technological innovation is a black box into which would enter various inputs and out of which would arise outputs in the form of commercializable technologies. Sometimes referred to as the "*Black Box of Innovation*."

Black holes — *See* innovation performance mapping.

Black space — *See* white space mapping.

Black swan events — *See* foresight.

Blindside curve — *See* S-curve.

Blowback innovation — *See* reverse innovation.

Boilerplate — In the innovation world, boilerplate is any text or code that can be used, and reused in new applications, without needing significant revision. This term is often used with reference to legal agreements, marketing documents, and even computer code (hence "*Boilerplate Code*").

Boom-bust cycle — *See* economic cycle.

Bootstrapping — *See* financing.

BOP innovation — *See* innovation at the bottom of the pyramid.

Bottom-up economic development — *See* top-down economic development.

Bottom-up innovation — *See* top-down innovation.

Brainstorming — *See* creative thinking models.

Brand — In business, a brand comprises the identity of an organization as distinguished from others, particularly from its competitors, and can serve to help distinguish an organization's products, processes, or services from those of its competitors. It can include tangible features, such as a name, logo, tagline, trademark, or other symbol or image. It can also include intangible features, such as how the organization is perceived by others and whether there is an implied promise to stakeholders associated with the brand. The manner in which a brand is perceived by stakeholders is sometimes termed "brand reputation." Although a brand is at least partly intangible, it is still given accounting treatment as a financial asset and the value of a brand is sometimes termed brand equity. *See also perceptual equity* and *trademark*.

Brand equity — *See* brand.

Brand reputation — *See* brand.

Breakthrough — A theory, concept, or application that, for the first time (in the world), overcomes a substantial barrier that had been holding back progress among some kind of regional, national, or international peer group. The related term "New-to-the-World Solution" refers to a breakthrough in business and/or innovation. *See also* Camelot Scenario and innovation barrier.

Breakthrough innovation — *See* disruptive innovation.

BRICK — Brazil, Russia, India, China, and South Korea. *See* BRIC nations.

BRIC nations — One of several acronyms that attempt to group together major countries that are in a similar stage of "emerging" economic development. BRIC (or "Big Four") stands for "Brazil, Russia, India, and China," BRICK (or BRICS)

	stands for "Brazil, Russia, India, China, and South Korea," and BASIC stands for "Brazil, South Africa, India, and China."
BRICS	Brazil, Russia, India, China, and South Korea. *See* BRIC nations.
Bridging organization	*See* intermediary organization.
Broker	*See* intermediary organization.
Building cycle	*See* Kuznets cycle.
Bush, Vannevar (1890–1974)	An American engineer and science administrator best known for his influence on government's role in the U.S. research, development, and innovation system. In addition to his own research and academic career, he also served on and/or led numerous national-scope research funding and advisory committees culminating in pioneering the role of science adviser to government. He became head of the U.S.'s World War II National Defense Research Committee (NDRC) and then the Office of Scientific Research and Development (OSRD). These government R&D organizations helped achieve numerous wartime technological innovations in such areas as radar, explosives, and nuclear weapons. Bush also developed the U.S.'s post-war science strategy, which included financial aid for discovery and applied research, and led to the creation of the U.S. National Science Foundation (NSF). References [739, 740].
Business accelerator	An organization, facility, and/or service that provide modest amounts of capital, mentorship, and/or other services to a business start-up or early-stage business in return for small amounts of equity. The time spent in an accelerator is small, often only a few months. *See also* business incubation and business incubator.
Business angel	Angel investor. *See* seed capital.
Business case	The essential basis for the market potential and/or the commercial viability of a product, process, or service. More details regarding the market and the anticipated revenues and expenses would be contained in the business model, while details on the whole business process from development through to sales would be contained in the business plan. *See also* business concept, business model, and business plan.
Business concept	An idea for commercializing a new product, process, or service that has been expanded into a description of the nature of the new product, process, or service; what is does; how it is different and better than what is already in the marketplace; how it would be delivered; and who would be the customers for it. Thus, a business concept is often referred to as being "*a bridge between an idea and a business plan*." *See also* product definition, business case, and business plan.
Business culture innovation	*See* administrative innovation.
Business cycle	*See* economic cycle.
Business development	The processes of identifying which business areas in a company can and should grow, developing strategies to achieve such growth, and it may also involve implementing such strategies. Business development builds on the positioning created through marketing and helps focus sales. It may also identify opportunities for strategic alliances, mergers and acquisitions or, conversely, opportunities for divestitures. The business development function usually also includes the process of identifying and nurturing customer relationships, in partnership with the sales function. *See also* marketing, sales, and distribution.

Business ecosystem — An organized network of organizations that collectively support a product, process, or service business. Also termed "*virtual cluster*." The anchor and/ or leader of a business ecosystem is sometimes referred to as the "*ecosystem leader*," "*keystone*," or "*platform leader*" [270]. The other ecosystem organizations are sometimes referred to as "*niche players*," or "*complementors*," whose products and services contribute to the productivity and outputs of the greater ecosystem. Examples: Walmart, Apple, and Mozilla have been ecosystem leaders for their respective business ecosystems. A business ecosystem is sometimes referred to as an *innovation ecosystem*, but this term is better used to refer to the broader meaning of the important entities involved in technological innovation (i.e. industry, academia, intermediaries, and government). *See also* innovation ecosystem.

Business growth planning — *See* Ansoff Matrix.

Business incubation — The provision of services and/or facilities in order to enable and/or accelerate the development and growth of new, usually small- and/or medium-size enterprises (SMEs), from the start-up phase through to some level of stand-alone maturity. The services may span a wide range from standard business services, such as legal, financial, recruitment, office facilities, and marketing, to custom services, such as access to angel- and venture-financiers and successful entrepreneurs, research, development, access to specialized facilities, prototyping, and demonstration. The goals in business incubation are usually to enable SME growth while reducing the associated risks, costs, and/or time involved. *See also* business accelerator, business incubator, and small- and/or medium-sized enterprise.

Business incubator — An organization, facility, and/or service in which significant amounts of capital, mentorship, management, education, infrastructure, and/or other services are provided to a business start-up or early-stage business, usually in return for significant amounts of equity. The goal is to provide enough support for entrepreneurs to survive the *valley of death* period between initial financing and significant sales of the new business' products, processes, or services. Depending on the nature of the business, this could take several months to several years. Also termed "*technology hatchery*." *See also* business incubation, business accelerator, cluster, research and technology park, and valley of death.

Business intelligence — (BI) *See* competitive intelligence.

Business logistics management — *See* value chain.

Business model — A description of the manner in which a business earns revenues and profits. Business models range from very simple, and brief descriptions to more complex, and lengthy descriptions resembling the business plans. *See also* business case, business concept, business plan, and commercialization plan.

Business model innovation — *See* strategic innovation.

Business model trap — *See* innovation barrier.

Business plan — A plan that describes the intended process for taking a new product, process, or service concept through the process of development, commercialization, and into an actual business enterprise. The business plan often evolves out of a *Commercialization Plan* and will typically cover

	descriptions of the overall project and value proposition; the new product, process, or service and its current development status; the *Market Analysis* and a marketing strategy; the *Intellectual Property*; an operations plan; a management plan, including descriptions of the key personnel; financial information and projections; the risk analysis, including how the principal risks are being managed; and any other special circumstances. *See also* business case, business concept, business model, and commercialization plan.
Business plan innovation	*See* strategic innovation.
Business process innovation	*See* administrative innovation.
Business R&D intensity	(Innovation indicator) An indicator of the R&D investments by and within the business sector is the ratio of business-enterprise funded R&D to gross domestic product (BERD/GDP). *See* innovation indicators and Tables 8.2–8.4.
Business structure innovation	*See* administrative innovation.
Buying hierarchy	A new product life cycle concept attributed to Windermere Associates (San Francisco, CA, USA). In this model, a new product goes through an evolutionary sequence comprising four phases: (1) functionality, in which the new product is able to compete based on a functionality not satisfied by competitors; (2) reliability, in which the new product is able to stay ahead of competitors based on market demand for reliability; (3) convenience, in which the new product competes by providing functionality, reliability, and superior convenience; and (4) price, in which the earlier attributes are all satisfied by competitors in the marketplace and competition comes down to price. *See also* product lifecycle curve in the entry for S-curve.
Buying unit	*See* Awareness, Trial, Availability, Repeat model.

11.4 C

Ca.	See *circa*.
Camelot Scenario	An imaginary scenario in which some kind of substantial barrier that had been holding back progress in business and/or innovation that does not exist. The creation of Camelot Scenarios is a brainstorming tool that is sometimes used in an attempt to find a solution or breakthrough. *See also* *breakthrough* and *innovation barrier*. Reference [737].
CapEx	*See* capital expense.
Capital	The assets, such as cash, investments, and property that a person or organization has available. In the context of innovation, this refers to the financial resources to support the research, development, and commercialization of a new product, process, or service. Debt capital refers

	to some form interest-bearing loan, often from a financial institution. Grant capital refers to some form of (usually) government grant assistance. Equity capital refers to some form of partial ownership in return for financial investment. Other examples of capital include sweat equity, seed capital, pre-venture capital, and venture capital.
Capital expense	(CapEx) A financial term for an expenditure for something that is intended to provide benefit to an organization for more than a year, as opposed to an "operating expense" (OpEx), which refers to expenditures for things that are used or done within an operating year. Accordingly, capital expenses are recognized on financial statements over time through annual depreciation amounts, while operating expenses are recognized within the year they are incurred. CapEx examples include major equipment and facilities. OpEx examples include research and development (R&D) and selling, general, and administrative (SG&A) expenditures.
Captive inventor	An inventor who works under a contractual arrangement that assigns intellectual property ownership to someone else, usually their employer.
Carve-out	A synonym for spin-off. *See* spin-out.
Cash flow	The difference between the amount of money flowing in during a specified period of time (often a month) and the amount flowing out during the same period. If the inflow of money is greater, it is referred to as *positive cash flow*; the converse is referred to as *negative cash flow*.
Catalytic innovation	*See* disruptive innovation.
Caveat innovator	A term coined by Richard Foster with reference to the perils of ignoring the implications of ignoring the plateauing nature of S-curves. According to Foster, the evolutionary approach "is doomed to fail" in the face of discontinuities brought about by competing technologies or product/ services. Reference [131]. *See Foster* and *S-curve*.
CBE	*See* circular business model.
CDA	Confidential disclosure agreement. *See* non-disclosure agreement.
CE	*See* circular business model or concurrent engineering.
Central chain-of-innovation	*See* chain-linked model.
Chain-linked model	An example of a non-linear fourth-generation *model* of the technological innovation process, which was proposed by Stephen Kline in 1985. In the chain-linked model, the innovation process begins with the identification of a potential market and then proceeds through the conception of a design (possibly but not necessarily involving research), development of the design, prototyping and testing, then redesigning and retesting, then production, marketing, and distribution. This principal pathway has been termed the "central chain-of-innovation." Supplementing and interconnecting with this principal pathway are a number of feedback loops including loops that feed-in knowledge and research results of various kinds such that there is not a single pathway but at least five major possible pathways from beginning to end (see in Figure 3.2). In the chain-linked model, research processes, for example, can feed most of the other elements in the process. Also termed *"linked-chain model"* and *"chain of innovation."* References [63, 67]. *See also* fourth-generation innovation model, non-linear innovation models, and generations of innovation.
Chain of innovation	*See* chain-linked model.
Chasm	*See* tipping point.

Chesbrough, Henry W. (1956–Present)
: An American business professor known to the innovation world for his 2003 book on *Open Innovation*. Although the activities now associated with open innovation were already being practiced, especially since the 1980s, Chesbrough's work helped coalesce them into an overall approach that could be incorporated into an organization's strategy. He also connected such processes as the acquisition of external technologies and other information (the inbound open innovation process) and the use of internal technologies and other information externally (the outbound open innovation process). References [160, 161].

Chief innovation officer
: *See* participative innovation.

Christensen, Clayton M. (1952–Present)
: An American business professor best known to the innovation world for his works on business strategy and technological innovation by businesses. He coined the phrases "Innovator's Dilemma" and "Technology Mudslide Hypothesis," among others. Reference [79].

CI
: *See* competitive intelligence.

Circa
: A Latin term meaning "*around*" or "*about*." It is frequently used to mean "*approximately*," particularly when referring to historical dates. There are many abbreviations for the term *circa*, including *c.* and *ca.*

Circular business model
: (CBM) This refers to a business model within the circular economy that is designed to reduce or eliminate waste and pollution, keep products and materials in use for as long as possible, and regenerates energy and natural resources. Unlike traditional linear models (take-make-waste), circular models create, deliver, and capture value by promoting strategies like sharing, leasing, reusing, repairing, refurbishing, and recycling materials and products, ultimately extending their lifespan and reducing the need for new resource extraction. For a case study example, see reference [648].

Circular economy
: (CE) This refers to an environmentally friendly economic model that is inherently restorative or regenerative, and employs approaches such as recycling, remanufacturing, reusing, and generally reducing the degree to which energy and natural resources are consumed. This is as opposed to the traditional economic model in which natural resources are consumed to create products that, once obsolete, are cast-off as waste.

Clinical trial
: The biomedical equivalent of a pilot or field test, in which a new, prospective medical or veterinary medical treatment is tested on live human or animal subjects. There are four standard phases of clinical trial: Phase 1 refers to testing for safety, Phase 2 refers to testing for the possibility of efficacy, Phase 3 refers to testing for the degree of efficacy, and Phase 4 refers to testing for long-term effects. The first three phases are generally considered to be part of the R&D process, while the fourth phase is usually only considered to be part of R&D if it results in an additional scientific or technological advance. Reference [485]. *See also* field pilot.

Closed innovation
: *See* open innovation.

Closed science
: *See* open science.

Clust-bun effect
: *See* clustering-bunching effect.

Cluster — (Innovation cluster) A group of organizations in fairly close proximity comprising companies with similar business interests and often supporting or coordinating organizations. These organizations normally share sufficient common interests that they become interconnected in formal and/or informal ways. The cluster theory of economic development holds that such clusters strengthen both entrepreneurship and innovation by providing a supportive, knowledge-rich, and resource-rich environment. For the same reasons, a cluster environment can be particularly attractive to start-up and small- and medium-sized enterprises. Clusters are normally led by industry but are frequently supported by government agencies and/or government funding. They are often engaged with intermediaries, including RTOs. Also termed *"innovation cluster," "cluster ecosystem," or "regional innovation cluster." See also* business incubator and innovative regional cluster.
(Technology cluster) A technology cluster is a grouping of two or more technologies that are closely related, especially as perceived by customers. Example: a household paper recycling service and a bottle/can recycling service. Rogers found that some businesses market clusters of technological innovations because they find more rapid consumer acceptance that way [77].

Cluster ecosystem — *See* cluster (innovation cluster).

Clustered Regularly Interspaced Short Palindromic Repeats — *See* CRISPR.

Clustering-bunching effect — A reference to the tendency for investment funds to flow to locations that have experienced bursts and/or clusters of inventions and/or technological innovations. Thus, the *clustering* of the inventions/innovations is viewed as leading to the *bunching* of investment funds, which intensify the business activities. Also termed *"clust-bun effect."* Reference [52].

Codified knowledge — *See* knowledge.

Cognitive innovation — The ideation, development, and deployment of improvements in ways of thinking and conceptual models. Example: lateral thinking. This is a form of non-commercial innovation. *See* innovation (non-commercial). *See also* creative thinking models and social technology [8].

Cognitive tacit knowledge — *See* social technology.

Collaborative compromise paradox — *See* innovation paradoxes (innovation process paradoxes).

Collaborative innovation — A form of open innovation in which two or more organizations work together on the development of innovation(s). Such organizations can include multiple customers, suppliers, and technology developer/providers. Advantages can include efficiency, and cost and risk sharing, plus the ability to leverage each other's technological and inventive capacities. Disadvantages can include difficulties maintaining cohesion and focus, and/or intellectual property issues. This is sometimes termed participative innovation, although the latter term can have other meanings as well. *See also* innovation ecosystem, participative innovation, and open innovation.

Collectors — *See* early majority.

Comets — *See* innovation performance mapping.

Commercial innovation	The meaning of innovation as originally defined by Schumpeter [15, 16]. *See* innovation. The term *commercial innovation* is sometimes used to distinguish from *non-commercial innovation*, which refers to any aspects of improving an organization's efficiency or effectiveness that manifest themselves in ways other than the introduction into the marketplace of commercial products, process, or services. Example: in some usage, the term *commercial innovation* is used to specifically exclude military innovations. *See also innovation (*20th–21st *centuries).*
Commercialization	The process of developing a new product, process, or service into a form that is market-ready and introducing it into the marketplace (product launch). The commercialization process generally follows the discovery and invention processes, if these are needed. *See also* business planning, commercialization plan, market analysis, product/process/service development, and technology.
Commercialization gap	*See* innovation gap.
Commercialization gap theory	*See* innovation gap theory.
Commercialization plan	A plan that describes the intended process for taking a new product, process, or service concept through the process of research, development, demonstration, and commercialization. A commercialization plan will typically cover descriptions of the overall project and value proposition; the new product, process, or service and its current development status; the *market analysis*; the *intellectual property*; the risk analysis, including how the principal risks are being managed; and the next steps in the product/process/service development plan. A commercialization plan is often later expanded into a *business plan*. *See also* business plan.
Commercialization valley of death	*See* valley of death.
Commercial readiness index	(CRI) A representation of the readiness of a mature technology, and an extension of the technology readiness level (TRL) scale, transcending TRL 9. The original Australian CRI system uses a scale ranging from 1 to 6 (*See* Table 4.4 and Figure 4.4). Reference [197]. *See also* technology readiness and technology readiness level.
Commitment trap	*See* innovation barrier.
Communication model	A term referring to, or a version of, the "*coupling model*" of the technological innovation process. *See* non-linear innovation models.
Compartmentalization	An approach to developing a complex device, or process, in which individual components are developed and tested before integrating them into the overall system [5].
Community of practice	(CoP) A CoP is a group of people that share a concern, problem(s), or interest in something, and who engage together on the topic. Principally bound by common interest they need not be geographically clustered. A CoP can exist within an organization. In the realm of open innovation, a CoP can exist among multiple organizations and can be effective in facilitating knowledge-sharing and problem-solving. *See* reference [677].
Competence-destroying innovation	*See* competence-enhancing innovation.

Competence-enhancing innovation	An innovation process that adds to and reinforces competencies, as opposed to "*competence-destroying innovation*," which supersedes previously existing competencies and may even make the latter obsolete. For example, incremental innovation is competence enhancing, whereas disruptive innovation is usually competence destroying. An example of competence-enhancing innovation is the electric typewriter over the mechanical typewriter, while an example of competence-destroying innovation is the transistor over the vacuum tube. Reference [6].
Competitive intelligence (CI)	Knowledge about developments and trends relevant to an organization's competitive position. Such knowledge could relate to discoveries, inventions, and/or technologies; products, processes, and/or services; partners, competitors, and/or suppliers; and customers and markets, for example. Competitive intelligence requires human analysis and judgement, and encompasses data, information, and knowledge (see also *Wisdom Hierarchy*). Competitive intelligence may refer to the process of producing such intelligence, or to the results themselves. Also termed "*business intelligence." See also social, technological, economic, environmental, and political analysis*.
Competitiveness	(Economics) The ability of an organization, country, or region to offer and supply products, processes, services, or jobs that effectively compete with others in the same market(s). The term has also been applied to specific regions or marketplaces, where it is used to refer to the extent to which such regions or marketplaces comprise competitive environments. *See also competitors, macroeconomics,* and *microeconomics*.
Competitiveness drivers	In making comparisons among countries' global competitiveness, the World Economic Forum (WEF) considers each county's stage of economic development maturity, using three broad categories of competitiveness driver and different weightings for calculating the Global Competitiveness Index (GCI) within each category. In the WEF categorization, *Stage 1*, or *Factor-Driven Economy*, refers to countries having GDP *per capita* of less than $2,000; *Stage 2*, or *Efficiency-Driven Economy*, refers to GDP *per capita* within the range $3,000 to $8,999; and *Stage 3*, or *Innovation-Driven Economy*, refers to GDP *per capita* of more than $17,000. *Economies in Transition* lie between Stages 1 and 2 or between Stages 2 and 3. Depending on the category, the weightings are adjusted in an attempt to make the GCI appropriate to a given country's degree of economic development and therefore the manner in which it competes globally. For example, a factor-driven economy competes mostly based on the strength of their "basic requirements" such as natural resources, infrastructure, and/or labour; an efficiency-driven economy competes mostly on the strength of their "efficiency enhancers" such as higher education and training, technological readiness, and market size; an innovation-driven economy competes mostly based on the strength of their "innovation and sophistication factors" such as capacity to innovate, value-added manufacturing, and value chains. Reference [491]. *See also* Global Competitiveness Index.
Competitor intelligence	*See* competitive intelligence.

Competitors	A description of the competitors in the marketplace for a product, process, or service, and also how the product is or will be differentiated from them. There are almost always competitors. If there are truly no competitors, then there is a good chance there is no market. *See also competitiveness, market analysis*. Reference [76].
Complementor	*See* business ecosystem.
Component innovation	A synonym for "*modular innovation.*" *See* architectural innovation.
Concept-push innovation	*See* design-driven innovation.
Concurrent engineering	(CE) A product development approach in which multidisciplinary functions work together through the entire process. *See also* Design for Excellence.
Confidential disclosure agreement	(CDA) *See* non-disclosure agreement.
Confidentiality agreement	*See* non-disclosure agreement.
Connectedness	(ICT) The availability and use of information and communications technology (ICT) to enable communications, information flows, and trade.
Connectedness Index	(ICT) In the area of ICT, the "Connectedness Index" is a blended indicator calculated based on such product and service factors as the market-ready supply of infrastructure, networks, and systems, demand, price, and usage. Example: The Conference Board of Canada Connectedness Index. *See* Innovation Indicators and Table 8.2. (Global business) In the area of international business and trade, a Global Connectedness Index is a blended indicator calculated based on such cross-border flows as trade, capital, information, and people. It is intended to indicate the breadth and depth of a country's interactions that cross its national borders. Example: DHL Global Connectedness Index (GCI) and McKinsey Global Institute (MGI) Connectedness Index.
Conservative	*See* late majority.
Consumer-innovators	(Innovators) Consumers that make inventions or modify existing products to create the first concepts or even the first prototypes of new technological innovations, especially in the area of products. Example: the skateboard was reportedly first developed and built by children for their own use by hammering roller skate wheel assemblies onto wooden boards. This is sometimes referred to as "user innovation," especially in the context of user modifications of current commercial products. Reference [85]. (Early adopters) The early adopters in the technological diffusion process, as described by the Rogers diffusion of innovation model. *See* Figure 4.8. These consumers are not actually innovators in the sense of making technological innovation happen, however.
Context-driven research	*See* modes of science.
Continuous innovation	*See* incremental innovation.
Convergent thinking	Part of creative problem solving. *See* creative thinking models.
Cooper, Robert G.	An American management scholar known for his contributions in product development and innovation management. He pioneered many groundbreaking discoveries in product innovation, including the Stage-Gate® Idea-to-Launch Process, which has been widely implemented in North American companies. References [223, 224].
Copyright	A legal protection for ownership of the original author(s) of published or unpublished artistic or literary creations, including computer software. The

specific rights and terms are determined by a given country's copyright law. *See also* intellectual property.

Core innovation — *See* incremental innovation.

Corporate immune system — A corporate cultural trait in which the internal "system" reacts against anything that is perceived to be a threat to the continuance of current norms and practises and/or a threat to the overall system. Although such a culture can help protect against things that could harm the organization, it can also repel new ideas and practises that could benefit the organization.

Corporate myopia — A corporate cultural trait that focusses on current "day-to-day" status quo business practises to the exclusion of potential or even prospective future business practises (hence the analogy with nearsightedness). Although such a culture can help with maintaining focus on the things that currently provide business value, it can also repel new ideas and practises that could benefit or even be critical to the organization in the future.

Cosmetic innovation — *See* incremental innovation.

Cost innovation — *See* reverse innovation.

Cost minimization — *See* late to market.

Country-first — *See* world-first.

Coupling model of innovation — *See* non-linear innovation models, third-generation innovation model. *See also* generations of innovation.

CPS — Creative problem solving. *See* creative thinking models.

Creative accumulation — *See* Schumpeter Mark I innovation.

Creative age or society — *See* technological ages.

Creative agglomeration — *See* Schumpeter Mark I innovation.

Creative problem-solving — *See* creative thinking models.

Creative thinking models — A number of different approaches to creative thinking, creative problem solving, and/or inventiveness have been developed, particularly since the 1940s. Listed below are some of the principal foundational approaches.

Brainstorming stands for "*storming a problem in a commando fashion*" and was coined by Alex Osborn. Osborn's brainstorming concept was aimed at groups and was designed to help get lots of ideas out into the open, avoid killing ideas with early criticism, and enable multiple ideas to be combined and/or lead to new ones. Also termed "*Think Up*" or "*Horizontal Thinking.*" References [170– 172].

Creative problem-solving (CPS) originated with Alex Osborn and Sidney Parnes and is famous for its use of "*divergent thinking*" and "*convergent thinking.*" Divergent thinking is open-ended, and wide-ranging thinking is used to create a broad set of options. Convergent thinking is solution-oriented thinking used to formulate a specific solution or approach to a problem using the results of divergent thinking. Also termed *Osborn-Parnes model.* References [171–173].

Parallel thinking and *lateral thinking* originated with Edward DeBono as creative counters to "*linear*" or "*vertical*" thinking in problem solving. *Parallel thinking* has to do with avoiding adversarial approaches in team-based creative thinking, focusing instead on more co-operative and constructive approaches. Parallel in this sense means having everyone on the team thinking in, broadly, the same direction. *Lateral thinking* refers to avoiding thinking about a problem "head-on" and/or in logical "step-by-step" fashion, and instead thinking about it in indirect and/or non-linear ways (sometimes referred to as *thinking out-of-the-box*). References [174, 175].

Synectics was developed by William J.J. Gordon. The name comes from the Greek word *synektiktein,* referring to the joining together of different, and possibly seemingly irrelevant, ideas in the process of creative problem solving by a group of people. Synectics originally referred to the study of problem-solving and invention by groups, but later evolved into a set of techniques that include what are now known as *creative problem-solving, brainstorming*, and *lateral thinking*. Synectics is also a trademarked name. *See also* Medici Effect. References [178, 179].

Theory of Inventive Problem Solving (TIPS, or TRIZ, the Russian acronym) was developed by Genrich Altshuller. TIPS is a systematic approach to invention based on a comprehensive analysis of the patented solutions to hundreds of thousands of previously solved inventive problems. Altshuller identified a series of approaches that, taken together, frequently lead to finding an inventive solution. Reference [72].

Spider diagrams are diagrams that use text and drawings and/or pictures to visually organize information. Spider diagrams, or at least their forerunner seems to have originated with the famous notebooks of Leonardo Da-Vinci. There are other variations of spider diagrams, including *idea sun bursts* and *mind maps*.

Creator — *See* early adopter *and* innovator.

CRI — *See* commercial readiness index.

CRI 1 — Commercial readiness index 1. There can be somewhat different definitions, but on a six-point CRI scale, CRI 1 generally refers to the stage at which the technology is commercially prospective but has not yet been commercially tested or proven. *See* Table 4.4 and Commercial readiness index.

CRI 2 — Commercial readiness index 2. There can be somewhat different definitions, but on a six-point CRI scale, CRI 2 generally refers to the stage at which the technology has passed a first, small-scale, commercial trial. *See* Table 4.4 and Commercial readiness index.

CRI 3 — Commercial readiness index 3. There can be somewhat different definitions, but on a six-point CRI scale, CRI 3 generally refers to the stage at which the technology has been commercially deployed, driven at least partly by market-pull. *See* Table 4.4 and Commercial readiness index.

CRI 4 — Commercial readiness index 4. There can be somewhat different definitions, but on a six-point CRI scale, CRI 4 generally refers to the stage at which multiple commercial deployments have occurred, possibly with some level of government support. *See* Table 4.4 and Commercial readiness index.

CRI 5 — Commercial readiness index 5. There can be somewhat different definitions, but on a six-point CRI scale, CRI 5 generally refers to the stage at which market-pull is driving broad commercial deployment and competition is emerging in the marketplace. *See* Table 4.4 and Commercial readiness index.

CRI 6 — Commercial readiness index 6. There can be somewhat different definitions, but on a six-point CRI scale, CRI 6 generally refers to the stage at which mature commercial deployment with established standards, regulatory acceptance, and performance track record. *See* Table 4.4 and Commercial readiness index.

CRISPR — (Clustered Regularly Interspaced Short Palindromic Repeats) A new (as of 2013) genome editing tool that enables accurate and rapid targeting, investigation, and splicing and editing of specific DNA sequences in a genome. Potential applications range from drug development, to disease treatment, to genetic trait modifications that could extent to food crops, animals, and humans. Also termed *"CRISPR-Cas9,"* for the first of several nucleases discovered. References [525, 526].

CRISPR-Cas9 — *See* CRISPR.

Critic — *See* early majority.

Critical technology events — (CTEs) Key advances in knowledge during the development and commercialization of a technology. CTEs have been defined as "ideas, concepts, models, and analyses that had a major impact on the development" of a particular innovation, and which led to significant improvements over the preceding technology [92].

Cross-license — A mechanism for exchanging technologies by which people or organizations grant royalty-free licenses to other people or organizations, in exchange for reciprocal rights to the latter's technologies. *See also license, licensing, and royalty-free license.*

Crowdfunding — *A relatively recent (i.e. since the 2000s) mechanism for raising early-stage (usually modest) capital for a company and/or commercialization process. In crowdfunding, a website typically serves as a platform that enables entrepreneurs to pitch their project(s) and interact with potential funders whose contributions could be any combination of donations, loans, equity, or future-product purchases. See also* seed capital, start-up capital.

CTEs — *See* critical technology events.

Cultification paradox — *See* innovation paradoxes (innovation process paradoxes).

Cumulative synthesis model — A model for the innovation process proposed by Abbott Payson Usher, by which a sequence of steps describes the process of creating something novel, practical, and useful (technological innovation). One formulation of these steps is: "*perception of the problem*," followed by "*setting the stage*" (i.e. acquiring the elements necessary for the solution), followed by the "*primary act of insight*" (in which the solution to the problem is found, followed by "*critical revision and development*" (in which the solution is made practical). Usher's model is a hybrid of the transcendentalist and mechanistic models in that somewhere in the process there is still a critical act of inspiration and/or insight. References [96, 97]. *See also* transcendentalist model and mechanistic model.

Current assets — *See* assets.

Customer intelligence — *See* competitive intelligence.

Customer-oriented innovation — This refers to maintaining a customer-orientation throughout most or all of the technological innovation processes. Also termed "*outcome-driven innovation.*" Reference [8].

Customers — A description of the target purchasers, and also of the ultimate end-users, of a product, process, or service. *See also* market analysis. Reference [76].

Customer value-chain analysis	(CVCA) An approach to product development in which, beginning with the product-definition phase, design teams identify key stakeholders (including customers and suppliers, all of whom are referred to as "customers" in this case), their interrelationships if any, and their role(s) in the new product's lifecycle. CVCA helps design teams to recognize diverse product requirements at an early stage of the development process. CVCA is an example of a Design for Excellence (DFX) tool. Reference [189]. *See also* Design for Excellence, voice of the customer.
CVCA	*See* customer value-chain analysis.
Cybernetic revolution	*A macroeconomic wave subsequent to the Industrial Revolutions that predominantly involves computer automation and self-regulating systems. Reference [340].*
Cycle time	*See* time to market.

11.5 D

Dark factory	An entirely automated factory would have so few people that the lights could mostly be left switched off, hence a "dark factory."
Data hierarchy	*See* Wisdom Hierarchy.
Data-Information-Knowledge-Wisdom Hierarchy	(DIKW Hierarchy) *See* Wisdom Hierarchy.
Davidow's law	Essentially a strategy that a company that is the first to release a product may be able to dominate the marketplace, especially if it is the first to obsolesce its own, previously dominating, product. Named for William H. Davidow, an executive at Intel in the 1970s and 1980s. Reference [234]. *See also* the Borg law.
Death valley	*See* valley of death.
Death valley curve	*See* valley of death.
Debt capital	*See* capital.
Debt financing	*See* financing.
Decision-making errors	Among the kinds of errors that can occur in a decision-making process are (i) judging that there is a risk, benefit, or opportunity when there is not (*Type I* or *false-positive error*) and (ii) judging that there is no risk, benefit, or opportunity, or failing to notice the same, when there is (*Type II* or *false-negative error*). In the innovation realm, a false negative, or Type II, error could involve being too skeptical or too negative about the prospects for a new technology or its market opportunity. *See also* technology readiness and technology acceptance model.
Deep learning	A form of machine learning, in which computers (including computers in robots) use massive data sets and neural networks to learn through experience. This represents a form of artificial intelligence. Also termed *"deep structured learning," "hierarchical learning,"* or *"deep machine learning." See also* singularity.
Delphi method	A foresight tool by which a group of experts in a field are surveyed for their opinions about a possible future situation or scenario, a summary of

the entire group's responses and rationales is provided as feedback, and then the group is surveyed again one or more times until something approaching a consensus, the *"expert consensus,"* is achieved. Originally developed by the RAND Corp. (Santa Monica, CA, USA) in the 1950s with military foresight in mind, the Delphi method has since been widely used in foresight activities of all kinds. *See also SWOT analysis* and *STEEPV analysis*.

Delta — (Business planning) The difference between the revenues that an organization is forecasting to generate and the revenues that they have targeted to generate.

Demand-induced innovation — *See* linear innovation models.

Demand innovation — *A synonym for demand-pull innovation. See* linear innovation models.

Demand-pull innovation — *See* linear innovation models.

Demographic cycle — *See* Kuznets cycle.

Demonstration — (1) In some usage, this is simply a physical demonstration to show that an invention basically works. Such demonstration could involve an engineering prototype, or at least a working model. Also termed *"proof of concept." See also* prototype.
(2) In other usage, this is a demonstration and/or test of a production prototype of a new or improved product, process, or service conducted at full-scale and under conditions of actual field, industrial plant, or market operation. With this definition, "demonstration" is usually the next step in commercialization beyond the engineering prototype and/or field test step. Sometimes termed *field demonstration, plant demonstration*, or *market demonstration. See* Figure 2.1. *See also* field pilot, production prototype, and prototype.

Deployment — *See* technology deployment.

Design-driven innovation — A form of technological innovation that is not "market pull" and also not quite "technology push" but rather *"concept push"* – meaning that the innovation relates to an entirely new product or service concept that was not previously imagined and which prospective customers did not previously know that they might want and/or need. Where they are successful, design-driven innovations are frequently breakthrough innovations. Also termed *"concept-push innovation."* Examples include the Wii entertainment system, which combined video gaming with physical activity and social interaction, and the Swatch, which transformed timepieces into fashion accessories. Reference [104]. *See also* disruptive innovation.

Design for all — *See* Design for Excellence.

Design for assembly — (DFA) *See* Design for Excellence.

Design for Excellence — (DFX) A suite of product development design programmes in which customer needs are identified at an early stage in the process. Some specific DFX tools include *"Design for Assembly"* (DFA), *"Design for Manufacture"* (DFM), *"Design for Manufacture and Assembly"* (DFMA), *"Design for Excellence"* (DFX), and *"Design for All."* References [187, 188]. *See also* concurrent engineering and customer value-chain analysis.

Design for Manufacture — (DFM) *See* Design for Excellence.

Design for Manufacture and Assembly — (DFMA) *See* Design for Excellence.

Design for Reuse	The including of previously designed components in a new product, process, or service. The term originated with reference to the use of previously designed code or data in the software engineering field, but is now used more broadly. Also termed *"design reuse."* Reference [741].
Design for X	(DFX) *See* Design for Excellence.
Design model	*See* production prototype.
Design patent	*See* patent.
Design reuse	*See* Design for Reuse.
Development	Experimental development. *See* research and development.
DFA	Design for Assembly. *See* Design for Excellence.
DFM	Design for Manufacture. *See* Design for Excellence.
DFMA	Design for Manufacture and Assembly. *See* Design for Excellence.
DFX	*See* Design for Excellence.
DHL GCI	DHL Global Connectedness Index. *See* Connectedness Index.
DHL Global Connectedness Index	(GCI) *See* Connectedness Index.
Diffusion	*See* technology dissemination.
Diffusion of innovations model	*See* Rogers.
Diffusion of technologies model	*See* Rogers.
Digital age	*See* technological ages.
DIKW Hierarchy	Data-Information-Knowledge-Wisdom Hierarchy. *See* Wisdom Hierarchy.
Direct indicators	*See* innovation indicators.
Dirigiste regional innovation system	*See* regional innovation system types, Table 5.1.
Discontinuity	*See* S-curve.
Discontinuous innovation	*See* disruptive innovation.
Discoverer	A person that has discovered or co-discovered a new addition to the world's body of knowledge. Discoveries are generally attributed to the first to document, publicly present, or publish their discoveries. *See also* discovery, innovator, inventor, and patent.
Discovery	A new addition to the world's body of knowledge. Such an addition could be developed through the formulation and testing of new theories or through the observation and study of new phenomena. One can discover physical things (such as previously undiscovered elements, species, or stars), ideas (such as new concepts that lead to new theories), or understanding (such as new and improved understanding the behaviour of natural phenomena). Discovery is different from invention and innovation. *See also* innovation, invention.
Discovery-push innovation	*See* linear innovation models.
Discovery research	*See* research and development.
Disembodied technology	*See* acquisition of technology.
Disruptive innovation	An innovation that displaces an earlier commercial product, process, or service and significantly changes (disrupts) a market niche, or even an entire market. Disruptive innovations are unexpected in the marketplace and are different from evolutionary or incremental innovations. Sometimes termed *breakthrough innovation, catalytic innovation, disruptive technology, discontinuous innovation, fundamental innovation, radical innovation, seismic*

innovation, technology-push innovation, transformational innovation, or *revolutionary innovation*. Disruptive innovation is usually very R&D intensive and may require discovery research. Examples: the electronic calculator made the slide rule obsolete; the electric typewriter made the manual typewriter obsolete. *See* Figures 2.4 and 2.5. *See also* innovation, design-driven innovation, evolutionary innovation, incremental innovation, and radical technological transition.

Disruptive technology — *See* disruptive innovation.

Distribution — In business, distribution refers to the process of making commercial products, processes, or services available to customers, either directly or through an intermediary. *See also* marketing, business development, and sales.

Distribution channels — A summary of the pathways available to get a product to the customers and/or end-users. *See also* market analysis. Reference [76].

Divergent thinking — Part of creative problem solving. *See* creative thinking models.

"Does it pencil?" — This phrase refers to the question of whether a new product or service concept has a sound business basis, meaning whether its cost and selling price will yield enough profit for it to be feasible. The phrase "*back of the envelop calculation*" is sometimes used with the same meaning (although this phrase can also refer to purely technical, approximate calculations). Reference [76].

Domestic product of industry — (DPI) A measure of the economic production of industry within a region, given by the gross domestic product (GDP) less the production value of non-industrial goods and services, such as those provided by governments, not-for-profit organizations, and academia. *See also* gross domestic product.

Downstream innovation — *See* upstream innovation.

DPI — *See* domestic product of industry.

Drucker, Peter (Ferdinand) (1909–2005) — An Austrian-born American management consultant and author. He is associated with the development of modern (post-1950) management principles. Among his influential writings are those on innovation, such as his 1985 book *Innovation and Entrepreneurship*.

Dual ladder — *See* technical ladder.

Dual-use technologies — Technologies developed for application in one sector that can be used or adapted for use in another sector. Examples include the use of defence or aerospace technologies in manufacturing or other sectors. In some usage, *dual-use technologies* specifically refer to defence technologies that can also be used for non-military applications.

Due diligence — The process of systematically researching and evaluating the accuracy of a statement or the nature and status of a person, group, or organization. Examples include conducting due diligence on a financial statement, the financial status of an organization, and the background of a potential new employee. The goal of due diligence is usually to ensure that the important information related to a potential decision is known before making that decision, in order to reduce the risk of making a poor decision. Where patents are concerned, due diligence sometimes refers to a legal principle that patent owners need to develop a product, process, or service out of the patented intellectual property in question, rather than simply use the patent as a barrier to prevent others from doing so.

Dynamically continuous innovation — *See* evolutionary innovation.

11.6 E

Early adopter — A category of technology adopter in Rogers' diffusion of innovation model. Early adopters have sometimes been referred to as "*consumer-innovators*," although the term refers to their role as early adopters and not as innovators. Also sometimes referred to as "*visionaries*" or "*creators*." *See* technology adoption life cycle. *See* Figure 4.8. *See also* Rogers, Everett M.

Early growth phase — *See* venture capital.

Early majority — A category of technology adopter in Rogers' diffusion of innovation model. Also sometimes referred to as "*pragmatists*," "*critics*," or "*collectors*." *See* Figure 4.8. *See* technology adoption life cycle.

Early modern innovation — *See* innovation (early modern).

Early-stage company — *See* start-up capital.

Early-stage technology — A technology concept that is beyond the speculative concept level and is either being developed into something feasible or has been demonstrated to be feasible. At this stage of development, the technology does not have to be practical, efficient, or cost effective. In terms of Technology readiness levels, a demonstrated proof of concept is beyond TRL 2 but may not yet fully satisfy TRL 3. *See also* proof of concept, speculative concept, and technology readiness level.

Earnings Before Interest, Taxes, Depreciation, and Amortization — (EBITDA) A financial measure intended to enable comparisons of companies' operating profitability by eliminating the effects of interest payments, tax jurisdictions, asset depreciation, and takeover activities. Some related terms include earnings before interest and taxes (EBIT), earnings before interest, taxes, and amortization (EBITA), earnings before interest, taxes, and depreciation (EBITD), and earnings before interest, taxes, depreciation, amortization, and restructuring or rent costs (EBITDAR).

Ease of entrepreneurship index — A blended indicator calculated based on such factors as barriers to competition, regulatory and administrative opacity, and administrative burdens. Example: The Conference Board of Canada Ease of Entrepreneurship Index. *See* innovation indicators and Table 8.2.

EBIT — Earnings before interest and taxes. *See* earnings before interest, taxes, depreciation, and amortization.

EBITA — Earnings before interest, taxes, and amortization. *See* earnings before interest, taxes, depreciation, and amortization.

EBITD	Earnings before interest, taxes, and depreciation. *See* earnings before interest, taxes, depreciation, and amortization.
EBITDA	*See* earnings before interest, taxes, depreciation, and amortization.
EBITDAR	Earnings before interest, taxes, depreciation, amortization, and restructuring or rent costs. *See* earnings before interest, taxes, depreciation, and amortization.
Eco-industries	As defined by the OECD and Eurostat, eco-industries produce products, processes, or services that measure, prevent, limit, minimize, or correct environmental damage to water, air, and soil, or otherwise deal with problems related to ecosystems. *See also* eco-innovation.
Eco-innovation	*See* ecological innovation.
Ecological innovation	The ideation, development, and deployment of improvements in environmental protection and/or contribute to sustainable development. Also termed "*eco-innovation*." Example: a new recycling programme. This is usually a form of non-commercial innovation, but technological innovation in this area occurs as well. Reference [8]. *See* innovation (non-commercial). *See also* eco-industries.
Economic cycle	The concept that, for various reasons, economies can cycle through periods of building, then rapid growth, then a plateauing and/or crisis, then a recession, and finally some kind of recovery or rebuilding. Economic cycle is sometimes referred to as *business cycle*, or *boom-bust cycle*. Several kinds of economic cycles have been identified, each with its own drivers, measures, and time frames (periods).
Economic innovation	The ideation, development, and deployment of improvements in economic models and processes. Example: Microcredit. *See also* innovation. Reference [8].
Economy in transition	*See* competitiveness drivers.
Ecosystem innovation	Technological innovation that is identified, developed, and commercialized through a multiorganization partnership. This is similar to an R&D alliance except that ecosystem innovation is aimed at developing and commercializing large technological innovation opportunities in a short time period, whereas R&D alliances are more focussed on the R&D components and over moderate to long periods of time. Reference [314]. *Also termed "multiparty innovation."*
Ecosystem leader	*See* business ecosystem.
Edison, Thomas A. (1847–1931)	An American inventor and businessman best known for his prolific invention record, which led to over a thousand patents and numerous innovations. He was a champion of what is now termed market-pull rather than technology-push innovation. In 1880, he became one of the first inventors to apply the principles of organized science and teamwork to the process of invention, working with many researchers and employees. He founded the Edison Electric Light Electric Company, which, following a number of mergers ultimately became the General Electric Company. References [54, 742].
Educational innovation	The ideation, development, and deployment of improvements in education models and processes. Example: E-learning. This is a form of non-commercial innovation. *See* innovation (non-commercial). Reference [8].

Efficiency-driven economy	*See* competitiveness drivers.
Efficiency enhancers	*See* Global Competitiveness Index.
Electronification of innovation	A reference to the increasing use of electronic tools, such as expert systems, simulation modelling, and even artificial intelligence in fifth-generation innovation practices. Reference [98]. *See also* generations of innovation, non-linear innovation models.
Embedded knowledge	Knowledge that was once held only by a small number of specialists and/or experts but which has become embedded more generally in an organization's products, processes, services, or practices. *See also* acquisition of technology.
Embodied technology	*See* acquisition of technology.
Emergent novelty	*See* innovation (20th–21st centuries).
Endless frontier research	*See* third mission.
Endless transition research	*See* third mission.
Endogenous Growth Theory	An economic growth theory that assumes economic growth is the result of internal (endogenous) rather than external (exogenous) factors. There are various versions of this theory, but the main internal growth factors are investments in human capital, technological innovation, and knowledge as they relate to productivity. As a result, the theory also gives importance to such things as industrial business strategies and government policies insofar as they affect the three principal drivers. Sometimes referred to as the "New Growth Theory." The Endogenous Growth Theory is different from external (exogenous) economic growth theories, such as the Solow-Swan Growth Model, which are based on the accumulation of physical capital and growth of the labour force, although both kinds of theories assume a strong connection between continuing technological innovation and overall economic growth. References [38–40]. *See also* Solow-Swan Growth Model and exogenous variable.
Endogenous variable	*See* exogenous variable.
End user	The ultimate user of a product, process, or service. Such a user may or may not have any significant expertise in the nature of the product, process, or service, and the term is sometimes meant to imply a specific lack of such expertise. The purchaser of a product, process, or service is not necessarily an end user, which is an important consideration in the development of marketing strategies.
Engineering prototype	A working model of a product or process that has been built carefully enough for use in testing and demonstration of important design parameters. An engineering prototype may not be built to full scale, but will be designed to be scaled-up. *See also* field test and prototype. Reference [76].
Engineering research	*See* scientific and engineering research.
Enterprise unit	A kind of business unit within an organization that has independent operations including its own strategy, business plan, management and financial system, and its own culture, marketing, and sales functions.

Nucleating and nurturing an enterprise unit can pave to way to it becoming a subsidiary, spin-out, or spin-off. The process of creating and nurturing enterprise units is sometimes termed internal corporate venturing (ICV).

Entrepreneur — A person that starts and manages a business, assuming all or most of the financial risks involved. *See* Reference [186]. *See also* intrapreneur.

Entrepreneurial university — A university that has incorporated a "third mission," beyond those of teaching and research, which is to engage with other organizations in the development of practical uses of new knowledge (*"Mode 2 Research"*). Such a university can play a larger role in an innovation ecosystem. The entrepreneurial university is frequently associated with research and/or technology parks, incubators, and/or start-up and spin-off companies. In some definitions (including that of OECD), the entrepreneurial university is also itself associated with an entrepreneurial culture and can play a role in helping to develop entrepreneurs. The evolution of the entrepreneurial university with its third mission has also been referred to as the result of the *"second academic revolution,"* the *first academic revolution* being the evolution of the research university, with its second mission, from the teaching university. References [262, 273]. *See also* modes of science, innovation ecosystem, triple-helix model, and quad-helix model.

Envelope S-curve — *See* S-curve.

Equity — A term referring to the fractional ownership of a business corporation. It is sometimes used to refer to a specific percent of ownership in a business, sometimes to a number of shares (or the value of a number of shares), and sometimes to the total number or value of shares of a business.

Equity capital — *See* capital.

Equity financing — *See* financing.

Era of ferment — A reference to the early stages in a product/process/service lifecycle, in which a new product is introduced to the marketplace, with associated business development, design improvements, and early to rapid sales growth as the product finds a successful market niche. The related terminology for the latter phases of the lifecycle is the *"Era of Incremental Change,"* in which sales plateau and then decline, as the market niche becomes saturated, and competition is mostly based on price. Reference [132].

Era of incremental change — *See* era of ferment.

Etzkowitz, Henry (1940–present) — An American sociologist known to the innovation world for his work on university-industry interactions and for his co-development (with Leydesdorff) of the triple-helix model curve, a sociological model that was developed to describe the roles and intersections of governments, universities, and industry in advancing knowledge-based economies and which by extension has been applied to the advancement of economies based on innovation. He was also the originator of the "entrepreneurial university" concept and co-founder of the Triple Helix International Conference Series (since 1996). *See* triple-helix model. *See* Figure 5.4. *See also* references [259, 261, 262].

Evolutionary innovation — An innovation that competes with an earlier commercial product, process, or service but does not significantly change (disrupt) a market, or even a market niche. Evolutionary innovations usually significantly help companies enhance their competitive positions in the marketplace. Also termed *"dynamically continuous innovation," "leap-forward innovation,"* or *"sustaining innovation."* Sometimes termed *"continuous innovation"* although this term is probably better suited to *"incremental innovation."* The terms *"architectural innovation"* and *"modular innovation"* are probably best considered to be forms of evolutionary innovation. Evolutionary innovation is usually very applied R&D intensive. *See* Figures 2.4 and 2.5. *See* architectural innovation; modular innovation. *See also* innovation, disruptive innovation, incremental innovation; Abernathy-Clark model; and Henderson-Clark model.

Ex ante — A Latin term meaning "from before." In business, the term *ex ante* usually refers to forecasts. For example, an *ex ante* estimate would be a forecast rather than a post-event, or "actual," estimate of financial or production results. In commercial innovation, the term is sometimes used to refer to things, activities, or processes that are dealt with before an event. For example, several collaborators and/or their activities could be synchronized *ex ante* (i.e. before) an event of some kind. The opposite of *ex ante* is *ex post*, meaning "from after."

Exclusive license — *See* license agreement.

Exit — *See* exit strategy.

Exit strategy — A plan for when and how to leave a position. The position could be an ownership position, such as in the ownership of a start-up company, and the timing could be triggered by the achievement of pre-set objectives or the occurrence of unexpected risks or losses beyond pre-set tolerances. The plan could be as simple as selling shares, or might be more complex, and it could involve a phased transition.

Exogenous variable — A variable whose origin is external to the system or model under consideration and which is independent of the system or model. Example: in classical economics (before the Schumpeter era), technological progress was generally assumed to be exogenous, meaning that it was assumed to be solely determined by non-economic forces. The opposite of exogenous is endogenous.

Experimental development — *See* research and development.

Explicit knowledge — *See* knowledge.

Explorers — (Innovation strategy) A characterization of one of four kinds of organizational approaches to innovation strategy. *"Explorers"* tend to be in emerging, rapidly expanding markets, or redefining their markets, but in either case to be in markets demanding speed and agility. They also tend to be focussed on customers, partners, and markets, generally for new insights and opportunities, users of rapid prototyping and testing of large numbers of relatively small ideas, and innovating through a diffuse, customer-facing process. Jaruzelski and Dehoff have referred to such organizations as *"Need Seekers"* [381]. *See* innovation strategy mapping.
(Technology adoption) A category of potential customer or technology adopter in a psychological model of technology adoption. *See also* technol-

ogy acceptance, technology acceptance model, Technology readiness index, technology adoption life cycle.

Ex post — *See ex ante.*

Ex situ — In science and engineering, the Latin term *ex situ* generally refers to an aspect of a reaction or process taking place away from where it normally occurs or was created. *See also in situ.*

External patent applications — (Innovation indicator) An indicator of the number of practical inventions with international potential that are being protected is the number of external patent applications per 10,000 personnel in the labour force of an economy. *See* innovation indicators and Tables 8.2 and 8.4.

11.7 F

Fab lab — *See makerspace.*

Factor-driven economy — *See* competitiveness drivers.

Fail better — *See* fail fast.

Fail cheap — *See* fail fast.

Fail early — *See* fail fast.

Fail fast — A common business mantra, especially with regard to product development, entrepreneurs, and start-up companies. Two key concepts associated with this phrase are: (1) not to fear failure but learn from it, and (2) that it is better to learn quickly if a new product, process, or service is not going to be successful so that it can either be improved or abandoned in favour of a new one. A goal could be to experience one or more small failures from which useful learnings can be obtained, increasing the probability of achieving a significant success, rather than risking a single large failure that might not be survivable. Also termed *"quick to fail."* Other nuances that build upon the same philosophy include: *"fail cheap," "fail small," "fail often," "fail early," "fail better," "fail well,"* and *"fail forward." See also* "works like" model.

Fail forward — *See* fail fast.

Fail often — *See* fail fast.

Fail small — *See* fail fast.

Fail well — *See* fail fast.

False negative error — *See* decision-making errors.

False positive error — *See* decision-making errors.

False summit — The phenomenon in a research or development process when a problem is solved and it is then discovered that there is another previously unrecognized problem yet to be solved before the real goal(s) of the process can be achieved. This is an analogy to a false summit in mountaineering, in which on ascent, it appears that the summit of a mountain is in sight, but once reached, it turns out that it was simply a ridge or another feature of the mountain that had prevented sight of the true summit. The *"false summit effect"* refers to the psychological impact of reaching and recognizing a false summit, which is usually some level of

	discouragement. In early-stage innovation processes, the false summit effect occurs quite often because, by definition, not enough is known about the technological solution or how to reach it until the development is fairly well advanced.
False summit effect	*See* false summit.
Fast follower	A person or organization that takes advantage of a new idea, technology, invention, or technological innovation (developed by a "first mover") and implements a copy or an adaptation of it. A *"fast follower"* strategy (also termed a *"second-to-market strategy"*) has a goal of being an early participant in the marketplace as a new product, process, or service evolves "up" its lifecycle trajectory. Example: in the videotape industry, Sony was a first mover with its Betamax technology, while RCA and Matsushita were fast followers with their VHS technology. *See also* adopt and adapt, first mover, late to market, market segmentation, and reverse engineering.
Fast innovator	*See* time to market.
Father of TRIZ	*See* Altshuller.
FBE	*See* fuzzy back-end.
Fear of innovation	Some public fears related to technological innovation are that it may create and set loose upon society monsters (the *"Frankenstein Hypothesis"*); job losses and unemployment due to any or all of mechanization, process efficiency improvements, and robotics (*"technological unemployment"*); technologies that are advanced and marketable but environmentally and/or socially unsustainable; the driving or determining by technology of a society's social structure and cultural values (*"technological determinism"*). Reference [52].
Feasibility study	An evaluation of a proposed initiative or business concept that is aimed at identifying strengths, weaknesses, opportunities, and threats (SWOT), the potential for success, and the resources needed to further develop the initiative or business concept. Feasibility studies are often used to help decide whether to proceed with an initiative or business concept. *See also* SWOT.
FEF	Front-end fuzziness. *See* fuzzy front-end.
FFE	*See* fuzzy front-end.
FFF capital	Friends, family, and fools capital. *See* seed capital.
Field demonstration	*See* demonstration.
Field pilot	A test of an engineering prototype of a new or improved product, process, or service under conditions of actual field, industrial plant, or market operation. This is usually the next step in commercialization beyond laboratory, proof-of-concept, or working model testing. Field pilots may be conducted at full commercial scale or at a reduced scale. Sometimes termed *field test, pilot plant, pilot test, plant pilot, plant test*, or simply *"pilot."* *See* Figure 2.1. *See also* clinical trial, demonstration, engineering prototype, prototype, and scale-up.
Field test	*See* field pilot.
Fifth-generation innovation model	Any of the advanced, non-linear models like those of the fourth generation, but to which have been added time- and cost-efficiency aspects, as in lean innovation. This is frequently associated with the

"system integration and networking (SIN) model." *See* Section 3.1 and Table 4.2. *See also* systems integration and networking model, generations of innovation, and non-linear innovation models.

Fifth-generation of technology foresight — *See* generations of technology foresight.

Fifth wave — *See* 5th wave.

Financing — The acquisition of capital. Some common forms in innovation are *bootstrapping*, in which funds are generated internally from other income, *debt financing*, in which funds are borrowed, or *equity financing*, in which funds are obtained through the sale of a share in ownership.

Firm-first — *See* world-first.

First academic revolution — *See* second mission, entrepreneurial university.

First-generation innovation model — Any of the linear and sequential technological innovation models that begin with discovery (basic) research and then evolve upwards through applied research and engineering development phases. These are often referred to as being "technology-push" or "supply-side" models. Although they have some value in terms of distinguishing broad driving forces in innovation, the processes themselves don't usually begin this way, nor are they very often sequential or linear. *See* Section 2.1 and Table 4.2. *See also* generations of innovation, linear innovation models, and technology-push model.

First-generation nanotechnology — *See* generations of nanotechnology.

First-generation of technology foresight — *See* generations of technology foresight.

First Industrial Revolution — *See* technological ages.

First mission — The first mission of a university is teaching. The second mission is research. The third mission of an "*entrepreneurial university*" involves participation by universities in working with governments and industry to enable economic growth (via innovation) and social progress. *See also* entrepreneurial university, modes of science, innovation ecosystem, triple-helix model, quad-helix model.

First mover — A person or organization that is first to market a new product, process, or service, or the first to establish itself in a given marketplace. Also termed *"first to market,"* or *"originator."* A "*market leader*" strategy (also termed a "*first-to-market strategy*") has a goal of being the first to introduce new products, processes, or services into the marketplace. Example: in the videotape industry, Sony was a first mover with its Betamax technology, while RCA and Matsushita were fast followers with their VHS technology. *See also* fast follower, late to market, market segmentation.

First to market — *See* first mover.

First wave — *See* 1st wave.

Fixed assets — *See* assets.

Foresight — A systematic process involving environmental and horizon (futures) scanning aimed at anticipating future events in order to be able to develop strategies to encourage, prevent, change, or simply manage the impacts of such events. Sometimes termed "futures research." The principal kinds of future events anticipated by foresight activities usually involve technological, economic, environmental, political, social, and/or ethical (TEEPSE) futures. Innovation foresight (compared with technology foresight) has to look further than the development of new technologies and into their commercialization (this is also termed "*innovation system foresight*"). Two terms that occur frequently in foresight processes and scenarios are "*wild cards*" and "*weak signals*." In foresight scenarios, wild cards are events that are perceived to be of low probability of occurrence but have potentially high impact should they actually occur. Also termed "*outliers*," "*black swan events*," or simply "*black swans*." In foresight scenarios, weak signals are unclear observables that provide a warning of the probability of possible future events, including wild card events. Reference [425].

Formal knowledge — *See* knowledge.

Formative stage company — *See* start-up capital.

Forward innovation — *See* reverse innovation.

Foster, Richard N. — An American engineer and business consultant known to the innovation world for his adaptation of the sigmoid S-curve to applied research and technology development and to product lifecycles. This is described in his 1986 book, *Innovation: The Attacker's Advantage*. The S-curve is sometimes termed as Foster curve. *See* S-curve. Reference [131].

Foster's curve — *See* S-curve (technology).

Fourth-generation innovation model — Any of the advanced, non-linear models of technological innovation that introduce parallel processes and are integrated across the organization, its customers, and its suppliers with multiple feedback loops. Examples are the "chain-linked model" and "integrated model." *See* Section 3.1 and Table 4.2. *See also* generations of innovation, non-linear innovation models, chain-linked model, and integrated model.

Fourth-generation nanotechnology — Molecular nanotechnology. *See* generations of nanotechnology and molecular nanotechnology.

Fourth-generation of technology foresight — *See* generations of technology foresight.

Fourth Industrial Revolution — *See* technological ages.

Fourth-pillar organization — An organization that works to enable and/or assist with the innovativeness and competitiveness of companies, by working with industry, government, and academia (which represent the other three "pillars"). Such organizations use these linkages to help companies develop and deploy new commercial product, processes, and/or services. Fourth-pillar organizations are usually either government-owned corporations or not-for-profit corporations. Examples include RTOs, industry associations, and economic development organizations. Fourth-pillar organizations are also examples of "*intermediary organizations*." Reference [274]. *See also* intermediary organization, knowledge-intensive business services, innovation ecosystem, and technology and innovation centre.

Fourth wave — *See* 4th wave.

Frankenstein Hypothesis — *See* fear of innovation.

Frascati manual	An Organization for Economic Co-operation and Development (OECD) document providing recommended methods for collecting and interpreting data on research and development (R&D). The counterpart to the Frascati Manual is the Oslo Manual, which is concerned with data on innovation. Reference [485]. *See also* Oslo Manual.
Friends, Family, and Fools capital	(FFF capital) *See* seed capital.
Front-end fuzziness	(FEF) *See* fuzzy front-end.
Frugal engineering	*See* frugal innovation.
Frugal innovation	This term generally refers to reducing the cost of an existing product, process, or service. Depending on the usage, there may be an implication that certain features have been removed for a particular market niche, that performance has been reduced, and/or that quality has been reduced. Also termed "*inclusive innovation.*" The terms "*Jugaad innovation*" (using meagre resources) or "*Gandhian innovation*" (affordable and sustainable) are sometimes used to describe Frugal innovation with respect to the Indian market. Part of the process involved is sometimes termed "*Frugal engineering.*"
FTE	*See* full-time equivalent.
Full-time equivalent	(FTE) A unit of employment that recognizes fractional employment in order to give a better representation of workforce capacity than, for example, only recognizing full-time employment. As a result, the numbers of FTEs in an organization, or assigned to a programme, may be very different from the number of people so employed or assigned. Example: the number of FTE researchers in an organization or assigned to a programme.
Fundamental innovation	*See* disruptive innovation.
Fundamental research	*See* research and development.
Funding gap	*See* valley of death.
Future pull	Any kind of creative thinking, planning, or product/process/service development that is influenced mostly by a sense or vision of future benefits, needs, or demands could be said to be influenced by "*future pull.*" This term is sometimes used to provide contrast with the terms "*technology push*" and "*market pull.*"
Futures research	*See* foresight.
Future technology myopia	*See technological myopia.*
Fuzzy back-end	(FBE) The final phase of new product/process/service development, when there is a very high degree of uncertainty about marketing and sales, and how best to manage these processes to achieve a commercially successful conclusion to the entire process. The term was coined to contrast with the "*fuzzy front-end*" phase. *See also* fuzzy front-end.
Fuzzy front-end	(FFE) The early phase of new product/process/service development, when there is a very high degree of uncertainty about the nature of the future product, whether it will be technically successful, what it will cost in terms of time and resources, and whether it will be likely to be commercially successful. The fuzzy front-end phase is typically from product conception through to the point where there is a validated product concept, before a decision is made to proceed with the product development phase. Also termed "*upfront phase.*" The uncertainties inherent in this phase are

sometimes referred-to as *"front-end fuzziness,"* or FEF. *See also* prototype, time to market, and fuzzy back-end.

11.8 G

GAAP — *See* Generally Accepted Accounting Principles.

Gamma release — *See alpha test.*

Gamma test — *See alpha test.*

Gandhian innovation — *See* Frugal innovation.

Gartner hype cycle — *See* technology hype-cycle.

Gated process — *See* idea-to-launch process.

GBAORD — Government Budget Appropriations or Outlays for R&D. *See* Government R&D expenditures.

GCI — (1) *See* Global Competitiveness Index.
(2) Global Connectedness Index. *See* Connectedness Index.

GDP — *See* gross domestic product.

GDP per capita — Gross domestic product per person, usually referring to a country or region within a country. *See also* competitiveness drivers.

Generally Accepted Accounting Principles — (GAAP) A set of standardized procedures and rules for recording and reporting corporate financial information.

Generations of innovation — Rothwell identified five generations of models of technological innovation, also termed the *"Rothwell models of innovation."* Reference [98, 126]. The first two are linear models of innovation, while the other three are non-linear models:

- First generation: technology-push model
- Second generation: market-pull model
- Third generation: coupling model
- Fourth generation: integrated model
- Fifth generation: systems integration and networking model

Open innovation, and/or the functioning of regional clusters of innovation have recently been bundled into the category of sixth-generation models of technological innovation, see Table 4.2.
See also linear innovation models, non-linear innovation models, and generations of technology foresight.

Generations of nanotechnology — The development and evolution of nanotechnology has been divided into four stages, or generations. First-generation nanotechnology refers to passive nanostructures created to carry out a single, specific task. This includes many formulated products having nanoscale components and products involving surface coatings. Second-generation nanotechnology refers to active nanostructures created to carry out more than one task, sequentially and/or in parallel. This includes nano-scale actuators, sensors, and drug delivery devices. Third-generation nanotechnology refers to the expected emergence of functioning nanosystems comprised of many (i.e.

	thousands of) interacting components. Fourth-generation nanotechnology refers to the expected emergence of nanosystems designed and constructed at a molecular level (molecular nanosystems). Such systems are predicted to work with hierarchical systems within systems, like living human or animal cells, and/or like nano-scale robots. *See also* molecular nanotechnology.
Generations of technology foresight	A set of descriptions of major advances in the evolution of technology foresight methodologies since the beginning of the 1990s. They are based in part on Rothwell's [98, 126] five generations of models of technological innovation. First-generation foresight focusses on the prediction and diffusion of technologies, and often involves the "technology-push" model of innovation. Second-generation foresight adds to this the marketplace and the activities of the participants in technology development, they be industry, intermediary, government, and/or academic. This level of sophistication often involves the "market-pull" model of innovation. Third-generation foresight adds interactions with society, and often involves a "coupling" model of innovation (such as the "chain-linked" model). Fourth-generation foresight adds to this the distributed roles of the involved parties in innovation systems and often involves the "integrated" model of innovation. Fifth-generation foresight adds to this structural and policy issues and often involves the *"system integration and networking model" model of innovation. References [*422, 424*]. See also* generations of innovation.
GEO	*See* geostationary orbit.
Geostationary orbit	*GEO. See* low earth orbit.
Geosynchronous orbit	*GSO. See* low earth orbit.
GERD	*See* gross domestic expenditure on R&D.
GIL	*See* Government Innovation Lab.
Ginarte-Park Index	An indicator of the relative strength of (level of protection provided by) a country's intellectual property laws. The index is based on the sum of five components of a country's patent laws, each of which is given a score between 0 and 5. The components are breadth of coverage, participation in international patent agreements, protection against loss, degree of patent enforcement, and length of patent term. References [487, 488]. *See also* patent and patent protection.
Global Competitiveness Index	(GCI) An indicator of a county's economic competitiveness that is calculated and regularly published by the World Economic Forum. The principal components of the GCI are "basic requirements," covering institutions, infrastructure, and the economic environment; "efficiency enhancers," covering higher education and training, the labour market, technological readiness, and market size; and "innovation and sophistication factors," covering business sophistication and capacity to innovate. Reference [491]. *See also* competitiveness drivers.
Globalized regional innovation system	*See* regional innovation system types, Table 5.1.
Global MNE	Global Multinational Enterprise. *See* multinational enterprise.
Globalization	*See* internationalization (product, process, or service).

Global Multinational Enterprise	(Global MNE) *See* multinational enterprise.
GNERD	*See* gross national expenditure on R&D.
GNP	*See* Gross National Product.
Gompertz curve	A technology S-curve model. The Gompertz logistic curve model relates the technology maturity (y) to time (t) as: $y = y_\infty \cdot e^{\alpha \cdot \exp(-\beta t)}$, where the maximum value for y is given by y_∞ and the shape of the S-curve is determined by the adjustable parameters α and β. In this model, the growth curve is not symmetrical about the inflection point. Instead, it sharply increases up to the inflection point, after which the rate of growth slows. *See also* S-curve and pearl curve.
Go-to-market strategy	A strategy by which an organization connects with, and delivers products, processes, and/or services to, its target customers.
GOVERD	*See* government intramural expenditure on R&D.
Government Budget Appropriations or Outlays for R&D	(GBAORD) *See* government R&D expenditures.
Government-financed GBAORD	*See* government R&D expenditures.
Government-financed gross domestic expenditure on R&D	(Government-financed GERD) *See* government R&D expenditures.
Government Innovation Laboratory	(GIL) Not a laboratory but a physical or virtual space in which government officials, sometimes with other stakeholders such as business or community people, meet to conceptualize, discuss, and propose innovative government practices, policies, and/or services. For an example of a virtual GIL, see Reference [743]. Innovation in this context is not technological innovation, but is a form of organizational and/or social innovation. *See also* innovation (non-commercial).
Government intramural expenditure on R&D	(GOVERD) Research and development expenditure made within the government sector. *See also* government R&D expenditures.
Government R&D expenditures	There are several ways to measure or estimate government spending on R&D. Two commonly used measures are *government-financed gross domestic expenditure on R&D* (government-financed GERD) and *government budget appropriations or outlays for R&D* (GBAORD). Government-financed GERD is simply the portion of GERD that is financed by government. GBAORD is obtained from government budgets by identifying all the budget items involving R&D and estimating their R&D content in terms of funding. GBAORD therefore represents government R&D spending intentions rather than actuals. *See also* government intramural expenditure on R&D, gross domestic expenditure on R&D, gross national expenditure on R&D, and Frascati manual. Reference [485].
Government R&D intensity	(Innovation indicator) An indicator of the R&D investments made by governments is the ratio of government budget appropriations or outlays for R&D to gross domestic product (GBAORD/GDP). *See* innovation indicators and Table 8.2.

Grant capital — *See* capital.

Grassroots regional innovation system — *See* regional innovation system types, Table 5.1.

Great surges — *See* Kondratieff waves.

Grey space — *See* white space mapping.

Gross domestic expenditure on R&D — (GERD) The total expenditure on R&D performed within a country or region during a given period. GERD includes R&D performed within the country or region and funded from abroad, but it does not include payments for R&D conducted outside the country or region. GERD is calculated by adding the expenditures in each of the four performing sectors: business, not-for-profit, government, and academic. *See also* gross national expenditure on R&D, Frascati manual, and government R&D expenditures. Reference [485].

Gross domestic product — (GDP) The total economic productivity of a region, such as a country, or of an industrial sector during a specified period of time, usually a year or quarter of a year. GDP is different from business sales in that it takes into account the sales value of the goods and services produced, less the value of the goods and services consumed during the production. Thus, GDP can potentially increase with increased production, increased sale prices, and/or reduced costs of production. The difference in GDP per capita between a country or region and some specified peer or peers is termed the "*prosperity gap*." *See also* domestic product of industry and macroeconomics. *See also* gross national product.

Gross national expenditure on R&D — (GNERD) The total expenditure on R&D financed by a country's institutions during a given period, including the financing of R&D performed outside of the country. GNERD is calculated by adding the expenditures in each of the four performing sectors: business, not-for-profit, government, and academic. *See also* gross domestic expenditure on R&D, Frascati manual, and government R&D expenditures. Reference [485].

Gross national product — (GNP) The total economic productivity of the people and companies of a country, regardless of where the actual production took place, during a specified period of time, usually a year. GNP may be lower or higher than GDP depending on how much of a country's production takes place outside of its borders and how much of other countries' production takes place within its borders. *See also* gross domestic product.

Gross rate of replication — *See* technological capacity.

Growth curve — *See* S-curve.

Growth model — *See* Solow-Swan Growth Model.

GSO — Geosynchronous orbit. *See low earth orbit.*

11.9 H

Hackerspace — *See* makerspace.

Hack lab — *See* makerspace.

Henderson-Clark model — An innovation model that defines forms of evolutionary innovation and distinguishes between an organization's product component knowledge and its architectural knowledge (i.e. knowledge of the linkages among product components). In the *Henderson-Clark model*, organizations having low degrees of both component and architectural knowledge are most likely to achieve modest incremental innovations at best, whereas those with high degrees of both component and architectural knowledge are most likely to achieve disruptive innovations. In between these extremes fall two categories of organizations. One category represents organizations having a low degree of component knowledge but a high degree of architectural knowledge, and therefore are most likely to achieve *architectural innovations*, in which improvements are made to the linkage(s) between the components in a product, but not to the components themselves. The other category represents organizations having a low degree of with architectural knowledge and a high degree of component knowledge, and therefore are most likely to achieve *modular innovations*, in which improvements are made to one of more of the components in a product, but not among the linkages between those components. The Henderson-Clark model has been used to explain how incumbent organizations in a marketplace can achieve substantial innovations that fall short of being disruptive. Reference [87]. *See* Figure 2.6. *See also* Incremental innovation; evolutionary innovation; disruptive innovation; Abernathy-Clark model; platform innovation; and radical technological transition.

HEO — High earth orbit. *See* low earth orbit.

HERD — Higher education expenditure on R&D.

Heroic theory of discovery (or invention) — *See* multiple discovery theory.

Hesitators — A category of potential customer or technology adopter in a psychological model of technology adoption. *See also* technology acceptance, technology acceptance model, technology readiness index, and technology adoption life cycle.

Heterodox paradigm — The heterodox paradigm was coined by Michael Storper in 1997 [258] to represent a new way of thinking about economic development (and innovation), in terms of regions, organizations, and technologies and their linkages and interdependencies (as embraced in the *"Holy Trinity" model*). Storper's heterodox paradigm is an example of *heterodox economics* coupled with technology development and sociology. Heterodox economics refers to approaches to economics that go beyond, or at least fall outside of classical or conventional economics, such as socioeconomics or eco-economics. *See also* "Holy Trinity" model.

High earth orbit — HEO. *See* low earth orbit.

High-tech — *See* high technology.

High technology — This usually refers to the technological sophistication of a new product, process, or service, although the term has also been applied to entire industries, sectors, and even regions. Technological sophistication, in this context, is often assessed based on direct (and sometimes also indirect) R&D intensity. Four categories of technological sophistication are often distinguished, based on some measure of R&D intensity, such as R&D investments divided by value-added and/or gross production values:

- High-technology (also termed *"high-tech"*), such as aerospace, pharmaceuticals, and instruments,
- Medium-high-technology (also termed *"medium-high-tech"*), such as chemicals, electrical equipment, and motor vehicles,
- Medium-low-technology (also termed *"medium-low-tech"*), such as refined petroleum products, metals and metal products, and shipbuilding, and
- Low technology (also termed *"low-tech"*), such as food, pulp and paper, and textiles.

Such categorizations are both generalized and relative. A "*high-tech*" industry can produce "*low-tech*" products, and *vice versa*. "*High-technology*" is sometimes used simply as a synonym for "*advanced technology*." Reference [20].

High technology complex — *See* innovation ecosystem.

Holding period return — *See* return on investment.

Holistic innovation management — A set of management of innovation processes in an organization that combines strategic, total, open, and collaborative innovation. In some of the literature, holistic innovation management is also taken to include aspects of sustainability [396]. In some ways, holistic innovation management represents an organizational, strategic version of the open innovation and ecosystem approaches discussed in Section 5.1.

Holland, Maurice (1891–1981) — An American engineer and research manager. He was the first head of Engineering and Industrial Research at the U.S. National Research Council and seems to have published the first description of the steps involved in the "*research cycle*" leading to innovation (in 1928). His work was the forerunner of the first linear model of innovation, advanced by Maclaurin in the 1940s. He was also the founder of the U.S. Industrial Research Institute (in 1938). References [58, 59]. *See also* Maclaurin.

"Holy Trinity" model — A regional innovation system model advanced by Michael Storper in 1997 to describe the roles and intersections of territories, technologies, and organizations in linking regions, technologies, and production capabilities into innovation systems that generate new products and economic development [258]. Storper termed this way of thinking about economic development (and innovation) the heterodox paradigm. The first-order interactions in this model would produce economic technological systems, regional technology systems, and regional productions systems, all of which combined would lead to a stream of new products entering the marketplace. *See also* heterodox paradigm and innovation ecosystem.

Horizontal thinking — A synonym for brainstorming. *See* creative thinking models.

Howitt, Peter (1946 – Present) — A Canadian economist who was co-awarded the 2025 Nobel Prize in Economic Sciences, in his case for having co-identified the prerequisites for sustained growth through technological progress (his co-awardees were Joel Mokyr and Philippe Aghion).

Hype-Cycle — *See* Technology Hype-Cycle.

Hyper-Innovation — (Pace of Innovation) The simplest definition of hyper-innovation is innovation conducted at a rapid pace. Depending on the specific usage,

there is frequently an associated connotation of rapid innovation that is driven, or at least enabled, by multiple interconnections among people, organizations, information technologies, and markets.
(Amount of innovation) In some contexts, the term hyper-innovation is used to mean "too much" innovation. Example: an organization that has been highly successful with one or more major technological innovations could become vulnerable to a culture shift towards hubris, undermining the kind of culture that originally enabled the innovation(s). *See also* hypo-innovation.

Hypo-innovation — This term generally refers to "too little" innovation. In this context, the prefix *hypo* means "less than normal." Example: an organization that has not produced significant technological innovations and is being outpaced by its competitors for this reason could be considered to suffer from hypo-innovation. *See also* hyper-innovation.

11.10 I

I18N — "I-eighteen letters-N" referring to a specific kind of internationalization. *See* internationalization (product, process, or service).

I-CAN — *See* Innoventures Canada Inc.

ICT — *See* information and communication technology.

ICT investment intensity — The ratio of ICT sector expenditures to gross domestic product in an economy. *See* innovation indicators and Table 8.2.

ICV — Internal corporate venturing. *See* enterprise unit.

Idea funnel — *See* product development funnel.

Idea generation — *See* ideation.

Idea sun bursts — *See* spider diagrams, *under* creative thinking models.

Ideation — The process of idea generation whether through divergent thinking, brainstorming, Camelot scenarios, or any other approach. Organized and/or facilitated ideation is sometimes used to originate ideas for new theories, applications, solutions, processes, or products. *See also* breakthrough and innovation project.

Idea-to-launch process — Any number of processes that have been developed to guide the managed evolution of an idea into a commercial enterprise, product, process, or service. Also termed "*gated process*." Examples of idea-to-launch processes for new product development in an organization include the Stage-Gate® process and the product and cycle-time excellence (PACE) process. *See* Stage-Gate® product development process, technology stage-gate process, and Figure 4.13.

ILO — *See* Industry Liaison Office.

Imitative innovation — *See Shanzhai* innovation.

Imitator — *See* innovator.

Impact multiplier — A number that quantifies the indirect effects of an economic impact. The principle is that an economic impact such as in increase in product sales will be multiplied in an economy through increased spending on the jobs, goods, and services needed to produce those products. Multipliers tend to be estimates because they attempt to account for indirect effects (sometimes termed *spillover effects*) that may not be directly or easily measurable. Multipliers are expressed as a ratio between a measure of a total effect and a measure of the direct effect that caused it.

Impacts — (Innovation) The ultimate consequences of the outputs from an organization or its employees, especially in the context of the mandate and/or mission of the organization. Examples: sales and adoption of a new product, process, or service (output) might result in significantly increased customer satisfaction, revenues, net income, jobs, exports, or economic activity. *See also* activities, inputs, outcomes, outputs, reach, and metrics.

Implementation — *See* technology deployment.

Inactive — *See* laggards.

Inbound innovation — *See* inbound open innovation.

Inbound open innovation — An aspect of the open innovation process in which the flow of technology or other information is into an organization from the outside. Example: licensing-in of technologies. Also termed *"inbound innovation." See also* open innovation and outbound open innovation.

Inclusive innovation — *See* frugal innovation.

Incremental innovation — An innovation that represents a small improvement, or a series of small improvements to an existing product, process, or service. Incremental innovations are often more or less continuous and may be required to maintain a competitive position in the marketplace. Sometimes termed *continuous innovation, core innovation, cosmetic innovation, market-pull innovation, or survival innovation*. Sometimes termed *sustaining innovation*, but this term is probably better suited to *evolutionary innovation.* Incremental innovation is usually very problem solving- and adaptation-intensive. *See Figures 2.4, 2.5, and 3.2. See also* innovation, disruptive innovation, evolutionary innovation, technology mudslide hypothesis.

Incrementalist strategy — *See* rationalist strategy.

Incremental-radical dichotomy — The dilemma faced by large, successful, incumbent companies in an established marketplace that tend to find it extremely difficult to implement disruptive innovations because, by their very nature, they require the creation of new technological knowledge and/or resources that render the old ones obsolete, and they involve the introduction of such radically new products, process, or services into the marketplace that they render the previous ones obsolete – including those of the innovating company. The dilemma is whether to risk disrupting a well-established and successful business, or line of business, by introducing a new product that may or may not be successful. Also referred to as the *innovator's dilemma. See also* Abernathy-Clark model and innovation paradoxes (innovation system paradoxes).

Incubation — *See* business incubation.

Incubator — *See* business incubator.

Indigenous innovation — (Self-determined) Innovation processes of any kind conducted for and/or by a group of people in a region or country for themselves. Also termed "*Zizhu chuangxin*," meaning self-determined innovation. Example: national innovation programmes.

(Indigenous people) Innovation, often social innovation, processes conducted for and/or by Indigenous people. A related meaning is social innovation processes that are informed and potentially influenced by the application of indigenous (traditional) knowledge as guided by elders, and hence the wisdom of ancestors.
(Global transfer) A process of making use of technologies transferred from other, more advanced economies, to develop new or improved technologies "at home." Example: China's 2006 indigenous innovation strategy. *See also Shanzhai* innovation.

Indirect indicators — *See* innovation indicators.

Individual contributor ladder — *See* technical ladder.

Induced innovation — An older term that is used with reference to drivers of innovation such as knowledge-induced (i.e. technology-push) innovation and demand-induced (i.e. market-pull) innovation. Such terminology was commonly used in the 1960s. Reference [83]. *See also* linear innovation models.

Industrial design — A grant of exclusive rights on the visual features of an original, manufactured item by a government. Such features generally include some combination of features of shape, pattern, or configuration, but not the method or materials used in the manufacturing. *See* invention, intellectual property.

Industrial innovation — *See* innovation (20th–21st centuries). *See also* Maclaurin.

Industry 4.0 — The fourth Industrial Revolution. *See* technological ages.

Industry Liaison Office — (ILO) A technology transfer and/or industry liaison office at an academic institution such as a university or polytechnic. The principal focus of such offices is usually to facilitate the transfer of the knowledge gained through discovery research to those that would undertake applied research and development, build links between academic researchers and industry, assist faculty members with patenting, and negotiate licensing agreements, particularly the latter. At universities, these are sometimes termed "*University-Industry Liaison Offices*" (UILOs).

Information age or society — *See* technological ages.

Information and communication technology — (ICT) A rapidly growing area that includes communication devices, applications, and services. Examples range from satellite sensing and communication systems, to network hardware and software, to devices such as radios, televisions, phones, and computers. ICTs formed the basis for key developments in the **fifth** Kondratieff wave of disruptive innovations. *See also* **5th** wave and Kondratieff waves.

Information-based economy — *See* knowledge-based economy.

Information economy — *See* knowledge-based economy.

Information hierarchy — *See* Wisdom Hierarchy.

Information society — *See* knowledge-based economy.

Infrastructural investment cycle — *See* Kuznets cycle.

Infrastructure of the information society — *See* Internet of Things.

In-kind — A phrase meaning "with something similar." In business this phrase usually refers to a non-monetary payment or investment such as with goods or services, in which case the in-kind contribution is usually assigned an equivalent monetary value.

Innovation

(Early modern) In the 1500s through 1800s, innovation meant "*introducing novel change*," particularly with regard to religious and/or political change. In these early times, the connotation of the term innovation was negative: as it was frequently meant to imply that the changes were unwanted, unnatural (apart from the natural order of things), revolutionary, and/or dangerous. In contrast, the terms "*reformation*" or "*restoration*" were frequently used to describe positive, moderate, natural-order-restoring changes. Reference [1].

(20th–21st centuries) From the 1930s to the present day, a technological, or commercial, meaning of the term has been widely used, in which innovation involves introducing novelty to the commercial marketplace through new products and services, and in which innovation has a positive, but still disruptive, connotation. This meaning was defined by Schumpeter in the 1930s as the conversion of ideas and knowledge into commercially successful products and services. Here, "successful" means meeting the needs of customers, in a way that encourages them to take up the new approach and have it diffuse through the marketplace. Whereas an invention is the first occurrence of an idea or concept for a new product or process, innovation involves taking it into commercial practise. In this sense, some innovations can be described as successfully commercialized discoveries or inventions. The terms "*technological innovation*,"[120] "*commercial innovation*," or "*industrial innovation*" are often used to distinguish the 20th–21st centuries, or Schumpeterian definition of innovation from the Early Modern definition. An early synonym, attributed to A.P. Usher [96] is "*emergent novelty*." References [15, 16]. *See also* discovery, invention, top-down innovation, and upstream innovation.

Several types of technological innovation have been defined based on the nature of the technology, hence *product innovation*, *process innovation*, and *service innovation*. Examples of process innovations are Henry Ford's automobile assembly line manufacturing process, and FedEx's overnight long-distance parcel delivery process. *Marketing innovation* and *organizational innovation* refer to improvements in the practice of marketing or the operations of an organization, respectively, in ways that ultimately translate into improved sales and/or margins for the products, processes, or services that the organization is selling in the marketplace. Reference [26]. Several types of innovation have been defined based on outcomes. *See* disruptive innovation, evolutionary innovation, and incremental innovation. *See also* adjacent innovation.

(Non-commercial) In the 21st-century usage, the earlier meaning of the term innovation as introducing novel change has again become widely used, but this time with a positive connotation, and it has been extended to realms beyond those of religion, politics, and technology, such as orga-

120 The term "*technological innovation*" came into use in the 1950s, representing a merger of the work of Maclaurin and Schumpeter and may have been coined by Maclaurin, who frequently referred to "technological change."

nizational processes, marketing, and social structures. An example is the ideation, development, and deployment of new and improved internal processes within an organization, even if these have little or no connection to commercialization or the marketplace. The terms "*non-commercial innovation*" and "*non-technological innovation*" are sometimes used to distinguish such non-Schumpeterian innovation. Thus, the OECD *Oslo Manual* definition of innovation would be taken to include both technological and non-commercial innovation. *See also* organizational innovation and Government Innovation Lab.

Innovation 2.0 — A variation on the "*system integration and networking model*" (SIN Innovation Model) that involves parallel development integrating horizontal linkages among communities, governments, industry, research organizations, and universities, with the aim of creating and deploying disruptive innovations that drive economic growth, improved quality of life, and reduced resource use and environmental impacts [112]. Example: a community's adoption of a smart traffic flow-adjusting system that responds to real-time traffic and air quality sensor readings [112]. *See also* non-linear innovation models.

Innovation accelerator — *See* business accelerator.

Innovation activities — *See* activities. *See also* inputs, outcomes, outputs, metrics, reach, and impacts.

Innovation and sophistication factors — *See* Global Competitiveness Index.

Innovation at the bottom of the pyramid — (BOP innovation) Technological innovation (usually products) developed in and targeting emerging economies. Also termed "*resource-constrained innovation*." *See also* reverse innovation.

Innovation barrier — Any challenge or obstacle to the successful process of innovation up to and including successful technology commercialization. Such barriers could be technological, financial, manufacturing, cultural, or regulatory, for example. Such barriers are more specific, however, than the general requirements to conduct R&D, experimental development, prototyping, pilot testing, demonstration, preproduction, financing, marketing, sales, and deployment. By definition, all innovation barriers have to be successfully identified and managed before innovation (commercialization) is possible. *See also* breakthrough and Camelot scenario.

Some common "traps" that can create barriers to innovation include the "*performance trap*" and the "*commitment trap*." The "*performance trap*" can occur when a company is not performing well and focusses on short-term tactics at the expense of long-term strategy (sometimes referred to as the "*short-term mindset trap*"), or when a company is performing well and focusses on its current successes at the expense of other opportunities (sometimes referred to as the "*business model trap*"). The "*commitment trap*" can occur when a company fails to commit to a new opportunity and extensive research, analysis, and/or testing inhibit or prevent progress, or when a company over-commits to an opportunity that no longer appears prospective, inhibiting, or preventing retreat. Reference [380].

Innovation black box — *See* black box.

Innovation bridge	A reference to the function of technological innovation in providing a bridge between the world of knowledge, new discoveries, and ideas on the one hand, and the marketplace of commercialized products, processes, and services, on the other hand.
Innovation cluster	*See* cluster (innovation cluster).
Innovation continuum	(Technology) A way of describing and/or mapping the progress of technology development illustrating the stages involved and the degrees of technology maturity that they represent. *See* S-curve (technology) and Figure 3.5. *See also* technology readiness. (Degree of innovation) A way of describing and/or mapping degrees of innovation and their impact on an economy. *See* S-curve (innovation) and Figure 3.7. Also termed *"innovation S-curve."*
Innovation-decision process	The process by which a potential customer evolves from first becoming aware of a technological innovation, to forming an opinion on it, to reaching a decision on whether or not to purchase it, to acting on the decision, to confirming the purchase decision. Rogers has summarized these as: (1) knowledge, (2) persuasion, (3) decision, (4) implementation, and (5) confirmation [77].
Innovation deficit	This generally refers to either a rate comparison or a return-on-investment comparison. In the former case, it is when the rate of introduction of new technological innovations is less than that of another specified organization, region, or country. In the latter case, it is when the rate of introduction of new technological innovations is less than what would be taken to be an acceptable rate of return given the investments made in support of innovation. In either case, this term is usually used subjectively. It is usually not innovation that was measured but some other measure, such as economic activity or patents granted, with an implicit assumption that the measure is directly linked to innovation.
Innovation diffusion model	*See* Rogers.
Innovation dissonance	The tensions that can arise among people and teams when an organization is trying to encourage and promote technological innovation and encounters any of the several so-called *innovation paradoxes* by which the drivers of innovation come up against elements of excess, competition, organizational momentum, or even strategy. Depending on the nature and degree, such tensions could be considered to be productive or unproductive. Reference [242]. *See also* innovation paradoxes (innovation process paradoxes).
Innovation-driven economy	*See* competitiveness drivers.
Innovation ecosystem	(General) An innovation ecosystem includes all of a region's public and private sector people and organizations whose activities include any or all of developing, enabling, producing, or diffusing innovations. These people or organizations constitute innovation ecosystem entities because of their capabilities and regardless of whether or not they actually interact with each other. Examples of innovation ecosystem entities include individual inventors and entrepreneurs, for-profit companies, not-for-profit companies, government organizations, and academic organizations. An *innovation system* is a network of innovation ecosystem people and/or

entities that interact with each other to develop, enable, and/or produce innovations. *See* Figure 5.5. *See also* business ecosystem, innovation system theory, innovation ecosystem models, innovative regional cluster, triple-helix model, quad-helix model, and quad model.
(National and regional) The term *national innovation system* (or national system of innovation, NSI) refers to a nation's innovation system, and *regional innovation system* to that of a region. A national or regional innovation system comprises the network of public and private sector innovation-related institutions and their mutual interactions. Reference [249]. Regional innovation systems have also been referred to as: *innovative milieux, high technology complexes, technopolis complexes*, and *technopoles*. Clusters can be considered a special kind of regional innovation system. *See also* regional innovation system types.

Innovation ecosystem entities — *See* innovation ecosystem.

Innovation ecosystem models — Any of several models of innovation ecosystems that focus on the interacting roles of multiple institutions in the process of creating technological innovations. The kinds of institutions considered vary with the model but generally include three or four of governments, universities, intermediary organizations, and industry. These models generally attempt to describe how discovery, knowledge-generation, invention, product development, market launch, and so on depend on interactions among institutional players, involving elements of control and/or incentives, knowledge generation, and wealth generation. Examples include *"Holy Trinity" model, triple-helix model, quad-model, quad-helix model*, and *N-tuple helix model*, among others. *See* Figures 5.4 and 5.5. *See also* Holy Trinity model, innovation ecosystem, quad-helix model, N-tuple helix model, and triple-helix model.

Innovation expenditures — The expenditures associated with innovation, including *research and development. See also* government R&D expenditures, gross domestic expenditure on R&D, gross national expenditure on R&D, activities, research and development.

Innovation foresight — *See* foresight.

Innovation funnel — *See* product development funnel.

Innovation gap — The gap between desired and actual innovation performance as measured or perceived by stakeholders of some kind. The innovation performance gap could apply to regions, countries, sectors, or companies, for example. The stakeholders could be governments, organizations, companies, leaders, employees, or constituents, for example. Also termed *"commercialization gap"* or *"technology gap."*

Innovation gap theory — A theory that a country, region, industry, or company that achieves new innovations (products, processes, or services successfully introduced into the marketplace) will have a competitive advantage over their competitors until the gap is closed. In the case of countries or regions the theory is sometimes extended by assuming that other countries or regions will have to import the new innovations until they can be replicated or substituted within their own regions. In this latter sense, an innovation gap can

	stimulate international trade. Also termed *"commercialization gap theory"* or *"technology gap theory."*
Innovation impacts	*See* impacts. *See also* activities, inputs, outcomes, outputs, metrics, and reach.
Innovation incubator	*See* business incubator.
Innovation indicators	Regions, countries, and organizations, in their efforts to measure and benchmark innovation performance, tend to use a combination of direct and indirect (proxy) indicators. Some of the indicators are "*lagging indicators*," meaning that they represent the past (such as measures of new products, process, and services already introduced into a marketplace). Other indicators are "*leading indicators*," meaning that they are used to represent possible future innovation performance (such as measures of resources and activities directed towards technological innovation). References [481, 482]. *See* Tables 5.1, 8.2–8.4.
Innovation in innovation	Reference coined by Etzkowitz to the broadening of the term innovation to include improvements in the way organizations interact in order to create the conditions for innovation and/or to enable innovation to occur [262]. *See also* entrepreneurial university, innovation ecosystem models, triple-helix model, and quad-helix model.
Innovation inputs	*See* inputs. *See also* activities, outcomes, outputs, metrics, reach, and impacts.
Innovation intermediary	*See* intermediary organization.
Innovation intermediation	*See* intermediary organization.
Innovation killer	Anything that contributes a barrier to creating innovation. Cynthia Rabe describes the ways in which having too much deep knowledge, or expertise, in an organization or team can be an innovation killer [744].
Innovation management	An organizational function that coordinates efforts to accomplish technological innovation. Such efforts could include planning, organizing, staffing, and/or leading processes involving, for example, market research, product research and development, manufacturing, and marketing. Innovation management is different from Technology Management. *See also* intellectual property portfolio mining and technology management.
Innovation metrics	*See* metrics. *See also* activities, inputs, outcomes, outputs, reach, and impacts.
Innovation models	*See* linear innovation models and non-linear innovation models.
Innovation outcomes	*See* outcomes. *See also* activities, inputs, outputs, metrics, reach, impacts.
Innovation outputs	*See* outputs. *See also* activities, inputs, outcomes, metrics, reach, and impacts.
Innovation paradoxes	The literature contains quite a few concepts of paradox inherent to innovation. These generally refer to either the pros and cons of pursuing technological innovation in the face of organizations or products/processes/services that are already successful ("*innovation system paradoxes*"), or else the challenges that may be inherent within the process of innovation itself ("*innovation process paradoxes*"). As such, they are usually more about risks than paradoxes in a literal sense, but the literature references are most often to "paradoxes." (*Innovation system paradoxes*) One system paradox could be called the "*success/failure paradox*," in which an organization's single-minded focus on "success" (and lack of tolerance for "failure") can blind it to the fact that

it actually needs to be able to tolerate, and even embrace failures in order to achieve success through technological innovation. Termed the "*innovation paradox*" by Farson and Keyes [240], who argue that it could be limiting or counter-productive to even think in terms of successes and failures. Another system paradox could be called the "*success paradox*," in which an organization becomes so successful, through technological innovation, that it becomes blinded to the need for more or new innovations, causing it to miss out on breakthrough innovations that, to its ultimate detriment are left to other, more nimble companies. Example: Nokia famously killed its smart phone because they were so deeply invested in "dumb phones." Termed the "*innovation paradox*" by Davila and Epstein [241]. Another version of this is the paradox an organization faces when it recognizes that developing a breakthrough innovation could lead it to future success but that the pathway could involve destroying its currently successful products, processes, or services. This is termed the "innovator's dilemma" by Christensen [79].

(*Innovation process paradoxes*) Some examples of innovation process paradoxes are included here: in the "*cultification paradox*," the single-minded pursuit of innovation is taken to such extremes that it overrides focus on other key activities such as sales and customer focus (i.e. too many pilots, too many specification changes, and too many new products) or that it causes changes to be made "for the sake of innovation" where there is no need for innovation at all (i.e. changing process that are already effective and efficient). In the "*collaborative compromise paradox*," multiple groups are brought together to collaborate on a complex innovation challenge but the competing needs and priorities of such groups lead to compromises that reduce their combined effectiveness, making the whole less than the sum of its parts, rather than more. In the "*invention within convention paradox*," inventions that could lead to innovations are developed but then cannot be developed further because the organization's operational processes are not flexible enough to accommodate changes. In the "*novelty paradox*," novelty is needed in product/process/service innovation, but a novel new product or a novel change to an existing product may not be perceived as an improvement by customers. Even worse, the introduction of a new or changed product is often accompanied by discontinuation of the previous one, which could lead to the loss of existing customers who are unable to purchase the old product and do not want the new one. In the "*Productivity Paradox*," investing in the development of a new product, process, or service with an expectation of increased productivity may lead to very positive results in other areas but either no increase, or even a loss in productivity itself. An example is the "*Solow computer paradox*." That increased investment in information technology could cause labour productivity to decrease instead of increase.[121] In the "*organizational paradox*," if an attempted technological innovation fails or performs poorly in the

121 In the early 1970s, many organizations began investing in computer systems with the expectation that they would enable labour productivity improvements (or even eliminate labour completely in some areas). However, although computers and information technology enabled many companies to

marketplace, or if an attempted organizational innovation fails, the impacts on the organization, such as supply chain problems or reduced spending budgets leading to reduced quality control (or reduced customer service), for example, could lead to the loss of existing customers. This could be considered to be "*negative innovation.*" References [8, 242]. *See also* innovation dissonance, negative innovation, and Solow computer paradox.

Innovation park — *See* research and technology park.

Innovation performance mapping — There are several systems for comparing the relative innovation performance and capabilities of organizations. Figure 7.5 shows an example in which organizations are rated as having both high innovation capabilities and performance (red giants), both low innovation capabilities and performance (black holes), high innovation capabilities but low innovation performance (comets), or low innovation capabilities but high innovation performance (shooting stars). The latter category includes organizations that have achieved innovation success(es) but which do not have the ability to sustain them. Reference [400].

Innovation pipeline — (I-Pipe) An idea pipeline, or gathering mechanism, intended to feed a new product development process. Generally attributed to Whirlpool. This is similar in concept to the idea-to-launch process or Product Development Funnel (Process). *See also* idea-to-launch process, product development funnel, Stage-Gate ® product development process, and technology stage-gate process. References [475, 476].

Innovation process paradoxes — *See* innovation paradoxes.

Innovation progression gap — *See* valley of death.

Innovation project — A project whose goal is to develop a new idea, discovery, or invention into an innovation. This could be an individual or group initiative.

Innovation reach — *See* reach. *See also* activities, inputs, outcomes, outputs, metrics, and impacts.

Innovation-related triple-helix — *See* triple-helix model.

Innovation sandbox — *See* sandbox.

Innovation S-curve — *See* S-curve (innovation).

Innovation snail — A term coined by Badulin for a technological product's development and business cycle, described as a linear series of 12 stages beginning with a product idea, and continuing through development, business planning, capitalization, production, market saturation, and then either declining sales or transitioning to another newer product. Badulin's sequence was drawn in an almost open-circular shape having the appearance of a snail's shell. Reference [64]. *See also* product lifecycle curve in the entry for S-curve, buying hierarchy.

Innovation strategy mapping — An approach to categorizing organizations' innovation approaches, having

---

achieve competitive advantage and market share, the desired labour productivity increases were largely unrealized, and in some cases, decreases were experienced.

four categories based on whether innovation is developed internally or externally, and how structured or informal the innovation process tends to be, *see* Figure 7.2. In this approach, the four categories are: "*explorers*," who tend to be in emerging, rapidly expanding markets, focussed on customers, partners, and markets generally for new insights and opportunities, users of rapid prototyping and testing of large numbers of relatively small ideas, and innovating through a diffuse, customer-facing process. "*Architects*," who tend to be in mature markets, with intensive capital and resource requirements and a centralized, structured approach to innovation, focussed on customers and competitors for new insights and opportunities, outsourcers of development and prototyping, and innovating through a top-down, formal process. "*Moonlighters*," who tend to be large companies with lots of technology and process expertise, that look inward to develop innovations based on internal ideas and know-how, focussed on a strong innovation culture that supports internal experimental and development project initiatives, including internal "moonlighting," willing to fund internally generated innovation projects, and innovating through bottom-up processes. "*Miners*," who tend to be very large companies with lots of technology and process expertise, very large internal business units, and strong silos, weak on internal innovation, necessitating the use of external innovation resources and/or organizations, focussed on finding innovation opportunities within the organization, and innovating through formal, centralized processes working through the external innovation organization. Reference [373].

Innovation system — *See* innovation ecosystem.

Innovation system entities — *See* innovation ecosystem.

Innovation system foresight — (ISF) *See* foresight.

Innovation system paradoxes — *See* innovation paradoxes.

Innovation system theory — The hypothesis that a nation's, or a region's, innovation performance depends upon the breadth and depth of their innovation ecosystem and on the quality of the relationships and interactions among the innovation ecosystem entities. A corollary is that entities like government departments and agencies, RTOs, universities and colleges, and businesses, and even entire industries in a country or region need to interact effectively with each other in order to stimulate and/or enhance system innovation performance. *See also* innovation ecosystem.

Innovation trap — *See* innovation barrier.

Innovation union — An initiative launched by the European Union's (EU) economic growth strategy "Europe 2020." The aim of the EU's innovation union is to improve Europe's technological innovation performance, in order to achieve economic growth and jobs growth.

Innovation voucher programme — Government programmes that are intended to help small- and medium-sized enterprises (SMEs) become more productive and innovative by developing and launching new products, processes, or services into the marketplace, while building stronger linkages with R&D providers (public and private). The vouchers are generally some kind of credit note for a

	certain amount of money that the SME can use to access expertise and technical solutions from R&D providers. Depending on the programme, the SMEs themselves may be required to contribute financially to the project as well. Examples of eligible services include applied research, development engineering, prototyping, product design and testing, intellectual property assessments and patenting, pilot and field testing, and market assessments and advice. Such voucher programmes have various names but they usually reference innovation, productivity, and/or competitiveness. Reference [307].
Innovative firm	An organization that has introduced an innovation (to the marketplace) during a specified period of time.
Innovative industrial cluster	*See* innovative regional cluster.
Innovative regional cluster	A geographically defined grouping of organizations of all sizes that comprise an industrial sector in which network relationships exist and are linked to associated academic, intermediary, and government organizations, and for which enhanced innovation performance have been attributed to the network as a whole. An "*innovative industrial cluster*" is similar, but is not geographically defined or constrained. *See also* innovation cluster and innovation ecosystem.
Innovative milieu	*See* innovation ecosystem.
Innovator	(Innovation) A person or organization that is the first to create an innovation, or at least is the originator of the critical conceptual elements of an innovation. Others making or adopting the same technological changes later are sometimes referred to as "*imitators*." The term innovator is often incorrectly applied to discoverers or inventors. *See also* discoverer, inventor, invention, and patent. (Technology diffusion) A personal or organizational category of technology adopter in Rogers' diffusion of innovation model. Also sometimes referred to as "*technologists*" or "*creators*." *See* Figure 4.8. *See* technology adoption life cycle.
Innovator's dilemma	A term coined by Clayton Christensen, referring to his observation that the practices that allow a company to become a leader in a mainstream market tend to cause them to miss opportunities offered by disruptive innovations, and that pursuing such disruptive innovations requires practices that could undermine the company's current business position. The dilemma then becomes whether to protect the company's current business position or risk it in pursuit of a new potential business. Reference [79]. Also referred to as the *incremental-radical dichotomy* in innovation. *See also* Abernathy-Clark model, innovation paradoxes (Innovation System Paradoxes).
Innoventures Canada Inc.	(I-CAN) I-CAN was a not-for-profit association of Canadian RTOs, some of whom were government organizations (Crown Corporations or Agencies), while others were private, not-for-profit corporations. It operated between 2008 and 2025. Membership has included the NRC, FP Innovations (British Columbia and Québec), InnoTech (Alberta), SRC, Industrial Technology Centre (Manitoba), Vineland Research (Ontario), Centre de Recherche Industrielle du Québec (CRIQ), National Optics Institute (INO, Québec), Research and Productivity Council (New Brunswick), Bio|Food|Tech (Prince Edward Island), and Research & Development Corp. (RDC, Newfound and

	Labrador). At the time of writing, I-CAN was in the process of rebranding as Canadian Association of Industrial Research and Technology Organizations (CAIRTO).
Inputs	(Innovation) Things that an organization or its employees receive or acquire in order to carry out a particular mission and/or achieve an objective. Example: contract or grant funding, new hires, purchases, or externally generated ideas or knowledge. *See also* activities, outcomes, outputs, reach, metrics, and impacts.
In silico	In science and engineering, the Latin-based term *in silico* generally refers to an experiment, calculation, or process simulation conducted by numerical (computer) simulation.
In situ	In science and engineering, the Latin term *in situ* generally refers to an aspect of a reaction or process taking place where it normally occurs, as opposed to moving it to some other place. Examples include studying a phenomenon where it occurs, in a reaction mixture, or in a process vessel. *See also ex situ*.
Institute of technology	*See* polytechnic.
Institutional innovation	*See* organizational innovation.
Intangible assets	*See* assets.
Intangible knowledge	*See* social technology.
Integrated circuit topographies	A grant of exclusive rights on the three-dimensional features of an original integrated circuit design by a government. Such features generally include various layers of metals, insulators, and semi-conductors in an electronic circuit, but not the actual functions of the circuit. *See* invention and intellectual property.
Integrated innovation	A coordinated, systems-engineering approach to innovation in which scientific, engineering, technological, business, and sociological aspects are all considered throughout the complete innovation process. This would include all of what are sometimes termed "*innovation activities*" plus the "*other innovation activities*." *See also* innovation, research and development, activities, other innovation activities.
Integrated model of innovation	*See* fourth-generation innovation model and non-linear innovation models. *See also* generations of innovation.
Intellectual capital	*See* technological capacity.
Intellectual property	(IP) Almost any definable intellectual creation including products, processes, services, literary and artistic works, symbols, names, images, designs (including logos), and other forms of know-how. Some forms of IP are protected by holding them in secrecy, such as concepts, discoveries, and *trade secrets*, which can include non-patentable inventions. Some forms of IP are legally protectable under intellectual property law, such as by *contract, copyright, patent, trade dress*, or *trademark*. *See* intellectual property rights. See also invention and technology.
Intellectual property mining	*See* intellectual property portfolio mining.
Intellectual property portfolio mining	(IP portfolio mining) The scanning and evaluating of intellectual property (IP), including but not limited to patents, held within an organization and/

	or external to an organization. The purpose is to identify IP that is dormant, underutilized, or that could be utilized in new ways, as part of an organization's innovation management programme. Classic examples are the identification of patented technologies from another country that are either not protected in the organization's home country or could be licensed for use in that country. Also termed "*intellectual property (IP) mining*," "*patent database mining*," or "*patent mining*."
Intellectual property rights	(IPR) Formal protection that is granted under a country's intellectual property law. Intellectual property rights that are protected by legislation include patents, trademarks, copyrights, industrial designs, integrated circuit topographies, and plant breeders' rights. Intellectual property can be held as a trade secret but, in this case, there is no legislative protection. *See* invention, intellectual property, patent, trademark, copyright, trade secret, industrial design, integrated circuit topographies, and plant breeders' rights.
Intelligent materials	*See* smart materials.
Interaction model	A term referring to, or a version of, the "*coupling model*" of the technological innovation process. *See* non-linear innovation models.
Interactive model of innovation	*See* linear innovation models.
Interactive regional innovation system	*See* regional innovation system types, Table 5.1.
Intermediary organization	(General) In general, an intermediary organization functions between an organization and some or all of its stakeholders by providing some kind of service, such as a programme function or technical assistance, for example. Brokers and dealers are sometimes referred to as technology market intermediaries (TMI). *See also* knowledge-intensive business services. (Innovation) In the technological innovation world, an intermediary organization is one that functions between industry and the marketplace in a way that supports technological innovation processes. Also termed "*bridging organization*," "*broker*," or "*innovation intermediary*" (and sometimes "*superstructure organization*" or "*knowledge processor*"). Example: a fourth-pillar organization such as an RTO. The specific functions provided by the intermediary organization could be almost anything that enables, assists, or leverages an innovation process, such as foresight, information/knowledge processing, research and development, testing, accrediting, and commercializing, for example. The process of conducting such activities is sometimes referred to as *innovation intermediation*. References [276, 277]. *See also* fourth-pillar organization, knowledge-intensive business services, innovation ecosystem models, research and technology organization.
Internal corporate venturing	*(ICV) See* enterprise unit.
International corporation	*See* multinational enterprise.
Internationalization	(Corporation or organization) The entry of an organization into operations that cross national borders. The term frequently refers to the entry of a company into sales of products, processes, or services in one or more international markets. (Product, process, or service) Designing a product, process, or service so that it can be sold in multiple countries, or such that it can be readily

adapted to be sold in multiple countries. This meaning of internationalization is also termed "*I18N*" (i.e. "I-eighteen letters-N"). The sub-process of actually adapting such a product, process, or service to a specific country, language, and/or culture is referred to as *localization*. This meaning of the localization is also termed "*L10N*" (i.e. "L-ten letters-N"). It follows that an internationalized product, process, or service should be fairly easy to localize. The process of internationalizing a product, process, or service and then localizing it for different specific countries, continents, or cultures is sometimes referred to as *globalization*.

Internationally advanced organization
An organization that stands out internationally by having globally unique, or at least globally uncommon, capabilities and/or products and services.

Internet of Things
(IoT) The networking of physical devices of all kinds ("*IoT devices*") such that they can collect and exchange data, allowing remote sensing and controlling across a network. Example: smart thermostat systems that can be remotely monitored and controlled via Wi-Fi. The Internet of Things has been called "the infrastructure of the information society" and could enable broad system controls such as the "*smart grid*" (a smart electric power grid) and "*smart cities*" (using a broad information and communication technology system to manage a city's infrastructure and systems). *See also* pervasive computing.

Intrapreneur
A person that acts entrepreneurially within a larger company. For example, an intrapreneur might develop, launch, and lead a new line of business or a new business unit within an organization, drawing upon the capacity of the larger organization but acting with greater risk-tolerance and nimbleness than is usual for the organization. *See also* entrepreneur. Reference [718].

In vacuo
In science and engineering, the Latin term *in vacuo* generally refers to a process or procedure conducted in a vacuum.

Invasionary technology
A technology that, upon commercialization, competes directly with one or more technologies that already dominate in the marketplace. Reference [76].

Invention
A new thing such as a new synthetic molecule, computer software program, device, or process. Some inventions are made public, some are described in materials that are copyrighted, some are held as trade secrets, and others are patented. A patentable invention is an invention that meets a patent system's requirements for novelty, usefulness, and significance. That is, to be patentable, an invention must be original enough not to be obvious to someone skilled in the field of the particular invention, more than a minor improvement on the prior inventions in the field, and it must be useful (it is usually described as a solution to a problem) – which is usually demonstrated by a practical application (*reduction to practice*) of the concept. Most inventions are minor and/or "obvious" improvements ("*sub-inventions*") and not patentable. Most patented inventions do not have significant commercial value, so it typically requires considerable R&D and a multitude of inventions to realize a patentable invention that does have significant commercial value.

	Invention is different from discovery and innovation. *See also* discovery, innovation, patent, and prior art.
Invention within convention paradox	*See* innovation paradoxes (innovation process paradoxes).
Inventive activity	Work that is aimed at creating inventions. Example: much of Thomas Edison's work was specifically aimed at creating practical inventions that could be commercialized. Reference [18].
Inventive potential	The total number of hypothetical inventions that could, in principle, be made by a member of an organization or society at a given point in time. Schmookler defined inventive potential as the number of inventions that a member of an organization or society could make at a given point in time with the talents that they have and the knowledge that anyone in the organization or society has available to them. Reference [18].
Inventor	A person that has created or co-created an invention. Where an invention is patented, there are rules for determining inventorship. *See also* discoverer, innovator, invention, patent.
Investment in knowledge	The ratio of total knowledge investments, including higher education, R&D, and software, to gross domestic product in an economy. *See* innovation indicators and Table 8.2.
In vitro	In science and engineering, the Latin term *in vitro* generally refers to a process or procedure conducted in an artificial environment such as a laboratory, as opposed to the "natural" setting. Example: growing biological cells in a laboratory culture, rather than within an organism (which is *in vivo*).
In vivo	In science and engineering, the Latin term *in vivo* generally refers to a process or procedure conducted in its natural environment, as opposed to in an artificial environment. Example: growing biological cells in a living organism, rather than in a laboratory culture (which is *in vitro*).
IoT	*See* Internet of Things.
IoT devices	*See* Internet of Things.
IP	*See* intellectual property.
I-Pipe	*See* innovation pipeline.
IP mining	*See* intellectual property portfolio mining.
IP portfolio mining	*See* intellectual property portfolio mining.
IPR	*See* intellectual property rights.
ISF	Innovation system foresight. *See* foresight.

11.11 J

J-curve	In general, a J-curve is any graph or diagram in which plotted values first decrease for some time but ultimately reach a minimum and rise thereafter. In business, business lines, and in technological innovation, J-curves usually refer to revenues initially being characterized by growing annual costs (negative revenue) as a business struggles to gain sales and market share, or as research, development, and commercialization costs are in-

curred, respectively. In successful businesses, business lines, or technological innovations, there is ultimately reached a tipping point beyond which sales revenues offset costs and ultimately total revenues become positive and growing. Such curves stereotypically approximate the shape of the letter "J." J-curves are also used in economics, with regard to trade balances for example. *See also* S-curve.

Joiner — *See* late majority.

Jugaad innovation — *See* frugal innovation.

Juglar cycle — An economic cycle having a period of about 7 to 11 years, reflecting oscillations in the demands on and levels of investments in production facilities. Within this cycle, businesses experience an expansion phase first, then a crisis, followed by a recession, and ultimately a recovery phase. *See also* economic cycle and Kondratieff waves.

Justified true belief — *See* knowledge.

11.12 K

KBE — *See* knowledge-based economy.

KETs — *See* key enabling technologies.

Key enabling technologies — (KETs) A small set of technologies that are expected to be the most important "building blocks" for future technological innovation across all industrial sectors. These technologies are expected to play critical roles in the evolution and sustainability of leading-edge economies. Examples: advanced materials, micro- and nano-electronics, nanotechnology, photonics, industrial biotechnology, and advanced manufacturing. Examples of industries for which KETs could lie at the heart of disruptive, or "game-changing," technological innovations include aerospace, agriculture, automotive, building construction, food, healthcare, mining and minerals, oil and gas, specialty chemicals, and textiles.
Reference [527].

Keystone — *See* business ecosystem.

Key technology — A mature technological innovation that has been widely accepted and adopted by customers, with essentially complete technology diffusion and market saturation. *See also* pacemaker technology; incremental innovation; evolutionary innovation; disruptive innovation.

KIBS — *See* knowledge-intensive business services.

KIS — Knowledge-intensive services. *See* knowledge-intensive business services.

Kitchin cycle — An older economic cycle (circa 1920s) having a period of about 3 to 4 years, reflecting how decision making in companies was affected by the time required for information to flow. This factor has become less important as the nature and pace of information flows has dramatically improved. In modern usage, the Kitchin cycle tends to refer more to the time to recognize that a market saturation has occurred, make decisions in response, adjust production and inventories, and/or wait for market demand to re-establish. *See also* economic cycle, Kondratieff waves.

KM — *See* knowledge management.

Know-how — *See* technology.

Knowledge	In the context of innovation, knowledge refers to facts, information, and understanding that have been acquired through education and/or experience, with reasonable assurance of their validity. This has also been referred to as *"justified true belief."* "*Tacit knowledge*" refers to knowledge acquired through thinking and experience in a specific context, comprises cognitive (beliefs and viewpoints) and technical (skills and knowhow) aspects, and is therefore difficult to transfer to others. "*Explicit knowledge*" refers to knowledge that can be accurately described, codified, and recorded and is therefore easy to store and transfer to others. Also termed *"codified knowledge"* or *"formal knowledge."* "*Knowledge codification*" refers to the conversion of tacit knowledge into explicit knowledge, permitting it to be accessed and used by others. *See also* acquisition of technology; knowledge-based economy.
Knowledge age or society	*See* technological ages.
Knowledge-based economy	(KBE) An economic model referring to a society whose economy is largely based on a combination of knowledge workers, the acquisition, creation, dissemination, and utilization of knowledge (embodied knowledge and codified knowledge), and the translation of such knowledge into learning, innovation, and economic development. In some usage, there is also an implication of broad access to and sharing of the knowledge (at least within the society under consideration). A knowledge-based economy is therefore more sophisticated than simply an "information-based" or "learning-based" economy (which only creates and disseminates information) or a "market-based" economy, and also different from a "political" economy. Also termed "*knowledge economy*" or, more broadly, "*knowledge society*" (in which economic health and growth can be translated into social health and growth). Similarly, an information-based economy is also termed "*information economy*" or "*information society.*" Reference [258]. *See also* knowledge.
Knowledge codification	*See* knowledge.
Knowledge economy	*See* knowledge-based economy.
Knowledge hierarchy	*See* Wisdom Hierarchy.
Knowledge-induced innovation	*See* linear innovation models.
Knowledge-intensive business services	(KIBS) Almost any kind of business service that involves or is strongly reliant on sophisticated technological and/or other professional knowledge. The former, technological KIBS, is sometimes referred to as T-KIBS (e.g. science, engineering, and R&D) while the latter, professional KIBS, is sometimes referred-to as P-KIBS (e.g. accounting, law, marketing). Sometimes simply referred to as knowledge-intensive services (KIS). In some cases, KIBS organizations act as intermediary organizations. Brokers and dealers are sometimes referred to as *technology market intermediaries* (TMI). *See also* intermediary organization.
Knowledge-intensive services	(KIS) *See* knowledge-intensive business services.
Knowledge management	(KM) The processes of any combination of identifying, developing, capturing, storing, using, sharing, and/or selling knowledge (embodied knowledge and/or codified knowledge). This can apply to an individual or to an organization. *See also* knowledge-based economy.

Knowledge processor — *See* intermediary organization.

Knowledge society — *See* knowledge-based economy.

Knowledge transfer — The process of capturing, organizing, and transferring knowledge from one part of an organization to another or from one organization to another. The term is often used to refer to kinds of knowledge that are historically and/or culturally embedded and difficult to articulate.

Knowledge turn — The amount of time required to develop an idea, concept, theory, or discovery into a commercialized innovation. Reference [737].

Kondratieff, Nikolai Dmitriyevich (1892–1938) — A Russian economist and student of free markets who is best known for his theory that major capitalist economies tend to go through 50 year cycles in which an economic depression (of about 10 years), would be followed by a period of technical advances (of about 30 years), followed by a period of economic uncertainty (of about 10 years). In modern usage, these cycles are usually taken to have periods of about 40 to 60 years, with the periods declining in magnitude over time. Eponyms include Kondratieff waves and K-waves. In some works, his name is translated as Kondratiev. *See also* Kondratieff waves.

Kondratieff waves — Waves or cycles through which major economies evolve. Nikolai Kondratieff proposed in 1925 that economics tend to go through 50 year cycles in which an economic depression (of about 10 years) would be followed by a period of technical advances (of about 30 years), followed by a period of economic uncertainty (of about 10 years) [338]. These cycles have at their heart periods in which a number of key technological developments (as well as social/political and economic) developments enable bursts of disruptive product or service innovations that fuel a sharp increase in industrial, and therefore, economic growth. The waves of change also carry with them the seeds of the next wave so the cycles repeat, apparently with periods of approximately 40 to 60 years. Synonyms include long waves, long cycles, supercycles, great surges, K-waves, and long economic cycles. Kondratieff is sometimes written "Kondratiev." Reference [343]. *See also* 1st through 7th wave, economic cycle, macroeconomics, technological ages, and Figure 6.3.

Kuznets cycle — An older economic cycle (dating back to the 1930s) having a period of about 15 to 25 years, reflecting demographic changes, usually due to immigration or emigration waves, and their effects on building construction. Also sometimes termed Kuznets swing, demographic cycle, building cycle, and infrastructural investment cycles. *See also* economic cycle and Kondratieff waves.

Kuznets swing — *See* Kuznets cycle.

K-waves — *See* Kondratieff waves.

11.13 L

L10N — "L-ten letters-N" referring to a specific kind of localization. *See* internationalization (product, process, or service).

Laggard	A category of technology adopter in Rogers' diffusion of innovation model. Also sometimes referred to as "*late mass*," or "*sceptics*," or "*inactives*." *See* Figure 5.2. *See* technology adoption life cycle.
Lagging indicators	*See* innovation indicators.
Late majority	A category of technology adopter in Rogers' diffusion of innovation model. Also sometimes referred to as "*conservatives*," "*joiners*," or "*spectators*." *See* Figure 5.2. *See* technology adoption life cycle.
Later stage company	*See* start-up capital.
Late to market	In terms of innovation strategy, a person or organization that takes advantage of a new idea, technology, invention, or technological innovation (developed by a "first mover") and implements a specifically low-cost adaptation of it. A "*cost minimization*" strategy (also termed a "*late-to-market strategy*") has a goal of producing a product, process, or service in the middle to late stages of a lifecycle trajectory. *See also* first mover, fast follower, and market segmentation.
Large-size enterprise	*See* small- and medium-size enterprise.
Late mass	*See* laggards.
Lateral thinking	*See* creative thinking models.
Law of the minimum	*See* Liebig's law of the minimum.
Leading-edge customer	One of several kinds of stereotypical people that can play a critical role in the technological innovation process. Engaging prospective customers throughout the innovation process is often identified as a means of improving the success rate. An idealized leading-edge customer might be an early adopter of technology, an experienced user of previous innovative products, able to provide forward-looking advice on desired new products or product features or benefits, a potential tester of prototypes, and a source of potential post-launch improvement ideas. Reference [98]. *See also* Rainmaker, product champion, and technological gatekeeper.
Leading indicators	*See* innovation indicators.
Lead users	Key customers that organizations consult for ideas that could be developed into new products, processes, or services. Sometimes termed luminaries. This is somewhat similar to the concept of a leading-edge customer.
Lean innovation	A technological innovation process that is systematic, focussed on the things that add value, and eliminates waste. This is by analogy with the principles of "*lean manufacturing*." References [126, 745].
Leap-forward innovation	*See evolutionary innovation.*
Learning society	*See* technological ages.
Learning-based economy	*See* knowledge-based economy.
LEO	*See low earth orbit.*
LEPEST analysis	Legal, Environmental, Political, Economic, Social, and Technological analysis. *See* Social, Technological, Economic, Environmental, and Political analysis.
Leydesdorff, (Louis André) Loet, (1948–present)	A Dutch sociologist known for his work in the sociology of communication and innovation. He co-developed (with Eztkowitz) the triple-helix model, a sociological model describing the roles and intersections of governments, universities, and industry in advancing knowledge-based economies, and which by extension has been applied to the advancement of economies

based on innovation. *See* triple-helix model. See also References [260, 266, 268, 738].

License — *See* license agreement.

License agreement — An agreement under which a person, group, or organization is entitled to make, use, or sell some form of intellectual property. Under an *exclusive license*, a licensee gains the sole right to such use, although there may be some kind of limitation such as geographic territory, technical field of use, product application, or time. Under a *limited or non-exclusive license*, a licensee gains a right – but not the sole right – to such use, although again there may be some kind of limitation. Payments made by a licensee for the right to use licensed intellectual property are called *royalties*. *See also* licensing, cross-license, and royalty-free licence.

Licensing — The legal process of contractually granting an intellectual property license. *See also* license agreement, cross-license, and royalty-free license.

Liebig's Law of the Minimum — Named for Justus van Liebig, a famous chemist of the 1800s, this "law" originally referred to Liebig's theory that the yield of an agricultural crop is proportional to the amount of the most limiting nutrient. Among other versions or adaptations of this theory are the "law" that the capacity of a barrel with staves of unequal length is limited by the shortest stave, and that growth is controlled by the amount of the most limiting (often the scarcest) resource. Reference [746].

Limited license — *See* license agreement.

Linchpin technology — A technology that in some fashion holds a more complex structure together. For example, a linchpin technology may be the key to commercializing certain other technologies, or it may transform the way other pre-existing technologies are used, or its deployment may require the invention or application of other technologies.

Linear innovation models — There are numerous models for the process of commercial innovation. The primary linear models are "*technology-push*" and "*demand-pull*." These are both unidirectional. In *technology-push*, science drives discovery, which in turn drives invention, which in turn drives the development and commercialization of products, process, or services into the marketplace. Since it is *technology-push*, most or all of this process is conducted without regard for the existence of market niche or need, so the ultimate product, process, or service may not be commercially successful. Examples include the mass-market personal computer (1977: Commodore PET™, Apple™, and Tandy TRS-80™), Sony Walkman™ portable audio cassette player (1979), and the Apple iPad™ tablet computer (2010). Technology-push is also termed "*discovery-push*," "*science-push*," "*supply-side driven*," or (in the older literature) "*knowledge-induced*." It is also termed the first, or the first-generation, model of innovation. In *market-pull*, marketplace opportunities drive the search for new product, process, or service concepts, for which innovators search the existing body of knowledge, inventions, and technologies. Since it is *market-pull*, the market niche or need is clear, but the search for appropriate knowledge, invention, and/or product conceptualization and development may not be successful. Examples include digital photo editing software (1980s) and mass-market digital cameras (1990s). Market-pull is also termed "*demand-pull*," "*needs-pull*," or (in the older literature) "*demand-induced.*" It is also referred to as the second, or the second-generation model of innovation. *See* Figure 2.1. *See also* non-linear innovation models and generations of innovation.

Linear thinking — *See* creative thinking models.

Linked-chain model — *See* chain-linked model.

Local innovation — *See* reverse innovation.

Localist regional innovation system — *See* regional innovation system types, Table 5.1.

Localization — *See* internationalization (product, process, or service).

Logistic curve — *See* S-curve.

Logistics management — *See* value chain.

Long cycles — *See* Kondratieff waves.

Long economic cycles — *See* Kondratieff waves.

Long waves — *See* Kondratieff waves.

"Looks like" model — *See* "works like" model.

Low earth orbit — (LEO) Earth orbits in the altitude range 160 to 2,000 km. Satellites in this range are generally used for earth observation and internet. In contrast, medium earth orbit (MEO) is in the altitude range 2,000 to 35,786 km, and generally used for positioning and navigation systems like GPS, and in the case of military satellites, for missile warning systems. Geostationary orbit (GEO), also termed geosynchronous orbit (GSO), is at an altitude of 35,786 km and is generally used for weather imaging and telecommunications. Some weather satellites operate at high earth orbit (HEO), which comprises altitudes above 35,786 km.

Low-tech — Low technology. *See* high technology.

Ludd, Ned — *See* Luddite.

Luddite — Generally, a person that is strongly opposed to mechanization, industrialization, automation, or even new technologies of any kind. The original Luddites were textile workers that, fearing job losses, destroyed textile machines in England in the Industrial Revolution of the early 1800s. They are thought to have taken their name from Ned Ludd, a textile worker that had destroyed such a machine in 1779. Modern usage frequently refers to a person that is strongly opposed, or resistant, to the adoption of new technologies of any kind (hence "*neo-Luddism*" and the *avoiders*, or *non-adopters*, in technology acceptance models). The "Luddite fallacy" refers to the false hope that avoiding technological advances can in some way ensure economic success and/or sustainability. *See also* technology acceptance model.

Luminaries — *See* lead users.

11.14 M

Maclaurin, (William) Rupert (1907–1959) — A New Zealand-born, American economic historian known for his development of the first model of the process of innovation (building upon the industrial research cycle work of Maurice Holland). His "*theory of technological change*" laid out a sequential series of steps beginning with research and ending with commercialization that is now known as the technology-push linear model of innovation. He also suggested a number of measures that could be used as indicators of progress in each element of the sequence. It may have been Maclaurin that coined the term "*technological innovation*." Reference [62]. *See also* Holland.

Macroeconomics — A sub-discipline of the field of economics that focusses on entire economies, as opposed to specific markets within an economy. Macroeconomics includes the structures, behaviours, and performance of economies, which may be regional, national, or broader. Macroeconomics is different from microeconomics, which focusses more narrowly on specific markets, although they are linked because the health of economies affects the actions of businesses and consumers and *vice versa*. *See* microeconomics. *See also* competitiveness, gross domestic product, and Kondratieff waves.

Main science and technology indicators — (MSTI) Indicators of the degree and nature of science and technology activities of countries, particularly those assessed for this purpose by OECD. The OECD MSTIs cover selected inputs, such as research and development (R&D) spending and numbers of researchers, activities, such as R&D activities by sector; outputs, such as patents; outcomes, such as exports or export market share; and impacts, such as balances of trade or payments and international trade, jobs, and GDP growth. Reference [483].

Makerspace — A form of intermediary comprising a community- or private-sector operated, collaborative workspace, typically providing access to, for example, 3D printers, laser cutters, CNC machines, and other fabrication tools and machines. Such spaces help potential inventors, developers, entrepreneurs, and start-ups to learn, experiment, create, develop, and prototypes [747, 748]. Synonyms include fab lab and tech shop. A makerspace case study is provided in reference [699]. Hackerspaces are usually similar but more focussed on computer electronics and software development (hence also termed "*hack lab*").

Management innovation — New ways of organizing work or work-related functions, practices, or processes, including the management function, within an organization. Also termed "*managerial innovation*." *See also* innovation and organizational innovation.

Mandate effectiveness ratio — (MER) An innovation metric pioneered by RTOs and used to help quantify the amount of incremental economic activity that results from their work helping businesses address their technological challenges and opportunities in the marketplace. The definition is:

MER = (incremental economic impact)/(core government investment in the RTO)

The timeframe is usually one year but should be specified. Also termed "*mission effectiveness ratio*." Reference [514]. *See also* innovation metrics.

Manufacturing development — *See* research and development.

Mark I innovation — *See* Schumpeter Mark I innovation.

Mark II innovation — *See* Schumpeter Mark I innovation.

Market analysis — In the commercialization of a product, process, or service, several elements have to be considered to understand the market potential. A typical sequence includes market identification, market size, customers, distribution channels, and competitors and product differentiation in the marketplace. Reference [76].

Market barrier — *See* innovation barrier.

Market-based economy — *See* knowledge-based economy.

Market demonstration — *See* demonstration.

Market development strategy — *See* Ansoff matrix.

Market identification — A summary of the customer needs that will be satisfied, the specific customers who will buy the product, process, or service, the reason(s) they will buy it, and whether the timing is right in the marketplace. *See also* market analysis. Reference [76].

Marketing — The processes of identifying a company's main differentiating features, developing the key messages that explain these differentiators and their relevance, and developing and implementing strategies to position the company in its appropriate marketplace(s). Such strategies may include such things as branding, stakeholder relations, advertising, events, and communications (print, internet, and social media), for example. The principal components of marketing are *market research and planning* and *market management*. Marketing helps to create visibility and profile (brand awareness) for a company, and sets the stage for business development and sales. *See also* business development, sales, distribution, and innovation barrier.

Marketing innovation — New ways of marketing products, processes, or services that have been successfully implemented and improve market penetration and/or market size. *See also* innovation and organizational innovation.

Market intelligence — *See* competitive intelligence.

Market leader — *See* first mover.

Market management — This includes advertising, promotion, and customer service related to a product, process, or service. This forms part of *marketing and sales*. *See also* market research and planning, sales, and distribution.

Market penetration strategy — *See* Ansoff matrix.

Market-pull innovation — *See* linear innovation models, second-generation innovation model, and incremental innovation. *See also* disruptive innovation.

Market segmentation — In terms of innovation strategy, a person or organization that takes advantage of a new idea, technology, invention, or technological innovation (developed by a "first mover") and implements a customized adaptation of it that is aimed at one or more niche markets. A "*specialist*" strategy (also termed a "*market segmentation strategy*") has a goal of serving niche markets with customized versions of the basic product, process, or service, usually later in the core technology's lifecycle trajectory. *See also* first mover, fast follower, and late to market.

Market readers — *See* architects and moonlighters.

Market research and planning — The evaluation of the market and the sales potential of a product, process, or service; market barriers; distribution channels; and identification of the target customers. This forms part of marketing and sales. *See also* market management, sales, and distribution.

Market size	A description of the target market and/or market segments for a product, process, or service; and also the size of those markets in numbers of units that can be sold. *See also* market analysis. Reference [76].
Mass customization	*See* product platform.
Material technology	A technology category comprising physical things, such as a tangible manufactured product. A process could be either a material or a social technology. *See also* technology and social technology.
Material testing agreement	(MTA) A legal contract created to protect information considered to be proprietary or otherwise business confidential when materials are transferred from one party to another for testing, evaluation, or further development. The parties to the contract agree not to disclose to others the information covered by the agreement. Also termed *"material transfer agreement," "plant material testing agreement," or for plant materials: "pMTA." See also* non-disclosure agreement, option agreement, and boilerplate.
Material transfer agreement	(MTA) *See* material testing agreement.
McKinsey Global Institute Connectedness Index	*See* Connectedness Index.
ME	Medium-sized enterprise. *See* small- and/or medium-sized enterprise.
Mechanistic model	A model for the innovation process in which innovation arises from an accumulation of numerous smaller steps or ideas that occur over a reasonably long period of time. *See also* cumulative synthesis model and transcendentalist model.
Medici effect	When exposure to a range of fields, disciplines, or cultures leads to a combination of existing concepts that sparks a creative new idea. The term "intersection" in the Medici effect refers to the making of multiple connections among the various fields of knowledge, philosophies, and approaches. The name Medici refers to a family in 15th-century Italy that funded creators from a wide range of disciplines, drew them to a single location (Florence), and helped trigger the European Renaissance in such fields as art, science, politics, literature, and architecture. Reference [180]. *See also* creative thinking models.
Medium earth orbit	(MEO) *See* low earth orbit.
Medium-high-tech	Medium-high-technology. *See* high technology.
Medium-low-tech	Medium-low-technology. *See* high technology.
Medium-sized enterprise	(ME) *See* small- and/or medium-sized enterprise.
MEO	Medium earth orbit. *See* low earth orbit.
MER	*See* mandate effectiveness ratio.
Messy fireworks innovation	*See* non-linear innovation models.
Mesotechnology	*See* microtechnology.
Metrics	(Innovation) Measures used to quantify and compare innovation activities, outcomes, outputs, reach, and/or impacts. These are usually distinct from, but may overlap with or supplement traditional business metrics such as production, sales, net income, and market share. Example: new patents applied for and/or issued. For some specific examples, see New Product Vitality Index and mandate effectiveness ratio. See also activities, inputs, outcomes, outputs, reach, and impacts.
MFP	*See* multifactor productivity.
MGI CI	McKinsey Global Institute Connectedness Index. *See* Connectedness Index.

Micro	A prefix used to signify one-millionth part of a unit, a multiple of 10^{-6}. The common symbol for micro is μ, as in μm for micrometre. *See also* nano.
Microeconomics	A sub-discipline of the field of economics that focusses on specific markets within an economy, as opposed to entire economies. Microeconomics includes the positioning and behaviours of businesses and consumers in specific markets. Microeconomics is different from macroeconomics, which focusses on economies as a whole, although they are linked because the actions of businesses and consumers affect economies and *vice versa*. *See* macroeconomics. *See also* competitiveness and gross domestic product.
Micro-entity	*See* small- and/or medium-sized enterprise.
Micro-sized enterprise	*See* small- and/or medium-sized enterprise.
Micro-, small-, and/or medium-Sized enterprise	*See* small- and/or medium-sized enterprise.
Microtechnology	A materials science term referring to materials or structures at a scale of about one micrometre (typically 0.1 to 100 μm), as opposed to the nanoscale (nanotechnology; 0.1 to 100 nm). The term "mesotechnology" has been used to describe the regime in-between these two, although others are still using this term as a synonym for microtechnoogy. *See also* nanotechnology and molecular nanotechnology.
Mind maps	*See* spider diagrams, *under* creative thinking models.
Miners	A characterization of one of four kinds of organizational approaches to innovation strategy. "*Miners*" tend to be very large companies with lots of technology and process expertise, very large internal business units, and strong silos. They also tend to be weak on internal innovation, necessitating the use of external innovation resources and/or organizations, focussed on finding innovation opportunities within the organization, and innovating through formal, centralized processes working through the external innovation organization. Jaruzelski and Dehoff have referred to such organizations as "*technology drivers*" [381]. *See* innovation strategy mapping.
Minnesota Mining and Manufacturing Co.	*See* 3M Inc.
Mission effectiveness ratio	*See* mandate effectiveness ratio.
Mission-oriented research	*See* research and development.
MNC	*See* multinational enterprise.
MNE	*See* multinational enterprise.
Mockup	A model built to accompany a *product definition* and used to illustrate a product or process, and sometimes also aspects of the environment in which it would be used. For example, a mockup for a process concept might include a model of some of the process vessels, pumps, and piping in which it would be used. Mockups are usually built to scale, but usually are not working models. *See also* prototype. Reference [76].
Mode 1	*See* modes of science.
Mode 2	*See* modes of science.
Model	*See* prototype.
Models of innovation	*See* linear innovation models and non-linear innovation models.
Modes of science	A social science term for the way scientific knowledge is produced. Mode 1 refers to knowledge produced by discovery research conducted within

specific disciplines (biology, chemistry, physics, etc.). Mode 2 refers to applied, mission-oriented research produced by multidisciplinary teams. Mode 2 research has also been connected to the "*entrepreneurial university*" and to university research in the triple-helix model. Also termed "*context-driven research*," or "*post-academic science*." *See* research and development. *See also* entrepreneurial university and triple-helix model.

Modular innovation — *See* architectural innovation.

Mokyr, Joel (1946–present) — A Dutch economist who was co-awarded the 2025 Nobel Prize in Economic Sciences, in his case for the theory of sustained growth through creative destruction (his co-awardees were Philippe Aghion and Peter Howitt).

Molecular engineering — *See* nanotechnology.

Molecular fabrication — *See* nanotechnology.

Molecular manufacturing — *See* molecular nanotechnology and nanotechnology. *See also* generations of nanotechnology.

Molecular nanotechnology — The organization of atoms and molecules from the nanoscale up, in order to produce structures and/or materials having precise, predictable properties. Synonyms include "fourth-generation nanotechnology," molecular manufacturing, molecular engineering, and molecular fabrication. The nanoscale structures created are sometimes termed nanoscale architectures or nanoarchitectures. *See also* nanotechnology and generations of nanotechnology.

Moonlighters — A characterization of one of four kinds of organizational approaches to innovation strategy. "*Moonlighters*" tend to be large companies with lots of technology and process expertise, that look inward to develop innovations based on internal ideas and know-how. They also tend to be focussed on a strong innovation culture that supports internal experimental and development project initiatives, including internal "moonlighting," willing to fund internally generated innovation projects, and innovating through bottom-up processes. Jaruzelski and Dehoff have referred to such organizations (and also to "architects") as "*market readers*" [381]. *See* innovation strategy mapping. *See also* bottom-up innovation.

MSTI — *See* main science and technology indicators.

MTA — Material testing (or transfer) agreement. *See* material testing agreement.

Multifactor productivity — (MFP) An industrial measure of the change in output per unit of a number of combined inputs (including labour, materials, and capital). Any change in MFP is due to factors other than the combined inputs, such as technology changes, change of scale efficiencies, production changes, management changes, and changes due to technological innovation). Although it is only one of these other factors, MFP is sometimes used as an indicator of the impact of changes due to technological innovation, and sometimes as an indicator of the amount of technological innovation. This could be for an industry or for an individual enterprise. Reference [489]. *See also* total factor productivity.

Multinational corporation — *See* multinational enterprise.

Multinational enterprise — (MNE) A corporation that owns and/or controls facilities, assets, and usually production, in at least one country beyond their home country. An MNE generally coordinates operations from a central (head) office. Some MNEs develop new products in one or more countries, then have them manufactured in other countries, and then have them distributed and sold in yet other countries, in order to optimize materials costs, labour, costs, and/or trade advantages. Also termed *"multinational corporation (MNC),"* *"transnational corporation,"* *"international corporation,"* or *"stateless corporation."* In some usage, Global Multinational Enterprises (Global MNEs) are distinguished from regionally focussed MNEs, the former having substantial sales in more than one or two continents and the latter having most of its sales in only one or two continents.

Multiparty innovation — *See* ecosystem innovation.

Multiplier — *See* impact multiplier.

Multiple discovery theory — The hypothesis that most scientific discoveries and inventions are made essentially simultaneously by multiple (independent) discoverers and inventors, respectively. Also termed *"multiple invention theory"* or *"simultaneous discovery (or invention) theory."* Commonly cited examples include the theory of the origin (evolution) of species and the formulation of calculus. Merton postulated that this could be the result of a combination of the prerequisite knowledge and tools becoming available and having a substantial number of people pursuing similar investigations, possibly in response to similar social and intellectual forces [73]. Epstein points out that, in addition, most inventions are improvements of previous inventions [74]. An opposing theory is the "heroic theory of discovery or invention," which is the hypothesis that the principal discoverers and inventors of most discoveries and inventions, respectively, are rare *"greats"* or *"geniuses."*

Multiple invention theory — *See* multiple discovery theory.

11.15 N

Nanoarchitecture — *See* molecular nanotechnology.

Nano — A prefix used to signify one-billionth part of a unit, a multiple of 10^{-9}. The common symbol for nano is n, as in nm for nanometre. *See also* micro.

Nanomaterial — Any material having one or more nanoscale dimensions (0.1 to 100 nm). Use of this term is also usually taken to imply that the material has different properties than it would if not for the nanoscale dimensions.

Nanoscale architectures — *See* molecular nanotechnology.

Nanotechnology — A rapidly growing area of materials science involving the design, characterization, manipulation, incorporation, and/or production of materials and structures in the nanoscale range (typically 0.1 to 100 nm), by any of a variety of physical and chemical methods. An important distinction is that these applications exploit the properties of the nanoscale components, distinct from bulk or macroscopic systems. There is a substantial overlap of scale between nanotechnology and microtechnology. Nanotechnology also encompasses the "nano" approach, or molecular nanotechnology, by which meant the precise, controlled assembly of structures up from the molecular scale that are well-organized, and with reproducible properties. *See also* microtechnology. Nanotechnology formed the basis for key developments in the sixth

Kondratieff wave of disruptive innovations and is one of the possible technological drivers of the seventh Wave of innovation. *See* 7th wave and Kondratieff waves.

National innovation system	(NIS) *See* innovation ecosystem.
National system of innovation	(NSI) *See* innovation ecosystem.
NDA	*See* non-disclosure agreement.
Need seekers	*See* explorers.
Needs-pull innovation	*See* linear innovation models.
Negative cash flow	*See* cash flow.
Negative innovation	The impacts of a failed innovation process that leaves an organization with reduced or less competitive product/process/service offerings in the marketplace than they had before commencing the innovation process. *See* innovation process paradoxes (innovation process paradoxes).
Neoclassical Growth Model	*See* Solow-Swan Growth Model.
Neo-Luddism	*See* Luddite.
Net perceptual equity	*See* perceptual equity.
Net rate of replication	*See* technological capacity.
Network regional innovation system	*See* regional innovation system types, Table 5.1.
New Growth Theory	*See* Endogenous Growth Theory.
New media age	*See* technological ages.
New product development funnel	*See* product development funnel.
New product development process	(NPD) New product development processes generally involve a process for identifying new product ideas, followed by a process for deciding which and how many of them should be further explored or developed, and a process for re-evaluating after each of a succession of development stages. An example of the sequence of stages is: identification of the problem or opportunity; investigation, idea generation, and solution identification; further investigation and proposal development; solution development and proof of concept; prototype testing, demonstration, and marketability testing; production, marketing, and commercial deployment; and post-project review. *See also* Stage-Gate® product development process, technology stage-gate process, and waterfall method.
New Product Sales Index	*See* New Product Vitality Index.
New Product Vitality Index	(NPVI) An innovation metric pioneered by 3M Inc. and used to help quantify the amount of a company's business and/or business growth originating from products that were introduced to the marketplace within a specific time frame (in 3M's case, the previous 5 years). The 3M definition is: NPVI = (sales from products introduced within the past 5 years)/(total sales) Also termed "*New Product Sales Index*" or "*Year 1 New Product Sales Index*" (where the time frame is 1 year). *See also* innovation metrics.
New-to-the-firm innovation	The adoption and implementation of a product, process, service, or function by an organization that had not previously done so. Example: adopting new management practices.
New-to-the-world innovation	The first introduction of an innovation to a marketplace.

New-to-the-world solution	*See* breakthrough.
NGO	*See* non-government organization.
Niche fusion	A technology deployment strategy involving the introduction of a new product, process, or service into multiple market niches in order to achieve a greater competitive advantage in the combined niches than would have been possible with only a single niche. Reference [737]. *See also* single niche strategy.
Niche innovation	A form of innovation in which the new product, process, or service displaces or renders obsolete the competing products in a small market niche, but not an entire market. As such, niche innovation falls between incremental and disruptive innovation. Reference [86]. *See* Figure 2.5. *See also* disruptive innovation, evolutionary innovation, innovation, incremental innovation, and Abernathy-Clark model.
Niche player	*See* business ecosystem.
NIH syndrome	*See* not invented here syndrome.
NIS	National Innovation System. *See* innovation ecosystem.
Non-commercial innovation	*See* innovation (non-commercial).
Non-disclosure agreement	(NDA) A legal contract created to protect information considered to be proprietary or otherwise business confidential. NDAs are used to enable parties to share confidential information with each other but otherwise maintain confidentiality. The parties to the contract agree not to disclose to others the information covered by the agreement. Also termed "*confidential disclosure agreement (CDA)*" or "*confidentiality agreement.*" *See also* material testing agreement. *See also* public disclosure and boilerplate.
Non-exclusive license	*See* license agreement.
Non-government organization	(NGO) A not-for-profit organization that is mostly or completely independent of government(s). The nature, purpose, means of revenue generation, and activities can vary widely although a not-for-profit organization is driven by purpose ("mission-driven") rather than the need to make profits and pay dividends to shareholders. Most NGOs have some kind of social and/or environmental purpose, and some are registered charities. The revenues for NGOs can be any combination of grants, donations, or revenue contracts, including fee-for-service contracts. Some NGOs perform contract services in the marketplace, whereas others are exclusively funded by donations. Not all NGOs are completely independent of governments, however, as some NGOs receive substantial portions of their revenues in the form of government grants and therefore may subject to indirect influence and/or control by government(s).
Non-linear innovation models	There are a number of non-linear models for the process of how innovations are created, focussed on the role of institutions in the process. The primary non-linear models are the "*coupling model,*" "*integrated model,*" and "*systems integration and networking model.*" The "*coupling model*" is sequential, drawing on technology-push and/or market-pull at each stage, but it is non-linear in that the various kinds of communication and interaction among contributors creates feedback loops. Over time, the term coupling was gradually superseded by terms such as communication, interaction, and chain-linked. The term coupling has also been used in

innovation with reference to the coupling of ideas in the minds of, for example, discoverers and inventors. The "*integrated innovation model*" involves parallel, rather than sequential development, with simultaneous R&D, prototype development, and manufacturing activities, for example. Integrated innovation processes encompass integrated development teams, engagement with suppliers and customers, and possibly engagement with partner organizations (horizontal collaboration). The "*system integration and networking model*" (SIN innovation model) involves parallel development integrating horizontal linkages with customers, suppliers, collaborators, marketers, and others throughout the process. The non-linear innovation models attempt to deal with the fact that technological innovation processes can be highly irregular, involving jagged lines of activity, and highly uncertain and they are sometimes referred to using metaphors such as "*fireworks*" or "*messy fireworks,*" hence the term "*messy fireworks innovation.*" "*Innovation 2.0*" is a variation on this theme. *See* Figure 3.4. *See also* linear innovation models, generations of innovation, innovation 2.0, and innovation ecosystem models. Reference [64].

Non-provisional patent application — *See* provisional patent application.

Non-technological innovation — Non-commercial innovation. *See* innovation (Non-commercial).

Not invented here syndrome — (NIH syndrome) A tendency of people, groups, or organizations to resist ideas, inventions, and/or innovations that have been developed by others. This frequently leads to rediscovering, reinventing, or redeveloping things that were already available elsewhere (often referred to as "*reinventing the wheel*"). The NIH syndrome is generally attributed to some combination of belief that external groups and/or capabilities are inherently inferior, a desire for control and/or credit, or an undue attachment to the *status quo*.

Novelty paradox — *See* innovation process paradoxes (innovation process paradoxes).

NPD — *See* new product development.

NPD funnel — *See* product development funnel.

NPVI — *See* New Product Vitality Index.

NSI — National System of Innovation or National Innovation System. *See* innovation ecosystem.

N-tuple helix model — An innovation ecosystem (sociological) model developed by Etzkowitz and Leydesdorff, and others, as an expansion of the triple-helix and quad-helix models, to include additional players, beyond governments, universities, intermediaries, and industry. References [266, 268]. *See also* innovation ecosystem models, triple-helix model, and quad-helix model.

11.16 O

OECD — Organization for Economic Co-operation and Development, Paris, France. *See also* Frascati manual and Oslo manual.

OEM — *See* original equipment manufacturer.

Open information — Information that is openly available, without the need for purchasing, licensing, or even significant interaction with the source of the information.

Open innovation	The terms *open innovation* and *closed innovation* refer to two extremes in approaches a company can make to developing and commercializing technologies into innovations. In closed innovation, a company conducts essentially all of its innovation activities internally, often in a dedicated R&D department, from idea generation through R&D and the subsequent activities such as experimental development, prototyping, pilot testing, demonstration, and pre-production. The main advantages of closed innovation are the ability to control the process and to potentially achieve dominating competitive advantage positions in the marketplace. The main disadvantages are cost and the fact that the company is limited by its internal idea generation and innovation capacity. In open innovation, a company uses external resources to conduct some or even most of its innovation activities, especially in the areas of idea generation and "*adopt and adapt*" technology acquisition. The main advantages of open innovation are that a company is not limited by its internal idea generation and innovation capacity and that it may be less expensive to adopt and adapt ideas and technologies from others than to develop them internally. Another advantage of open innovation is that it lends itself to partnering with other organizations in the innovation ecosystem. The main disadvantages are that competitor companies can usually access the same external capabilities and may be able to exploit them in ways that are better, faster, and/or cheaper, making it difficult to achieve a competitive advantage in the marketplace. Although the foregoing represent extreme positions, many companies use a hybrid strategy and leverage internal innovation capabilities with external ideas, technologies, and other capabilities. For a variation on open innovations, see permissionless innovation. *See also* collaborative innovation, innovation ecosystem, open science, inbound open innovation, outbound open innovation, and open-source innovation.
Open science	The classical approach an organization or individual can make to conducting discovery research in science. In open science, scientists engage in a two-way exchange of ideas, data, and knowledge by collaborating with one another without charge, and by publicly disclosing (presenting and publishing) without charge and by applying only the restriction that users publicly acknowledge their sources of information and collaboration. This approach has some similarities with the concept of *open innovation*. Conversely, an example of closed science would be where a scientist makes use of publicly available ideas and information but then holds their own ideas, knowledge, and other results gained from their work confidential and does not release them publicly. This happens in some industrial R&D environments. *See also* open innovation.
Open-source innovation	A kind of open innovation in which the process, source, and outcome of the innovation are open as opposed to private. Example: open-source software. *See* open innovation.
Operating expense	(OpEx) *See* capital expense.
OpEx	Operating expense. *See* capital expense.
Opportunity cost	The cost, usually meaning the loss of potential gain, that could be realized from alternatives when a particular opportunity or course of action is chosen instead.

Option agreement — An agreement, often a legal contract, between two parties that grants one party the right to buy or obtain (or sell or dispose of) something at a specified price and at a specified future time. Example: an option agreement could be used to enable someone to try and/or evaluate a technology for a certain period of time before having to decide whether or not to buy or license it. *See* material testing agreement. *See also* boilerplate.

Organizational innovation — The ideation, development, and implementation of new and improved internal processes within an organization. Such new organizational processes should have some kind of efficiency or productivity benefit, even if they are internal and have little or no connection to commercialization or the marketplace. Also termed "*institutional innovation*," although sometimes the term institution is meant more broadly. Examples: administrative, management, marketing, political, and strategic innovation. *See also* innovation and paradigm innovation.

Organizational paradox — *See* innovation process paradoxes (innovation process paradoxes).

Original equipment manufacturer — (OEM) A company that manufactures a part, or even a sub-system, that is integrated into another company's product. A similar part or sub-system made by someone else and not part of the original product is often referred to as "*aftermarket.*"

Originator — *See* first mover.

Osborn-Parnes model — A synonym for creative problem solving. *See* creative thinking models.

Oslo manual — An Organization for Economic Co-operation and Development (OECD) document providing recommended methods for collecting and interpreting data on innovation. The OECD definition of innovation includes both *commercial innovation* and *non-commercial innovation*. The counterpart of the Oslo manual is the Frascati manual, which is concerned with data on research and development. Reference [26]. *See also* commercial innovation and Frascati manual.

Other innovation activities — Any or all of the scientific, engineering, technological, commercial, and financial steps – other than R&D – that are needed to achieve innovation. That is, the non-R&D steps needed to get products, processes, or services into the marketplace. Reference [26]. *See also* Oslo manual.

Outbound innovation — *See* outbound open innovation.

Outbound open innovation — An aspect of the open innovation process in which the flow of technology or other information is from inside an organization to the outside. Example: licensing out of technologies. Also termed "*outbound innovation*." *See also* open innovation and inbound open innovation.

Outcome-driven innovation — *See* customer-oriented innovation.

Outcomes — (Innovation) The initial consequences of the things (outputs) that are produced by and/or happen as a result of activities conducted by an organization or its employees. Examples: the results of research and development activities. *See also* activities, inputs, outputs, reach, metrics, and impacts.

Outliers — *See* foresight.

Outputs — (Innovation) Things that are produced by an organization or its employees. Examples: the communication, publication, and dissemination of the results of research and development activities, or the products, processes, and/or services produced and sold by a business. *See also* activities, inputs, outcomes, reach, metrics, and impacts.

11.17 P

Pacemaker technology	An early-stage technological innovation that has not yet been widely accepted and adopted by customers and whose applications, usefulness, and market potential are not yet well known. In some usage, the term is applied to the specific case for which the emerging technology is judged to have high potential for future applications. *See also* key technology, incremental innovation, evolutionary innovation, and disruptive innovation.
PACE process	*See* product and cycle-time excellence process.
Paid-up license	*See* royalty-free license.
P&ID	*See* piping and instrumentation diagram.
Paradigm innovation	Similar to organizational innovation but broader, potentially encompassing any improvements in organizational (usually company) modus operandi, behaviour, and strategy [10]. *See also* organizational innovation.
Paradoxes of innovation	*See* innovation paradoxes.
Parallel thinking	*See* creative thinking models.
Participative innovation	In *closed innovation*, this refers to engaging others within an organization, beyond those officially in R&D, in innovation processes. Such organizations may appoint a Chief Innovation Officer, or the like, to lead and coordinate such processes on behalf of the organization. *See also* open innovation. In *open innovation*, this refers to engaging others – both inside and outside of an organization, and beyond those officially in R&D, in innovation processes. Such organizations may appoint a Chief Innovation Officer, or the like, to lead and coordinate such processes on behalf of the organization. This meaning of the term *participative innovation* is essentially the same as collaborative innovation. *See also* collaborative innovation and open innovation.
Patent	A grant of property rights on new, useful, and non-obvious inventions by a government. The grant of a patent excludes, for a certain number of years, others from making, using, selling, offering to sell, or importing the invention in the specific country granting the patent. For this reason, many inventions are patented in multiple countries. Patents are generally awarded to the "first to file" a patent application rather than the "first to invent," although some countries allow a grace period before filing is required. For this reason, *provisional patent applications* are sometimes filed. The most common patent type is termed *utility patent*, which covers processes, machines, manufactured items, or composition of matter. Other types include the *design patent* (for designs related to manufactured items) and the *plant patent* (for certain kinds of new invented or discovered plants). *See* invention, intellectual property, patentable invention, and provisional patent application. *See also* prior art.
Patentable invention	*See* invention and patent.

Patent applications — (Innovation indicator) An indicator of the total number of practical inventions being protected in an economy is the number of patent applications registered per 10,000 personnel in the labour force. *See* innovation indicators and Tables 8.2–8.4.

Patent database mining — *See* intellectual property portfolio mining.

Patent family — A group of patents that have been granted in multiple countries on the same invention. Example: OECD Triadic Patent Family, which is a group of patents on the same invention that have been granted by the European Patent Office (EPO), United States Patent and Trademark Office (USPTO), and the Japan Patent Office (JPO). The Triadic Patent Family concept was adopted by OECD in order to provide an indicator of technological change at the international level.

Patent mining — *See* intellectual property portfolio mining.

Patent pending — A term used to indicate that a patent application has been filed but not yet determined. That is, it has not yet been allowed, disallowed, or abandoned. *See* patent.

Patent protection — (Invention) Patent protection for an invention means that, once patented, an invention cannot be commercially made, used, or sold by anyone else without permission. *See* patent.

(National system) Patent protection in a specific country refers to the degree of protection that a specific country provides for a patented invention and/or the relative strength of a country's intellectual property laws compared with others. *See* Ginarte-Park index.

Pearl curve — One of the simplest of the technology S-curve models, the Pearl logistic curve model relates the technology maturity (y) to time (t) as: $y = y_\infty / (1 + \alpha \cdot \exp^{-\beta t})$, where the maximum value for y is given by y_∞, and the shape of the S-curve is determined by the adjustable parameters α and β. *See also* S-curve and Gompertz curve.

Pencil — *See* "does it pencil?"

Per capita — In innovation literature, the Latin term *per capita* is generally used to mean "per person." Example: GDP *per capita*.

Percentage of patents with foreign co-inventors — (Innovation indicator) An indicator of the degree of connectedness and collaboration among a country's inventors and those in other countries is 100 times the ratio of the number of patents having foreign co-inventors to the total number of patents issued in a country. *See* innovation indicators and Table 8.2.

Perceptual equity — The value of a brand, product, process, service, or enterprise as perceived by prospective customers or some other key stakeholder group. For example: in the case of customers, a high degree of perceptual equity for a particular product can contribute to a willingness to try the product, pay a premium price for it, and/or remain loyal to its brand versus those of competing products. *Net perceptual equity* refers to the difference between the positive perceptions about a brand or product and the negative such perceptions. *See also* brand.

Perfectibility hypothesis — (Innovation) The theory that a new technology in any particular field is perfectible, that continuing developments cause it to approach perfection relatively quickly leaving diminishing returns for future potential advances,

	and that as a result, the technological maturity plateaus (i.e. the inventive potential becomes exhausted). This hypothesis has been used to attempt to explain technology S-curves. A counter-argument is provided by Schmookler (Reference [18]). *See* S-curve.
Performance trap	*See* innovation barrier.
Peripheral innovation	An innovation involving an aspect that is secondary to the central aspect of an innovation. For example, an innovative applicator bundled with a new cosmetic product can add value to the new product by creating differentiation in the marketplace while adding customer interest, acceptance, and even value. Lee and Kang [639] provide a case study example.
Permissionless innovation	A form of open innovation in which an organization can invite the public to conceive and submit new ideas, solutions to problems, analyses of data, and/or solutions to opportunities in return for some kind of incentive. Chesbrough and Van Alstyne refer to permissionless innovation "as a complement to traditional research and development." *See also* open innovation. Reference [158].
Pervasive computing	The commonplace inclusion of computers in consumer products to enable them to communicate information to and from remote systems via the internet. Also termed "*ubiquitous computing*." The devices themselves are sometimes termed IoT devices (for Internet of Things devices). An example is a smart metre replacement for a traditional electric power metre. *See also* Internet of Things.
PEST analysis	Political, Economic, Social, and Technological analysis. *See* Social, Technological, Economic, Environmental, and Political analysis.
PESTEL analysis	Political, Economic, Social, Technological, Environmental, and Legal analysis. *See* Social, Technological, Economic, Environmental, and Political analysis.
PESTLE analysis	Political, Economic, Social, Technological, Legal, and Environmental analysis. *See* Social, Technological, Economic, Environmental, and Political analysis.
PFD	*See* process flow diagram.
Phase 1–4 trial	*See* clinical trial.
Phase-gate model	*See* Stage-Gate® product development process.
Phase-gate process	*See* Stage-Gate® product development process.
Pillars of innovation	*See* principles of innovation.
Pilot	*See* field pilot.
Pilot plant	*See* field pilot.
Pilot test	*See* field pilot.
Pioneers	A category of potential customer or technology adopter in a psychological model of technology adoption. *See also* technology acceptance, technology acceptance model, technology readiness index, and technology adoption life cycle.
Piping and instrumentation diagram	(P&ID) A diagram for a chemical manufacturing process that shows how the processing equipment and control instrumentation are interconnected. *See also* field pilot and process flow diagram.
P-KIBS	Professional KIBS. *See* knowledge-intensive business services.

Plant breeders' rights	A grant of exclusive commercial rights on the production and sale of the seeds of an original plant variety by a government. The plant variety has to be uniform, stable, and different from other varieties. Although commercial rights are protected, this does not extend to the production and/or growth of the variety for personal use. *See* invention and intellectual property.
Plant demonstration	*See* demonstration.
Plant patent	*See* patent.
Plant pilot	*See* field pilot.
Plant test	*See* field pilot.
Plant material testing agreement	(pMTA) *See* material testing agreement.
Platform innovation	When a new core technology, based on a technological breakthrough, leads to multiple new products, processes, or services. Example: the technological breakthrough of electronic imaging enabled an entire generation of digital camera products to be introduced whose core technology was very different from the previous celluloid film technology. It is possible to have a platform technology and create modular and/or architectural innovations within it, as long as the core technological breakthrough remains. Example: in digital data recording, magnetic tapes and a succession of smaller and smaller-sized floppy disks involved different materials and components but were all based on the same breakthrough core technology (magnetic-head recording). *See also* Henderson-Clark model.
Platform leader	*See* business ecosystem.
PLC	Product lifecycle. *See* S-curve.
pMTA	Plant material testing agreement. *See* material testing agreement.
Polytechnic	An institute of technology that provides advanced education (and degrees) and/or vocational training (and diplomas). Some polytechnics also conduct advanced scientific and engineering research, and some conduct applied research.
Political economy	*See* knowledge-based economy.
Political innovation	New government policies, programmes, or legislation that help achieve social objectives. These could include improving service delivery or efficiency, or both. *See also* innovation and organizational innovation.
Pontin's First Rule of Innovation	"Any sufficiently radical invention seems ridiculous to most people when they first encounter it" [75].
Pontin's Second Rule of Innovation	"The first attempt to commercialize an invention almost never succeeds" [75].
Porter, Michael Eugene (1947–present)	An American economist and business strategist best known to the innovation world for his work in business competitiveness at the industrial and national levels. Two of his most famous books are probably *Competitive Strategy* (1980) and *The Competitive Advantage of Nations* (1990). *See also* Reference [749].
Positive cash flow	*See* cash flow.
Post-academic science	*See* modes of science.

Post-industrial age	*See* technological ages.
Post-modern age	*See* technological ages.
Practice	*See* technology.
Pragmatist	*See* early majority.
Pre-venture capital	Intermediate-stage, usually moderate-level capital for a commercialization process (frequently termed *start-up capital* for a new young company). The capital is typically raised through informal or formal investors, internal organizational funds, or government grants or loans. Pre-venture capital is sometimes enough to develop a technology from an engineering prototype or production prototype to the *qualified production prototype* stage of product development, possibly even limited production and a first introduction to the marketplace. By this point, the market analysis and commercialization plan have probably evolved into a *market strategy* and *business plan*. *See also* capital, sweat equity, seed capital, start-up capital, and venture capital.
Principles of innovation	Any of a wide range of authors' selections (especially in the business and government policy literature) of key philosophies, attitudes, environments, or practices that are advocated as being conducive to encouraging, accelerating, or improving the success rates in innovation. There does not seem to be a single generally accepted set of such principles. Also termed "*pillars of innovation*" or "*tenets of innovation.*" *See also* innovation, research and development, activities, other innovation activities, and ten Cs.
Prior art	All of the publicly available knowledge related to an invention that existed in the world prior to the date that the invention was made (or, depending on a given country's patent laws, prior to the date of filing of a patent application on the invention). This includes the knowledge represented in previous patents, but also that contained in previous patent applications, publications, and sometimes even presentations. Prior art is one of the things assessed in judging the patentability of an invention. *See also* invention and patent.
Process	*See* technology.
Process benchmarking	*See* benchmarking.
Process flow diagram	(PFD) A flowsheet diagram for a chemical manufacturing process that shows the principal processing equipment and the process flows that are intended to occur. *See also* field pilot, piping, and instrumentation diagram.
Process innovation	*See innovation.*
Product	*See* technology.
Product and cycle-time excellence process	(PACE process) A gated process and management system for the development of new products, processes, or services, developed by Michael McGrath. Reference [750]. *See also* idea-to-launch process, Stage-Gate® product development process.
Product champion	One of several kinds of stereotypical people that can play a critical role in the technological innovation process. A product champion is someone that can champion the innovation process throughout its stages of development with a clear goal of producing a successfully launched new product, process, or service in the marketplace. Regardless of the kind of organization, this person has to be able to guide the project through the "*valley of death*" and all other obstacles in its path. Having a product

champion can greatly increase the success rate. *See also* Rainmaker, leading-edge customer, and technological gatekeeper.

Product definition — A basic summary of a new product, process, or service including a description of what it is, a preliminary design, and a description of how it would be used. *See also* business concept and prototype.

Product development funnel — Regardless of an organization's innovation development model, there will be a stage at which a number of potentially good ideas have been identified, but this number needs to be reduced in order to devote the most resources to the ideas with the best chance of being successful. Many organizations use the idea of a product development funnel, or hopper, in which many new ideas are fed in and there is a process for deciding which and how many of them should be further explored or developed. This can be done over several stages, rather than just picking one from the funnel, and an organization may develop specific criteria for assessing the ideas at any given stage and deciding which ones to advance and which to either discard or move back to the front-end of the funnel. Ultimately, only a few, or even just one (in a small organization), may be selected for final development and commercialization. Also termed *"product development hopper," "product development pipeline," "innovation funnel," or "idea funnel."* *See also* Stage-Gate® product development process and technology Stage-Gate process.

Product development hopper — *See* product development funnel.

Product development pipeline — *See* product development funnel.

Product development process — *See* new product development process.

Product development strategy — *See* Ansoff matrix.

Product diversification strategy — *See* Ansoff matrix.

Product innovation — *See* innovation.

Productivity paradox — *See* innovation process paradoxes (innovation process paradoxes).

Production prototype — A full-scale, completely operational, model of a product that has been built to mirror a mass-produced unit, but which has been custom built as a one-off or in very limited quantity. A production prototype is used to demonstrate performance requirements, including operation, safety, and durability. Also referred-to as a design model. *See also* prototype. Reference [76].

Production technology — *See* technology.

Product lifecycle curve — *See* S-curve.

Product platform — A group of components (hardware and/or software) that form the core of a range (family) of products. This enables a range of customized products to be developed for different markets and/or customer segments ("mass customization"). It also enables more sophisticated, higher-end products, termed *"augmented products,"* to be built on the core product platform. Example: the printer-cartridge platform underlying Hewlett Packard ink-jet printers.

Product qualification prototype — *See* qualified production prototype.

Product technology — *See* technology.

Professional ladder — *See* technical ladder.

Profit forecast — *See* Awareness, Trial, Availability, Repeat model.

Pro forma — A Latin phrase, which in the innovation world is usually used with the meaning "as a matter of form." It is used as an adjective or noun to describe a standardized type of document. For example, a "*pro forma* report," or a "*pro forma* financial statement." In the financial sector, "*pro forma* statements" often refer to corporate reporting that emphasizes current and/or projected figures, and as such, may not comply with Generally Accepted Accounting Principles (GAAP).

Project SAPPHO — In the 1960s and 1970s, an investigation named "Project SAPPHO" attempted to discover the differences between successful and unsuccessful innovations by comparing pairs of successful and unsuccessful technological innovation attempts for which both members of each pair were aimed at the same market. SAPPHO stands for Scientific Activity Predictor from Patterns with Heuristic Origins.

Proof — *See* demonstration, proof of concept, and prototype.

Proof of concept — A physical demonstration that a technology concept can be made to work and is no longer a speculative concept. At this stage of development, the technology does not have to be practical, efficient, or cost effective. In terms of technology readiness levels, a demonstrated proof of concept satisfies TRL level 3. *See* Figure 2.1. *See also* speculative concept and technology readiness level.

Prosperity gap — The difference in gross domestic product per capita (GDP per capita) between a country or region and some specified peer or peers. *See also* gross domestic product.

Prototype — In the technical development of a product, process, or service, various kinds of prototypes, or models, are used for illustration, demonstration, and/or to provide technical data to aid in the process. A typical sequence includes a mockup, working model, engineering prototype, production prototype, and qualified production prototype. *See* Figure 2.1. References [76, 186]. *See also* demonstration, scale-up, and "works like" model.

Provisional patent application — A patent application that has been specifically filed as a "provisional application for patent" with the USPTO. Such an application allows filing relatively quickly and inexpensively, and without making specific patent claims, to establish an early filing date and to enable one to use the term "patent pending." However, a provisional patent application is only valid for 1 year, within which time one must either abandon the application or else file a full, non-provisional patent application. *See also* invention, intellectual property, patent, and patentable invention.

Proxy indicators — *See* innovation indicators.

Public disclosure — Any communication of information or technology to a person or people that have not agreed to keep the information confidential. Under some laws, public disclosure must be in writing whereas under others, it includes oral discussions or presentations, and/or "showing" or demonstrating a technology. Where patentability is concerned, public disclosure generally has to be substantial enough that it teaches or enables a party to duplicate an invention. *See also* non-disclosure agreement.

Purchasing unit — *See* Awareness, Trial, Availability, Repeat model.

Pure research — *See* research and development.

11.18 Q

Quad-helix model — An innovation ecosystem (sociological) model developed by Etzkowitz and Leydesdorff and others as an expansion of the triple-helix model to include intermediary organizations, hence: governments, universities, intermediaries, and industry. In the quad-helix model, intermediary organizations would be the primary incubators for many technology-based start-ups and most SMEs. Also termed "*quadruple-helix model*." References [264, 266]. *See* Figure 5.5. *See also* innovation ecosystem models, innovation ecosystem, triple-helix model, N-tuple helix model, and quad model.

Quad innovation models — The addition of a fourth kind of institution to any of the three-component innovation models can improve their sophistication and explanatory power. *See* quad model and quad-helix model. *See* Figure 5.5. *See also* innovation ecosystem models and triple-helix model.

Quad model — An innovation ecosystem (sociological) model developed by Wilson in about 2003. The quad model is like a "Holy Trinity" model to which a fourth segment has been added, comprising resourceful organizations that can act as catalysts, intermediaries, and/or entrepreneurs (hence the term "quad leaders"). The quad model's four kinds of institutions are: governments, research institutions (universities and RTOs), non-government organizations (NGOs), and industry. Another "*quad*" adaptation of the Holy Trinity model is: governments, universities, intermediaries, and industry. References [264, 265]. *See also* innovation ecosystem models, innovation ecosystem, triple-helix model, quad-helix model, and N-tuple helix model.

Quadruple-helix model — *See* quad-helix model.

Qualified production prototype — A full-scale, completely operational sample of a product that has been built using a limited production run to demonstrate that it meets design standards. Qualified production prototypes are also used to demonstrate that they meet industry or regulatory standards. Also termed "*product qualification prototype*." *See also* prototype. Reference [76].

Quick to fail — *See* fail fast.

11.19 R

Radical innovation — *See* disruptive innovation.

Radical technological transition — When the basic technology in a product, process, or service radically changes both the components in a product and the linkages between

	those components (i.e. the basic architecture). Examples: the technological change in ocean-going ships from sails to steam engines and the technological change in power shovels from steam to gasoline engines. *See also* disruptive innovation and Henderson-Clark model.
Rainmaker	One of several kinds of stereotypical people that can play a critical role in the technological innovation process. A Rainmaker is someone with highly developed creative abilities, who is good at problem identification and problem solving, and who can serve as a catalyst within a team. Also termed "*wizard*." Such a person could be engaged at any point in a new product, process, or service development process (not just at the new idea and invention stages) and can greatly increase the success rate. The Rainmaker Index refers to a "creative potential" scale that is based on the Myers Briggs Type Indicator (MBTI) personality-type indicator system and that highly ranks people who score highly in the personality preferences intuitive (N) and thinking (T) in the MBTI system. *See also* leading-edge customer, product champion, and technological gatekeeper.
Rainmaker Index	*See* Rainmaker.
R&D	*See* research and development.
R&D alliance	*See* ecosystem innovation.
R&D expenditure	*See* government R&D expenditures, gross domestic expenditure on R&D, and gross national expenditure on R&D.
R&D intensity	(Innovation indicator) An indicator of the R&D investments from all sectors in an economy is the ratio of gross domestic expenditure on R&D to gross domestic product (GERD/GDP). *See* innovation indicators and Table 8.2.
R&D personnel	(Innovation indicator) An indicator of the total number of people directly involved in R&D in an economy is the number of researchers involved in R&D per 10,000 personnel in the labour force. *See* innovation indicators and Tables 8.2–8.4.
R&D system	Research and development system. *See* innovation ecosystem.
Rate of replication	*See* technological capacity.
Rate of return	*See* return on investment.
Rate of technological progress	*See* technological capacity.
Rate of technology dissemination	*See* technological capacity.
Rationalist strategy	One of two traditional approaches to strategy and strategic planning: "rationalist" and "incremental." In essence, the rationalist approach consists of attempting to scan and understand the surrounding environment, develop a strategy taking the environmental scan into account, and implementing the strategy. This approach is similar to classical military strategy and builds upon the classical organizational SWOT (strengths, weaknesses, opportunities, and threats) approach to planning. In contrast, the incremental approach consists of developing and implementing tactics intended to take the organization in the desired general strategic direction, evaluate their effectiveness, and adjust and repeat as necessary. This kind of approach has also been referred to as "trial and error" strategy.
RCI	*See* resource-constrained innovation.

Reach	(Innovation) The extent of individuals, groups, or markets that are made aware of and/or connected to product, process, service, or programme outputs. Distinctions are sometimes drawn between potential customers for these outputs (i.e. primary targets or primary customers) and potential partners, suppliers, or agents. *See also* activities, inputs, outcomes, outputs, metrics, and impacts.
Recombinant innovation	Innovation resulting from the adaptation of previous ideas or concepts. An example is when an idea is adopted from one product, process, service, sector, or country, then adapted and deployed in another. *See also* adopt and adapt.
Red giants	*See* innovation performance mapping.
Reduction to practice	The translation of a new idea, discovery, or invention into a demonstrable practical application. Having achieved a reduction to practice is one of the differences between an invention and a patentable invention. *See also* invention and working model.
Regional innovation cluster	*See* cluster (innovation cluster), innovative regional cluster.
Regional innovation hub	*See* cluster (innovation cluster) and innovative regional cluster.
Regional innovation system	*See* innovation ecosystem and regional innovation system types.
Regional innovation system types	The character of regional innovation systems varies widely depending on the nature of the region and the depth of its pool of public and private innovation resources. A simplified taxonomy for regional innovation systems is illustrated in Table 5.1. Reference [253].
Re-invention	(Business) Changing a company's business strategy, business model, products, or marketing approach in order to maintain or increase its competitiveness. (Customer) The degree to which an innovation is changed or modified by a user in the process of its adoption and implementation. Examples: a genetically modified seed would be difficult for a customer to modify, but a car is relatively easy for a customer to modify.
Research and development	(R&D) Any or all of discovery research, applied research, and experimental development. *Discovery research* is experimental and/or theoretical investigation undertaken to acquire new knowledge and/or understanding of facts and/or phenomena – *without* any particular use or application in mind. Synonyms for discovery research include *basic research, fundamental research, pure research, upstream research,* and *upstream studies. Applied research* is experimental and/or theoretical investigation undertaken to acquire new knowledge and or understanding of facts and/or phenomena – but directed *with* a specific use or application in mind. Synonyms for applied research include *mission-oriented research* and *strategic research. Experimental development* is systematic work, drawing on existing knowledge gained from research and/or practical experience, aimed at producing new or improved materials, products, processes, systems, or services. Synonyms for experimental development include *advanced technology development, engineering development, manufacturing development*, or simply *development. See* Figure 2.1 and Table 2.1. *See also* Frascati manual, activities (innovation), modes of science, whole product R&D. Reference [485].

Research and development system — *See* innovation ecosystem.

Research and technology organization — (RTO) A government-owned corporation/agency or private not-for-profit company that is primarily focussed on developing and deploying practical technologies that address commercial marketplace problems or opportunities, and spanning multiple sectors of an economy. RTOs differ from academia, mainstream government, for-profit companies, and even from mainstream not-for-profit companies. They have "public good" missions and are not primarily profit-driven. They work in the public interest but on market-pull issues, generally using a business-like approach, and frequently with a secondary, socio-environmental agenda. One of the most important functions of an RTO is to help business enterprises access, absorb, adapt, deploy, and exploit new technologies in order to enhance businesses' ability to innovate and therefore their competitiveness and sustainability. Also termed "*applied research organization*." There are two principal associations of RTOs, the World Association of Industrial and Technological Research Organizations (WAITRO) and the European Association of Research and Technology Organizations (EARTO) [296, 297].

Research and technology park — Real estate developments that provide infrastructure and a "like" environment for organizations with a strong interest in research, development, and/or commercialization of technology. Generally located adjacent, or in close proximity, to a major university Business Accelerator campus, research and technology parks generally comprise units of the local university itself, RTOs, industry, government agencies, SMEs, and even start-up companies. Some research and technology parks provide business services and even business incubation programmes. Also termed "*innovation park*," "*research park*," "*science park*," or "*technology park*." *See also* business incubator.

Research park — *See* research and technology park.

Resource-constrained innovation — (RCI) Innovation (usually technological product innovation) developed in and for emerging economies in a context characterized by financial and technological resource constraints. Also referred to as "*innovation at the bottom of the pyramid*." *See also* reverse innovation.

Responsible innovation — *See* sustainable innovation.

Return — *See* return on investment.

Return on assets — *See* return on investment.

Return on capital employed — *See* return on investment.

Return on equity — *See* return on investment.

Return on innovation — (ROInn) A measure of the financial return on investment in innovation for one or more (usually specified) innovations, within a specified time period. Generally, a simple return on innovation would be the revenues derived from sales of a new product, process, or service less the costs associated with manufacturing, marketing, selling, and shipping, or providing, with the difference being divided by the costs associated with ideating, developing, and commercializing the innovation. Other measures could involve market results such as market share or new market penetration. Also termed *"return on innovation investment"* (ROI2 or ROII). *See also* return on investment.

Return on innovation investment — (ROI2) *See* return on innovation.

Return on investment — (ROI) An indicator of the effectiveness of an investment, given by the formula:
ROI = (Gain from the investment − Cost of the investment)/(Cost of the investment)
The ROI is usually expressed either as the simple quotient above or as a percentage. It is also sometimes used to compare the effectiveness of several different investments. Unless the investment is sold, then the gain will refer to a specific time period, which should be clearly stated. For a given investment, the actual ROI value can be quite sensitive to the choices one makes in what to include in the gain and what to include in the costs, so the best practice is to include the complete basis for the calculation. ROI is sometimes termed *holding period return*, *rate of return*, *return*, or *yield*. A common time period is 1 year, for which the ROI is sometimes termed the *annual return*. Finally, as an indicator of investment performance, ROI is different from return indicators based on organizational size or operations, such as *return on equity*, *return on assets*, or *return on capital employed*. *See also* return on innovation.

Reverse engineering — The process of determining how a product, process, or service works in order to design a competitive product. Example: by reverse engineering the IBM PC, competitors were able to develop and produce similar products that could be sold at lower prices. *See also* fast follower.

Reverse innovation — The process by which innovations that have been developed to meet the needs of customers in developing nations are adapted (sometimes mostly by repackaging) for marketing and sales as low-cost innovations in developed nations. These are usually products rather than processes or services. The aspects developed in the developing nation(s) could be any or all of the product idea conception, the product research and development, the product design, or the initial market targeting. Example: GE's portable electronic medical instruments. Reverse innovation, in this sense, is as opposed to *Forward Innovation*, meaning the practice of starting with products that have been developed for developed nations and then adapting them (usually by removing features and costs) to produce innovations for marketing and sales in undeveloped nations. The first phase of reverse innovation is sometimes called local innovation. References [153, 154, 155]. Also termed *"blowback innovation," "cost innovation,"* or *"trickle-up innovation."* See also innovation at the bottom of the pyramid.

Revolutionary innovation — *See* disruptive innovation.

Richard's curve — *See* S-curve.

RIS — *See* regional innovation systems and regional innovation system types.

Rogers diffusion of innovation model — *See* Rogers.

Rogers diffusion of technology model — *See* Rogers.

Rogers, Everett M. (1931–2004) — An American sociologist and author best known in the innovation field for his theory and book on the diffusion of technologies in the marketplace (i. e. the diffusion of technological innovations). He used a normal distribution (i.e. bell curve) to describe the progress of technology diffusion by different adopters, which he characterized as *innovators* (the first few adopters), *early adopters* (the next few adopters), *early majority, late majority*, and *laggards*. Rogers' "*diffusion of innovation model*" is sometimes referred to as the innovation (or technology) diffusion model or the diffusion of innovations (or technologies) model. Reference [77]. *See* Figure 4.8. *See also* technology adoption life cycle, technology dissemination, and technology readiness.

ROI — *See* return on investment.

ROI2 — Return on innovation investment. *See* return on innovation.

ROII — Return on innovation investment. *See* return on innovation.

ROInn — *See* return on innovation.

Routine innovation — *See* sub-invention.

Royalties — Payments made by a licensee for the right to use licensed intellectual property. *See* license agreement and royalty-free license.

Royalty-free license — An intellectual property license requiring no royalty payments. Such licenses are sometimes granted for free, sometimes as part of a bundle with another product or service, and sometimes following an up-front, one-time payment. Sometimes referred to as a *paid-up license. See also* cross-license, license agreement, and licensing, royalties.

RTO — *See* research and technology organization.

11.20 S

Sales — A company's sales function builds on the stage set by marketing and on the strategies developed by business development, to create exchanges of value for a company's products, processes, or services. The sales function helps identify, create, and maintain relationships with clients, customers, and/or users in order to maximize sales. Sales also includes the closing, or securing, of sales deals, which may be through exchanges of value or the signing of contracts. The process of identifying and nurturing customer relationships is usually conducted in partnership with the business development function. *See also* marketing, business development, and distribution.

Sales and distribution — This includes the sale process and sales force, and the distribution channels related to a product, process, or service. *See also* market management and market research and planning.

Sales forecast — *See* Awareness, Trial, Availability, Repeat model.

Sandbox — Traditionally, this is a separate, usually multi-disciplinary, team in an organization that is charged with the task of exploring new ideas. In some cases, a short-list of ideas is handed off to other parts of the organization, while in others, select new ideas are developed as far as the prototype stage. The term "sandbox" comes from the imagery of people coming together, in a specific place or environment, to "play" with new toys. The sandbox concept has also been applied to linkages among organizations by which people can interact and collaborate on new, pre-commercial, ideas for products, processes, or services. Also termed *"innovation sandbox."* Another connotation of the term innovation sandbox is any idea generation and nurturing activity that involves free-form exploration and experimentation. Reference [751].

SAPPHO — *See* Project SAPPHO.

Saturation curve — *See* S-curve.

Scale-up — Once a prospective new product or process has been shown to work at the laboratory-scale, or as a small-scale working model, a next step in its development is to design and test ("scale-up") a full-scale or intermediate-scale model. Depending on the nature of the product or process, a working model at larger scale might need to look and behave quite differently from the original. For example, in the case of a mechanical device, some components might have to be made with different dimensions or from different materials in order to function properly at larger scale. In the case of a chemical process, the nature and geometry of the vessels, mixers, and other equipment might have to be quite different from their smaller-scale counterparts in order to properly scale-up critical process variables, such as shear and residence time, for example. *See also* field pilot and prototype.

Sceptic — *See* Laggards.

Schmookler, Jacob (1917–1967) — An American economist known to the innovation world for his work studies of significant practical inventions and their adoption by industry. His work led him to identify (in 1966) the role of market demand (*"demand-pull"*) as a driver for the invention process and what is now known as the *"market pull"* model[122] of technological innovation. He argued that although technological progress had been previously treated as an exogenous variable in economic models, it is actually primarily an endogenous variable and primarily driven by market demand. References [18, 100]. *See also* linear innovation models.

Schmookler innovation — In modern usage, this refers to market-pull innovation, although Schmookler's actual work emphasized the role of economic conditions in driving the creation of practical, patented inventions. Schmookler's work covered the period from the Industrial Revolution to about 1950. A follow-up evaluation by Scherer in the 1970s, which covered more recent technological advances in industry, reached the same broad conclusion. *See* market-pull innovation. References [18, 100].

Schumpeterian Growth Theory — Schumpeterian innovation. *See* innovation (20th–21st centuries) and Schumpeter Mark I innovation.

Schumpeterian innovation — *See* innovation (20th–21st centuries) and Schumpeter Mark I innovation.

Schumpeter innovation — *See* innovation (20th–21st centuries) and Schumpeter Mark I innovation.

122 The concept of market-pull seems to also have originated from the 1950s work of Carter and Williams [22, 23].

Schumpeter, Joseph Alois (1883–1950) — An Austrian economist and political scientist who is possibly best known for his theory of creative destruction as a necessary component of sustainable and/or growing economies. He also coined and/or championed the concepts of entrepreneurship and technological innovation as the agents of creative destruction. It also seems to have been Schumpeter that coined the term "Kondratieff waves." Reference [15, 16]. *See also* Kondratieff waves.

Schumpeter Mark I innovation — Schumpeter's original description of economic growth involved "creative destruction" in the marketplace driven by the introduction into the marketplace of "game-changing" new products, processes, or services by entrepreneurs or inherently entrepreneurial companies. The creative destruction involved the innovative companies succeeding over and/or replacing previous dominant companies that did not innovate. This has been termed Schumpeter Mark I innovation. In later work, Schumpeter recognized another mode of innovation, in which dominating companies could maintain their competitive position by themselves introducing "game-changing" new products, processes, or services into the marketplace. This has been termed Schumpeter Mark II innovation. Such innovation-based activities by already dominating (usually mature and large) companies have sometimes been referred to as "creative accumulation," or "creative agglomeration," referring to the maintenance of competitive position by building an evolving and (usually) broadening portfolio of innovative product, processes, or services. From a broad market perspective, the Mark I and II innovation pathways are usually viewed as being complementary. *See also* innovation (20th–21st centuries). References [15, 16].

Schumpeter Mark II innovation — *See* Schumpeter Mark I innovation.

Science park — *See* research and technology park.

Science-push innovation — *See* linear innovation models.

Scientific and engineering research — Scientific research is the systematic investigation of hypotheses and theories (proposed explanations of things based on available facts and understandings) with the goal of generating new knowledge and understanding of things and phenomena. Engineering research is conducted somewhat similarly, but with the goal of generating new knowledge and understanding of how to make and/or control things. Some descriptions of technological results from scientific and engineering research are given in Table 2.1. *See also* research and development.

Scientific revolution — *See* technological ages.

Scientometrics — The use of bibliometrics, that is, the study of the published literature, to measure the quality and impacts of scientific research.

S-curve — **Science.** The first technological application of the S-curve concept appears to have been in 1845 by Pierre François Verhulst, who used it to illustrate stages in the development of yeast fungi colonies: a phase of initial growth and achievement of critical mass, a phase of rapid (approximately exponential) growth, a plateau-like phase of maturity, and a phase of decline due to depletion of the system's resources. Since then, S-curves have been used to describe many other kinds of phenomena, sometimes involving four phases as in the example just given and sometimes with only the first three phases (i.e. without the decline phase). Synonyms include growth curve, logistic curve, saturation curve, sigmoid curve, and Verhulst curve. *See also* Pearl curve and Gompertz curve.

Technology. With a slight re-statement of the phases, S-curves can be used to describe the development and lifecycles of new technologies. The technology S-curve concept was introduced by Richard Foster in 1986 to help R&D managers. It involves a linear model of technology development, with overlays to illustrate how technologies get displaced by new ones. The typical phases are: (1) a phase of hypotheses, experimentation, addressing fundamental issues, and trial and error; (2) a phase of rapid maturing of the new technology as the fundamental issues become resolved, new approach(es) begin to take shape, individual advances begin to cluster, and/or a breakthrough occurs; and (3) a plateau as the technology matures and physical limitations are reached. The vertical axis of a technology S-curve is usually technology maturity and/or performance, while the horizontal axis is usually some kind of measure of cumulative research and development effort over time, summarized as R&D effort, investment, or time (*see* Figure 3.5). Synonyms include saturation curve, sigmoid curve, innovation continuum, technology growth curve, technology lifecycle, technology S-curve, Foster's curve, blindside curve, and Richard's curve. *See also* Pearl Curve, Gompertz Curve, and technology readiness. References [129, 131, 133].

Innovation. With another re-statement of the phases, S-curves can be used to describe degrees of innovation and their impact on an economy. Figure 3.7 provides an illustration comparing incremental, evolutionary, and disruptive innovation. Also termed "*innovation continuum*" (degree of innovation).

Technology diffusion. A way of mapping the progress of technology diffusion in the marketplace. If technology adoption statistics are plotted cumulatively, then an S-curve usually results. Such a representation demonstrates the plateauing of technology adoption by the time a technology is adopted by what Rogers termed the laggards. *See also* technology adoption life cycle and technology readiness.

Product, process, or service life cycle. Again, with a slight re-statement of the phases, S-curves can be used to describe the life cycle of a product, process, or service. For a new product, the typical phases are (1) the introduction of a new product to the marketplace, with associated business development and early sales; (2) a phase of rapid sales growth as the product finds a successful market niche; (3) a sales plateau as the market niche becomes saturated; and (4) a phase of declining sales as the market-pull declines and/or competing products displace it in the marketplace. Each of the above phases can be considered in terms of factors supporting the product's development and factors acting against the product's development. Synonyms include saturation curve, sigmoid curve, logistic curve, Foster's curve, blindside curve, and product life cycle (PLC) curve. References [129, 131, 133]. *See also* J-curve, Pearl curve, technology hype-cycle, buying hierarchy.

In both the technology and product cases, the displacement of older technologies by newer ones, or of older products by newer ones, can be represented by families of S-curves. There is generally a gap, or transition phase, called a discontinuity (or technological discontinuity) between the plateauing of one S-curve and the rapid rise of a successive S-curve. Following this period of discontinuity, the newer technology or product has overtaken the earlier one. Within a given type of product, process, or service, a family of two or more S-curves, taken together, is sometimes referred to as an envelope S-curve. *See* Figure 3.10.

Second academic revolution — *See* second mission and entrepreneurial university.

Second generation innovation model — Any of the linear and sequential technological innovation models that begin from an identified market need. These are often referred to as being "market-pull" or "demand-side" models. Although they have some value in terms of distinguishing broad driving forces in innovation, the processes themselves are very often neither sequential nor linear. *See* Section 2.1 and Table 4.2. *See also* generations of innovation, linear innovation models, and market-pull model.

Second-generation nanotechnology — Molecular nanotechnology. *See* generations of nanotechnology.

Second generation of technology foresight — *See* generations of technology foresight.

Second Industrial Revolution — *See* technological ages.

Second mission — The second mission of a university is research. The first mission is teaching. The evolution of universities to have a second mission has been referred to as the result of the "first academic revolution" from the teaching university. The third mission of an "*entrepreneurial university*" involves participation by universities in working with governments and industry to enable economic growth (via innovation) and social progress. *See also* entrepreneurial university, modes of science, innovation ecosystem models, triple-helix model, and quad-helix model.

Second to market — *See* fast follower.

Second wave — *See* 2nd wave.

Seed capital — Early-stage, usually modest capital for a company and/or commercialization process that is typically raised through informal investors, internal organizational funds, or government grants. In some cases, it comes from "*angel investors*" ("*business angels*"), family, friends, and entrepreneurs (sometimes termed "*friends, family, and fools*," or FFF capital). Seed capital is sometimes enough to develop a technology from a mockup or working model stage to the engineering prototype or production prototype stage of product development and to develop a formal market analysis and a formal commercialization plan. *See also* capital, crowdfunding, sweat equity, pre-venture capital, and venture capital.

Seismic innovation — *See* disruptive innovation.

Self-determined innovation — *See* Indigenous innovation.

Selling, general, and administrative expenditures — (SG&A expenditures). *See* capital expense.

Sell-off	A synonym for spin-off. *See* spin-out.
Service innovation	*See* innovation.
Service mark	*See* trademark.
Seventh wave	*See* 7th wave.
SG&A expenditures	Selling, general, and administrative expenditures. *See* capital expense.
Shanzhai innovation	A reference to the development of technological products, such as electronics, for which the appearance and/or the internal designs are imitative of, or even copies of, another company's products. The new products may or may not be of lower quality or lower price than the ones being adapted, imitated, or copied. Also termed *"imitative innovation." See also* Indigenous innovation.
Shooting stars	*See* innovation performance mapping.
Short-term mindset trap	*See* innovation barrier.
Sigmoid curve	*See* S-curve.
Simultaneous discovery (or invention) theory	*See* multiple discovery theory.
Single niche strategy	A technology deployment strategy involving the introduction of a new product, process, or service into a single market niche in order to achieve a greater competitive advantage than would have been possible if introduced into multiple market niches. Reference [737]. *See also* niche fusion.
SIN innovation model	Systems integration and networking model of innovation. *See* non-linear innovation models. *See also* generations of innovation.
Singularity	The point in time at which truly intelligent machines come into being. The original concept of Vernor Vinge referred more specifically to the point in time at which machines with greater than human intelligence come into being. Also called the technological singularity. Reference [535]. *See also* deep learning.
Sixth-generation innovation model	Any of the advanced, non-linear models like those of the fourth or fifth generation, but expanded to embrace the concept of "open innovation," with the possibilities of internal and/or external idea generation and sourcing, and internal and/or external pathways to market, and so on as advocated by Chesbrough in the early 2000s [154, 158, 160]. In addition, sixth-generation innovation is often considered to involve regional innovation clusters. *See* Sections 4.2 and 5.1 and Table 4.2. *See also* generations of innovation, non-linear innovation models, and open innovation.
Sixth wave	*See* 6th wave.
Skeptics	A category of potential customer or technology adopter in a psychological model of technology adoption. *See also* technology acceptance, technology acceptance model, technology readiness index, and technology adoption life cycle.
Skunkworks	Generally, the workings of a small team of people, often within an organization's research and development department, that are fairly unconstrained, unstructured, and aimed at testing and developing radical ideas. Some skunkworks are officially supported and may be well funded, whereas others are completely unofficial and may not have any explicit

funding, and in either case, they may be highly secret. Example: Lockheed Aircraft Corporation's World War II "*Skunk Works*" project that developed the XP-80 Shooting Star jet fighter in 1943.

SLEPT analysis — Social, Legal, Economic, Political, and Technological analysis. *See* Social, Technological, Economic, Environmental, and Political analysis.

Small- and/or medium-size business — (SMB) *See* small- and/or medium-sized enterprise.

Small- and/or medium-sized enterprise — (SME) Business enterprises that are smaller than a specified number of employees (e.g. 500), and/or have annual revenues of less than a specified value (e.g. $75 million), as distinguished from large-size enterprises. Some jurisdictions further distinguish between medium-, small-, and even micro-sized enterprises based on specific employee and/or revenue numbers. Example: in some usage, a micro-enterprise (or micro-entity) is one having up to 10 employees. Due to a combination of their large numbers and typically entrepreneurial natures, SMEs are often referred to in discussions of innovation and economic growth. Sometimes referred to as *micro-, small-, and/or medium-sized enterprise*, or *small- and/or medium-sized business (SMB)*.

Small-sized enterprise — (SME) *See* small- and/or medium-sized enterprise.

Smart city/cities — *See* Internet of Things.

Smart grid — *See* Internet of Things.

Smart manufacturing age — *See* technological ages.

Smart materials — Smart materials are those that sense and react to any of mechanical, thermal, chemical, electrical, and/or magnetic stimuli. Smart materials are sometimes subdivided into passive smart materials (sensors; sensing only), active smart materials (actuators; sensing and reacting), and very smart materials (sensing, reacting, and adapting). Examples include switchable surfactants, smart dyes, smart textiles, and self-healing materials. Also termed "*active materials*." The related term intelligent materials usually refers to smart materials that also have the ability to self-control or self-regulate. *See also* self-healing materials, smart dyes, and smart textiles. *See* Reference [752].

SMB — Small- and medium-size business. *See* small- and medium-size enterprise.

SmE — Small-sized enterprise. *See* small- and/or medium-sized enterprise.

SME — *See* small- and/or medium-sized enterprise.

Social innovation — The ideation, development, and deployment of new and improved solutions, processes, or practices within society in ways that have no connection to commercialization or the marketplace. These could take place in such areas as education, healthcare, or social services, for example. A more accurate term would be *social invention*. *See also* innovation.

Social invention — *See* social innovation.

Social internal rate of return on R&D — *See* social return on investment on R&D.

Social rate of return on R&D — *See* social return on investment on R&D.

Social return on investment in R&D — (SROI on R&D) An indicator of the net benefits to society of investment in applied research and development. The SROI on R&D is usually expressed

	as the increased value produced for society associated with the investment divided by the amount invested. The value component needs to be specified; it can comprise such things as social value, economic value, environmental value, or a combination of these. In the case of social or environmental value, some kind of financial proxy is usually developed in order to be able to complete the calculation. Example: one form of SROI on R&D is net incremental industrial productivity achieved within a specific time period divided by investment in applied R&D over the same time period. Also termed *"social rate of return on R&D" or "social internal rate of return on R&D."*
Social, Technological, Economic, Environmental, and Political analysis	(STEEP analysis) A summary of environments and factors that could be assessed and considered as part of making business decisions, including those regarding technology commercialization and product, process, or service opportunities. Additional factors that are sometimes considered include legal, ethical, demographic, ecological, and/or regulatory. This has led to a host of acronyms including LEPEST, PEST, PESTEL, PESTLE, SLEPT, STEEPLE, STEEPLED, STEER and STEP. *See also* competitive intelligence and STEEPV analysis.
Social, Technological, Economic, Ecological, Political, and Values-Based Analysis	*See* STEEPV analysis.
Social technology	A technology category comprising practical knowledge, such as the knowledge of how to construct something. Such knowledge can be intangible, such as knowledge that has to be transferred by demonstration and observation (tacit knowledge), or tangible, such as any form of codified knowledge. Within tacit knowledge are sometimes distinguished *"technical tacit knowledge"* (how to do things) and *"cognitive tacit knowledge"* (facts, structures, models, and beliefs) [737]. A process could be either a material or a social technology. *See also* knowledge, technology, and material technology.
Socio-institutional innovation	The ideation, development, and deployment of new and improved internal processes within a community organization. Such new organizational processes should have some kind of efficiency or productivity benefit, even if they are internal and have little or no connection to technological innovation. Example: producer co-operatives. This is a form of non-commercial innovation. *See* innovation (non-commercial). Reference [8].
Soft innovation	Innovation that involves form, i.e. aesthetics, rather than function. This is a legitimate form of innovation as long as it results in a commercially successful product, process, or service.
Solow computer paradox	The risk, or "paradox," that increased investment in information technology could cause labour productivity to decrease instead of increase. In the early 1970s, many organizations began investing in computer systems with the expectation that they would enable labour productivity improvements (or even eliminate labour completely in some areas). However, although computers and information technology enabled many

	companies to achieve competitive advantage and market share, the desired labour productivity increases were largely unrealized, and in some cases, decreases were experienced. This is an example of the *"productivity paradox," see* innovation process paradoxes (innovation process paradoxes). *See also* Solow, Robert.
Solow Growth Model	*See* Solow-Swan Growth Model.
Solow, Robert (Merton) (1924–present)	An American economist known for his work in macroeconomics, and especially for his theory of economic growth (also termed the *"Neoclassical Growth Model"*). In 1987, he won the Nobel Prize in Economic Sciences for his analysis of economic growth. Solow's model of economic growth is termed the "Solow Growth Model" or the "Solow-Swan Growth Model" (having been independently developed and published in the same year – 1956 – by both Solow and Trevor W. Swan). It is also referred to as the "Solow-Swan Neoclassical Growth Model." The Solow-Swan model separates the contributions to economic growth into increases in inputs (labour and capital) and technological progress, and predicts that sustainable economic progress requires labour – enhancing technological progress in order to increase output without needing more labour or capital. In his 1957 paper, Solow calculated that about four-fifths of the growth in U.S. output per worker was attributable to such technological progress. Reference [35]. *See also* references [36, 37].
Solow-Swan Growth Model	An economic growth model. The Solow-Swan model separates the contributions to economic growth into increases in inputs (labour and capital) and technological progress, and predicts that sustainable economic progress requires labour – enhancing technological progress, in order to increase output without needing more labour or capital. In his 1957 paper, Solow calculated that about four-fifths of the growth in U.S. output per worker was attributable to such technological progress.[123] Also termed *"Solow Growth Model"*, or *"Solow-Swan Neoclassical Growth Model." See also* Solow and Endogenous Growth Theory. References [36, 37].
Solow-Swan Neoclassical Growth Model	*See* Solow-Swan Growth Model.
Specialist strategy	*See* market segmentation.
Spectator	*See* late majority.
Speculative concept	During the process of invention, various concepts will be developed that could link knowledge and/or discoveries to some kind of application. Such speculative concepts may or may not be feasible or even physically possible. At this stage of development, the technology does not yet have to be shown to be feasible. In terms of technology readiness levels, a documented speculative concept satisfies TRL level 2. *See also* proof of concept and technology readiness level.
Speed to market	*See* time to market.
Spider diagrams	*See* creative thinking models.
Spillover effect	*See* impact multiplier.

123 Also termed *"exogenous technological change."*

Spin-in | In a “spin-in,” a company creates a start-up initiative to develop and produce a kind of product, process, or service that the parent company needs or expects to need in the future. The parent company usually then assists with the nucleation and growth process. If the start-up succeeds and its products, process, or services still align with the parent's current or future needs, it is usually absorbed. If it succeeds but no longer aligns with the parent's strategy, it can be spun-off. *See also* spin-off and spin-out.

Spin-off | *See* spin-out.

Spin-out | In a “spin-out,” a company allows outside investors to acquire minority ownership of a segment of it business (such as a division, business unit, or subsidiary). Such a segment is usually not a part of the company's core business(es), but of sufficient strategic interest that it is not completely divested (in contrast to a spin-off). In a “spin-off,” a segment of a company (such as a division, business unit, or subsidiary) is completely divested and becomes either a separate company or a part of a separate company. Synonyms for spin-off include sell-off, split-off, and carve-out. *See also* spin-in.

Split-off | A synonym for spin-off. *See* spin-out.

SROI on R&D | *See* social return on investment in R&D.

Stage 1 economy | *See* competitiveness drivers.

Stage 2 economy | *See* competitiveness drivers.

Stage 3 economy | *See* competitiveness drivers.

Stage-Gate® product development process | A Stage-Gate® process is a project management tool by which a project progresses through a series of stages, each one of which is followed by a gate. Each gate triggers a review of accomplishments and decision as to whether the project may proceed to the next stage. The gate evaluations are normally completed with reference to a set of pre-determined success criteria and can involve many kinds of input. The tool can be used as an aid to objectively and consistently manage efficiency and risk. References [223, 224]. *See* Figure 4.13 for an illustration. Also termed: “*phase-gate model*,” “*phase-gate process*,” and “*technology stage-gate process*.” *See also* new product development process and Waterfall method.

Start-up capital | Intermediate-stage, usually moderate-level capital for a new young company (frequently termed pre-venture capital for a commercialization process). Sometimes termed “*A-Round Capital*.” The capital is typically raised through informal or formal investors, internal organizational funds, or government grants or loans. By this point, there will be a *market strategy* and *business plan*, but most of the development costs still lie ahead. Two stages are sometimes distinguished. *Early stage* (or formative stage) refers to companies beginning operations, and probably having a product or service in testing or pilot production, but which are not yet at the stage of commercial manufacturing and sales. *Later stage* refers to companies beginning commercial manufacturing and sales, but before any initial public offering (IPO). *See also* capital, crowdfunding, sweat equity, seed capital, pre-venture capital, and venture capital.

Start-up company | A new, or relatively new, company that is still in an early stage of development and may not yet have well-developed products, services,

	markets, or sales. For this reason, start-ups often need external financing. *See also* business accelerator and business incubator.
Stateless corporation	*See* multinational enterprise.
STEEP analysis	*See* Social, Technological, Economic, Environmental, and Political analysis.
STEEPLE analysis	Social, Technological, Economic, Environmental, Political, Legal, and Ethical analysis. *See* Social, Technological, Economic, Environmental, and Political analysis.
STEEPLED analysis	Social, Technological, Economic, Environmental, Political, Legal, Ethical, and Demographic analysis. *See* Social, Technological, Economic, Environmental, and Political analysis.
STEEPV analysis	A foresight evaluation process that involves considering possible future scenarios in terms of six points of view. The acronym STEEPV refers to social, technological, economic, ecological, political, and values-based points of view, and the idea is to evaluate a given situation or scenario from each of these six perspectives, as an aid to planning. *See also* Social, Technological, Economic, Environmental, and Political (STEEP) analysis, SWOT analysis, Delphi method.
STEER analysis	Social, Technological, Economic, Ecological, and Regulatory analysis. *See* Social, Technological, Economic, Environmental, and Political analysis.
STEP analysis	Social, Technological, Economic, and Political analysis. *See* Social, Technological, Economic, Environmental, and Political analysis.
Strengths, Weaknesses, Opportunities, and Threats	See SWOT analysis.
Strategic innovation	A kind of *organizational innovation*, strategic innovation refers to developing and implementing strategies for creating major and/or disruptive technological innovations (products/processes/services) that are intended to significantly improve an organization's performance versus its competitors. Ideally, such new strategies would create new value for the organization and for its customers, but it could be either. Such organizational innovations are probably the least common because they usually involve dramatic changes in how an organization operates. Accordingly, such changes have to be led, or at least supported, by CEOs. Example: the strategy change at General Electric under CEO Jack Welch in the United States in the 1980s. Another example is provided by Amazon.com, whose business strategy changed the breadth of product availability, distribution, and promotion, compared with traditional bookstores, while also reducing overhead costs and prices. Also termed "*business model innovation*." References [32, 182]. *See also* organizational innovation and innovation.
Strategic research	*See* research and development.
Strategy innovation	*See* strategic innovation.
Sub-invention	An older term for an invention that would have been obvious to one skilled in the art. A sub-invention is therefore not a patentable invention but nevertheless represents something new and potentially useful. Sub-inventions are often created by skilled practitioners in a field, as incremental improvements or adaptations of existing processes or things. This has also been referred to as "*routine innovation*." [18]. *See also* invention.

Success/failure paradox	*See* innovation paradoxes (innovation system paradoxes).
Success paradox	*See* innovation paradoxes (innovation system paradoxes).
Supercycles	*See* Kondratieff waves.
Superstructure organization	*See* intermediary organization.
Supply chain management	*See* value chain.
Supply-side driven	*See* linear innovation models.
Survival innovation	*See* incremental innovation.
Sustainability-related triple-helix	*See* triple-helix model.
Sustainable innovation	Innovation that improves an organization's sustainability performance (whether measured in terms of economic, environmental, or social aspects, but increasingly the latter two). Also termed *"sustainability-oriented innovation (SOI)"*. This is very similar to *"responsible innovation,"* which can be defined as innovation that takes into account social, environmental, and/or ethical values or aspects.
Sustaining innovation	*See* evolutionary innovation.
Swan, Trevor W	*See* Solow, Robert.
Sweat equity	Any unpaid work that is contributed by a person, group, or organization to the commercialization process. Sweat equity is sometimes enough to develop a technology to the *working model* stage of product development and to develop a high-level market analysis and a high-level commercialization plan. *See also* capital, seed capital, pre-venture capital, and venture capital.
SWOT analysis	An evaluation process for assessing a real or imagined, current or future situation or scenario and its potential impacts. The acronym SWOT refers to strengths, weaknesses, opportunities, and threats, and the idea is to evaluate a given situation or scenario from each of these four perspectives, as an aid to planning. For example, when faced with a new possible market situation, an organization might assess its own strengths and weaknesses in the context of that situation and consider the relative merits of avoiding or pursuing the potential opportunity or defending or fleeing from the potential threat. *See also* STEEPV and Delphi method.
Synectics	*See* creative thinking models.
Synthetic innovation	Technological innovation that involves using existing knowledge and/or ideas in new ways (i.e. without involving research).
System integration and networking model of innovation	*(SIN model) See* fifth-generation innovation model and non-linear innovation models. *See also* generations of innovation.

11.21 T

Tacit knowledge	*See* knowledge and social technology.
TAM	*See* technology acceptance model.
Tangible knowledge	*See* social technology.
Technical innovation	*See* technological innovation.
Technical intelligence	*See* competitive intelligence.

Technical ladder
: A career development pathway, available in some organizations, by which professionals can remain in a professional/technical career path as an alternative to the traditional management path. The idea is to motivate and retain experienced professional/technical staff that do not want to go into management by offering them additional career stages that either offer theoretically equivalent status and rewards or at least increasing status and rewards. The existence of two career development pathways in an organization is sometimes referred to as having a "dual ladder." Also termed "*professional ladder*" or "*individual contributor ladder.*"

Technical revolution
: *See* technological ages.

Technical tacit knowledge
: *See* social technology.

Technological ages
: Technological ages (or revolutions) represent periods of major technological, economic, and societal changes. At least four Industrial Revolutions have been broadly identified. *See also* Kondratieff waves and Figure 6.3.

The first Industrial Revolution (mid-1700s to mid-1800s). Some of the principal changes included transitions from wood to coal power and hand to machine production methods (particularly in the textile industry), the advent of chemical and iron manufacturing processes, then increasing use of steam power and the development of machine tools, and the transition to high-volume printing via the steam-powered printing press. The first Kondratieff wave lies within this period.

The second Industrial Revolution (mid-1800s to mid-1900s). Some of the principal changes included the transition from coal to oil power, the advent of electricity and the telegraph and telephone, mass production, the internal combustion engine, highways, and rapid transportation. Also known as the technological revolution. Most of the second and all of the third Kondratieff waves lie within this period.

The third Industrial Revolution and Post-Industrial Age (mid-1900s to present). Some of the principal changes have included transitions from analogue to digital technology (particularly digital computers and communications), the advent of the Internet, information technology, mass media, and globalization. Also known as the digital age, new media age, information age, or the post-modern age. Most of the fourth and all of the fifth Kondratieff waves lie within this period.

Several Post-Industrial Revolutions (ages or societies) have been identified, beginning with the Scientific Revolution (with key developments including electricity, chemistry, and the chemical industry) and the Technical Revolution (with key developments including automobiles, mass production, and the petrochemical industry), followed by the Information Age, the Knowledge Age, and the Creative Age. An *Information Society* creates and disseminates information. Although this was not new, the growth of ICT massively increased data production and dissemination, and the Internet has made global connectivity possible and rapid. However, the production and communication of information alone does not necessarily lead to knowledge creation. A *Knowledge Society* develops, processes, shares, and uses knowledge to improve economic, social, and/or environmental conditions. Knowledge societies are by nature also *learning societies*, and embrace the concept of lifelong learning. Reference [267]. A *creative society* comprises people that are developing their natural creative talents and energies, and combining the power of information, knowledge, and creativity into an important economic force. Reference [169].

The fourth Industrial Revolution (Industry 4.0) and smart manufacturing (future). Some of the principal changes being predicted for a future *"age"* include smart manufacturing, in which supply chains, logistics, production, and product life cycle management are all interconnected in systems that are much more intelligent, adaptable, and dynamic that they are at present, with the ability to improve through self-optimization and autonomous decision-making. The sixth and seventh Kondratieff waves are in the future.

Technological capacity — The amount of technological knowledge in an organization, region, or country, divided by the size of the labour force, or the population. Also termed *"intellectual capital."* The rate of growth of an organization's, region's, or nation's technological capacity depends on the rate at which new technology is produced (the *"rate of technological progress"*) and the rate at which technology is disseminated (the so-called *rate of replication*). Since some new technologies are incremental additions whereas others replace earlier technologies that are thus made obsolete, it is sometimes appropriate to distinguish among the *"net rate of replication"* and the *"gross rate of replication."* Reference [18]. *See also* technology.

Technological determinism — *See* fear of innovation.

Technological discontinuity — *See* S-curve.

Technological gatekeeper — One of several kinds of stereotypical people that can play a critical role in the technological innovation process. A technological gatekeeper is someone that, on the one hand, acts as a conduit to external knowledge by maintaining a watching brief on relevant parts of the science and technology world, staying on top of the literature and technical conferences, and being knowledgeable of and/or actually linked to key external science and technology individuals (experts). On the other hand, this person acts as the communicator of such information to others involved in the technological innovation process. Such a person could be engaged at any point in a new product, process, or service development process (not just at the new idea and invention stages), ideally throughout the entire process, and can greatly increase the success rate. *See also* Rainmaker, product champion, leading-edge customer.

Technological innovation — *See* innovation (20th–21st centuries). *See also* Maclaurin.

Technological myopia — *When management overestimates the improvement potential of existing technology, particularly given the amount of time it can take to develop a new innovation. In the extreme, technology development that is meant to address a particular challenge or opportunity may be no longer relevant by the time the development is completed because it simply took too long ("future technology myopia").*

Technological product and process innovation — (TPP innovation) A technological innovation involving a product or process. *See* innovation (20th–21st centuries).

Technological progressiveness — A simple lagging indicator of technological innovation performance for commercial enterprises or entire industries, by which their performance in terms of having introduced "important" new or improved products or processes into the marketplace within a certain period of time is rated as being high, medium, or low. Introduced by Maclaurin in 1955. References [62, 105].

Technological revolutions — *See* technological ages.

Technological sophistication — *See* high technology.

Technological unemployment — *See* fear of innovation.

Technological valley of death — *See* valley of death.

Technologist — *See* innovator.

Technology — A broad term representing the knowledge of how to effectively use a product, process, or service (*know-how*), how to conduct or control a manufacturing activity (*practice, process, or production technology*), or how to manufacture, use, or consume something (*product technology*). In this context, a *tool* can be considered to be a product. Each of these kinds of technology can be protected as intellectual property, and each can be commercialized. Technologies are sometimes distinguished as being material technology (physical things) or social technology (intangible knowledge) although some kinds of technology, such as a process, could be either. *See also* acquisition of technology and technological capacity.

Technology acceptance model — (TAM) A psychological model developed to assess and describe how the features of a technology influence people's views of that technology in terms of perceived usefulness and perceived ease of use. Such views influence the extent to which a technology will be used by customers, if at all. A related, complementary model is the technology readiness index (TRI), and there is also an integrated technology readiness and acceptance model (TRAM) that combines the TAM and TRI models to provide a single means of categorizing customer readiness to accept a new technology. In terms of psychological readiness of customers to accept and/or embrace new technologies, some literature makes reference to market segmentation in this context, a simple version of which would be to distinguish among *explorers* (the first few adopters), *pioneers* (the next few adopters), *skeptics*, *hesitators*, and finally the *avoiders* (the non-adopters). This has some analogy to the categorizations in Rogers' Technology Adoption Life Cycle. *See also* acceptance threshold, technology readiness, technology readiness index, technology readiness level, technology adoption life cycle, acceptance threshold, and Luddite.

Technology acceptor capacity — The ability of a company or industry to accept, adapt (if necessary), and adopt new technologies. Also termed "*absorptive capacity.*"

Technology adoption life cycle — The stereotypical pattern of *technology diffusion* in which a new product, process, service, or idea becomes adopted and spreads according to the sociological characteristics of different adopters. Rogers [77] originally used a normal distribution (i.e. bell curve) to describe these different adopters as *innovators* (the first few adopters), *early adopters* (the next few adopters), *early majority*, *late majority*, and *laggards*. *See* Figure 4.8.

See also technology dissemination. *See also* technology hype-cycle, and technology readiness.

Technology and innovation centre — (TIC) An organization that focusses on the production, dissemination, and transmission of knowledge, aimed at new-technology exploitation by companies that results in economic value creation. This is essentially the same as a fourth-pillar organization.

Technology balance of payments — The net technology transactions (in purchasing power parity U.S. dollars) per 10,000 personnel in the labour force. This represents the balance of sales versus purchases of technology (such as patents, licenses, designs, trademarks, and trade secrets) in an economy. See innovation indicators and Table 8.2.

Technology commercialization — *See* commercialization.

Technology dependency ratio — The number of non-resident patent applications divided by the number of resident patent applications.

Technology deployment — The processes involved in getting a new technology in place and operating properly in its intended working environment. Such processes generally include construction and/or installation, configuration, testing, and/or troubleshooting, and making any necessary final modifications. Also termed "*technology implementation*."

Technology diffusion — *See* technology dissemination.

Technology diffusion model — *See* Rogers.

Technology dissemination — The transfer, adoption, and use of technology by multiple users. Dissemination is the way in which technology (and therefore innovation) spreads from its developer and first implementer and/or its first customer, to different customers, industries, markets, and regions. Two modes of technology dissemination are distinguished: *technology transfer* and *technology diffusion*. *Technology transfer* refers to the transfer of technology from a specific person or organization to another and occurs for initial and/or limited technology dissemination. *Technology diffusion* refers to the transfer of technology from a person or organization to another where either party may be unaware of the identity of the other, and occurs when technology is becoming widely disseminated. *See also* technology adoption life cycle and technology readiness.

Technology drivers — *See* miners.

Technology foresight — *See* foresight.

Technology gap — *See* innovation gap.

Technology gap theory — *See* innovation gap theory.

Technology growth curve — *See* S-curve.

Technology hatchery — *See* business incubator.

Technology hype-cycle — A product life cycle model in which a new product, process, or service advances through an unrealistic peak in market expectations, then sales decline until improvements and/or greater realism in the market prevails, after which the product/process/service matures in the market and then completes its life cycle. Also termed "*Gartner Hype Cycle*." A chart or graph illustrating these stages is referred to as a hype-cycle curve or Gartner

	Hype-Cycle Curve. *See* Figure 3.11. *See also* technology adoption life cycle and S-curve.
Technology implementation	*See* technology deployment.
Technology life cycle	*See* S-curve.
Technology management	An organizational function that coordinates efforts to manage the technologies involved in conducting the organization's functions. Such efforts could include technology strategy, technology foresight, technology road mapping, technology portfolio management, and technology maintenance. Technology management is different from "*innovation management.*" *See also* innovation management.
Technology market intermediaries	(TMI) *See* knowledge-intensive business services and intermediary organization.
Technology mudslide hypothesis	A phrase coined by Clayton Christensen, referring to the difficulties companies face in trying to cope with the "relentless onslaught of technology change" in their respective markets as being like "*trying to climb a mudslide raging down a hill. You have to scramble with everything you've got to stay on top of it, and if you ever once stop to catch your breath, you get buried.*" This assumes that all of the competing companies in a market niche are constantly having to move "upward" in terms of innovations or else they will slide "downhill" and lose market share. Reference [79]. *See also* incremental innovation.
Technology park	*See* research and technology park.
Technology-push innovation	*See* linear innovation models, disruptive innovation, and first-generation innovation model. *See also* incremental innovation.
Technology readiness	There are two very different kinds of technology readiness: the readiness of a technology for deployment in the marketplace, and the readiness of customers to accept and/or pay for the new technology. The issues determining the former are primarily technical and are illustrated by, for example, the technology readiness level (TRL) scale. The issues determining the latter are primarily psychological and are illustrated by, for example, the technology readiness index (TRI) and the technology acceptance model (TAM) scales. *See* technology readiness index, technology readiness level, and technology acceptance model.
Technology readiness index	(TRI) A psychological model developed to assess and describe how people's personalities influence the degree of acceptance and use of a new technology. The TRI represents an attempt to balance factors that would contribute to technology acceptance (optimism about the technology and personal association with innovativeness), versus factors that would inhibit technology acceptance (discomfort caused by the technology and insecurity caused by mistrust of the technology). Such views influence the extent to which a technology will be used. The original TRI model comprises 36 elements, while a more recent "refined" TRI model comprises 16 elements. *See also* references [220, 221]. A related, complementary model is the technology acceptance model (TAM), and there is also an integrated technology readiness and acceptance model (TRAM) that combines the TAM and TRI models to provide a single means of categorizing customer readiness to accept a new technology. The TRAM

approach meshes the personality-related features of TRI with the system-related features of TAM. In terms of psychological readiness of customers to accept and/or embrace new technologies, some literature makes reference to market segmentation in this context, a simple version of which would be to distinguish among *explorers* (the first few adopters), *pioneers* (the next few adopters), *skeptics*, *hesitators*, and finally the *avoiders* (the non-adopters), This has some analogy to the categorizations in Rogers' Technology Adoption Life Cycle. *See also* acceptance threshold, technology readiness, technology readiness level, technology acceptance model, and technology adoption life cycle.

Technology readiness level — (TRL) A representation of an assessment of the maturity of an evolving technology. Different sectors and different countries use different TRL scales but the general philosophy is the same and they are each based on a linear model of innovation while at the same time recognizing that innovation is seldom linear, not all development cycles are the same, and the development of any particular technology may skip some readiness levels. The United States, Canada, Australia, and the European Commission, among others, use similar scales ranging from 1 to 9 (*see* Table 4.3 and Figure 4.4). Research and technology organizations work across the full TRL span but generally focus on developing technologies from about TRL 2 or 3 through to about TRL 7 or 8. Seven- and four-point scales are also in common use. Possibly the simplest TRL scale is TRL 1: discovery, TRL 2: development, TRL 3: demonstration, and TRL 4: deployment. *See* references [195–197]. *See also* proof of concept, speculative concept, technology readiness, Technology readiness index, technology acceptance model, commercial readiness index, technology adoption life cycle, and technology-translation gap.

Technology S-curve — *See* S-curve.

Technology Stage-Gate™ process — (TechSG process) Like the traditional Stage-Gate® process but designed specifically for technology development programmes, and having more emphasis on managing through the development steps lying within the fuzzy front-end (FFE) of the process. Reference [229]. *See also* idea-to-launch process and Stage-Gate® product development process.

Technology transfer — *See* technology dissemination.

Technology-translation gap — The gap between a technology that is at an early stage of development (i.e. at the discovery, technology concept, or invention stage) and the product development stage (from which a prototype could be developed). In terms of technology readiness levels, this would be the gap between a technology at TRL 2–3 and at TRL 5–6. *See also* technology readiness level.

Technopole — *See* innovation ecosystem.

Technopolis complex — *See* innovation ecosystem.

Technostructure — The management cadres of corporations that, taken together, are sometimes viewed as being the planning and decision-making agents of advanced capitalist economies. From this point of view, the technostructure strongly influences, guides, and/or determines economic development in an economy. Reference [52].

Technovation — A *portmanteau* or mixed word derived from the words technological and innovation. In the world of technological innovation, it is usually used to

	represent the process of technological innovation. Caution should be exercised, however, as it is also frequently used with other meanings, such as in educational and mentoring programmes, for example.
TechSG process	*See* technology Stage-Gate™ process.
Tech shop	*See* makerspace.
TEEPSE futures	Technological, economic, environmental, political, social, and/or ethical future possible events. *See* foresight.
Ten Cs	The "Ten Cs" for successful implementation of technological innovations, as described by Rothwell, are "*effective Communications to gain Consensus for Change, Champions to sustain Continuous Commitment to Change, a Culture that is Customer Centred.*" Reference [98]. *See also* principles of innovation.
Tenets of innovation	*See* principles of innovation.
Ten/five rule	*See* 10/5 rule.
Term sheet	A document that sets out terms and conditions of a proposed contractual agreement between two parties around the sale of a specific technology, line of business, or business. A term sheet can be binding or non-binding. The term sheet is used to guide the preparation of a final contractual agreement that is binding. *See also* boilerplate.
Tesla, Nikola (1856–1943)	A Serbian-born American engineer and inventor best known to the innovation world for his development of the rotating magnetic field that forms the basis of most alternating-current machinery. He also developed the three-phase system of electric power transmission and an induction coil (the Tesla Coil) used to produce high-voltage, low-current, high-frequency alternating current in early radios and some medical devices.
TFP	*See* total factor productivity.
The Borg law	Essentially a strategy that a "*laggard*" company should, or must, leapfrog the industry leader by absorbing and then extending the leader's advantage, in order to compete effectively in a marketplace. Example: Microsoft's adoption of the features of *Netscape* and subsequent improvements to better position *Internet Explorer*. Reference [234]. *See also* Davidow's law.
The chasm	*See* tipping point.
Theory of inventive problem solving	(TIPS, or TRIZ) A systematic approach to the process of invention developed by Genrich Altshuller in the 1960s, based on a comprehensive analysis of the patented solutions to hundreds of thousands of previously solved inventive problems. He identified a series of approaches that, taken together, were key to the solution of the majority of these problems. Altshuller called his approach the "*Theory of Inventive Problem Solving*" (TIPS, or TRIZ, the Russian acronym). The TRIZ approach involves several groups of methods that can be used to look at a problem in ways that, either individually or in combination, frequently lead to finding an inventive solution. *See also* Altshuller and creative thinking models. Reference [72].
Theory of the firm	Any of several economic theories that attempt to explain the nature of corporations, including their existence, behaviour, structure, and their relationship to other corporations and the marketplace. Most such economic theories date back to the mid-1900s, although one of the first

	was Adam Smith's agency theory and his description of the efficiency of corporations (in his book *The Wealth of Nations,* 1776).
Thinking out-of-the-box	A synonym for lateral thinking. *See* creative thinking models.
Think up	A synonym for brainstorming. *See* creative thinking models.
Third generation innovation model	Any of the early non-linear, interactive models of technological innovation that couple market needs with technological opportunities. These models are sequential, but with multiple feedback loops. An example is the "coupling model." *See* Section 3.1 and Table 4.2. *See also* coupling model, and generations of innovation, non-linear innovation models.
Third-generation nanotechnology	Molecular nanotechnology. *See* generations of nanotechnology.
Third generation of technology foresight	*See* generations of technology foresight.
Third Industrial Revolution	*See* technological ages.
Third mission	Participation by universities in working with governments and industry to enable economic growth (via innovation) and social progress has been referred to as "*Mode 2 Research*" and the "third mission" (after teaching and research) of an "*entrepreneurial university*." For universities that have transitioned to a "third-mission" strategy, reference is sometimes made to a shift from an "*endless frontier*" of discovery (Mode 1) research as an end unto itself to an "*endless transition*" in which discovery research is translated into applied (Mode 2) research and development, and thence into commercial deployment and use. Reference [261]. *See also* entrepreneurial university, modes of science, innovation ecosystem models, triple-helix model, and quad-helix model.
Third wave	*See* 3rd wave.
Thought leader	A person that demonstrates leadership in the communication of leading futuristic, strategic, and/or systems thinking to key stakeholders in a field of endeavour. Thought leaders are frequently advocates for and/or leaders of change in a field and are generally perceived to be "*innovative*" in the broadest sense of the term.
TH theory	*See* triple-helix model.
TIC	*See* technology innovation centre.
Timeboxing	A project or time-management technique in which a fixed time period ("time box") is set for the completion of a project, usually including the delivery of project deliverables. This tool is sometimes used in lean innovation and lean manufacturing.
Time to market	(TTM) In general, the elapsed time between origination of the concept for a new product, process, or service, and the point at which it is available for sale in the marketplace. Depending on the organization and the sector(s) in which it operates, the definition of the starting time can vary, such as when an idea is recorded, or when a product development project is approved or staffed. Ambiguity about the definitive start time has, on part, led to the term "*fuzzy front-end*." Similarly, the definition of the ending time can vary, such as when a first unit is shipped or purchased, or when a specified production rate has been achieved. TTM can be considered to be a product development process metric. Also termed "*cycle time*," "*rate of innovation*," or "*speed to market*." The term "*fast innovator*" refers to a

	person, team, or organization that is able to achieve reduced and/or short times to market. *See also* Fuzzy Front-End.
Tipping point	In general, a "*tipping point*" refers to a transition point, beyond which some kind of "*critical mass*" has been reached, and an idea, trend, social behaviour, or product sales level spreads and/or increases very dramatically. Also referred to as a "*turning point*." Example: in epidemiology, the tipping point is when an infectious disease spreads beyond the capacity of any localized efforts to bring it under control. In innovation, it is often described in terms of Rogers' diffusion of innovation model, as the transition between the stage of early adoptions (characterized by the innovators and the early adopters) and the next two large waves of adoptions (characterized by the early and late majorities). In this context, it is sometimes referred to as "*the chasm*." *See* Figure 4.8. *See also* J-curve.
TIPS	*See* theory of inventive problem solving.
T-KIBS	Technological KIBS. *See* knowledge-intensive business services.
TMI	Technology market intermediaries. *See* knowledge-intensive business services, intermediary organization.
Tool	*See* technology.
Top-down economic development	An approach to economic development by which government tries to "pick winners," whether in terms of choosing particular applied R&D pathways, products, processes, or services, companies, or even clusters. This is typically done through the direct provision of grants or subsidies. This approach is generally viewed as being inefficient and/or ineffective. Conversely, industry-led approaches ("*bottom-up economic development*"), that are supported by government, are generally viewed as being more efficient and effective, especially in the case of innovative regional clusters.
Top-down innovation	The process of seeking-out, identifying, and evaluating potential market opportunities and then challenging the organization to come up with and develop concepts for innovative new products, processes, or services that could align with the selected market opportunities. *Bottom-up innovation*, on the other hand, refers to the process of originating and developing concepts for new business products, processes, or services and then evaluating them for market potential. *See also* innovation and upstream innovation.
Total factor productivity	(TFP) An economic measure of the contribution of technological innovation to labour productivity. In this case, labour productivity is divided into three principal components representing contributions from educational attainment, capital input, and total factor productivity (TFP), where TFP represents the influence of technological innovation. Also termed "*Solow's residual*." Reference [322]. *See also* multifactor productivity.
Total operating expense	*See* capital expense.
TPP innovating firm	An OECD term for an organization that has implemented new or significantly improved technological products, processes, or services. The acronym "TPP" stands for "technological product and process." Reference [26]. *See also* innovation (20th–21st centuries).
TPP innovation	Technological product and process innovation. A technological innovation involving a product or process. *See* innovation (20th–21st centuries).

Trade dress	The physical appearance of a product and/or its packaging. If the physical appearance of a product is unique, unusual, and/or widely recognized by the public, then its trade dress may be legally protected in the countries within which they are registered, and specific registration may not be needed. *See also* intellectual property.
Trade secret	Any kind of technological or business know-how that is protected from competitors by keeping it confidential. *See also* intellectual property and acquisition of technology.
Trademark	A word, phrase, name, design, symbol, logo, or device associated with a business brand, product, or process, and which is used for differentiation from other brands or products in the same general line of business. The term *service mark* is sometimes used when a business service is meant. Registered trademarks and service marks are legally protected in the countries within which they are registered. *See also* brand and intellectual property.
Trademarks	(Innovation indicator) An indicator of the total number of new products and services being protected in an economy is the number of trademarks registered per 10,000 personnel in the labour force. *See* innovation indicators and Tables 8.2–8.4.
Transcendentalist model	A model for the innovation process in which innovation arises from an instantaneous act or inspiration of genius. *See also* cumulative synthesis model and mechanistic model.
Transformational innovation	*See* disruptive innovation.
Transilience	A term coined by Abernathy and Clark [86] to describe the potential influence of a technological innovation on an organization's prior technological knowledge and resources, and its influence on the competitive marketplace. The potential influences are represented as quadrants in a diagram termed a *transilience map*, which denotes "regular (i.e. incremental), niche, "revolutionary" (i.e. evolutionary), and "architectural" (i.e. disruptive) innovations. *See* Figure 2.5.
Transilience map	*See* transilience.
Transnational corporation	*See* multinational enterprise.
TRI	*See* technology readiness index.
Triadic patent family	*See* patent family.
Triadic patents issued	(Innovation indicator) An indicator of the number of practical inventions with international potential that are being protected is the number of patents in triadic patent families issued per 1 million people in the population of a given economy. Triadic patent families are groups of patents that have been granted in multiple (three or more) countries on the same invention. *See* innovation indicators and Tables 8.2–8.4.
Trialability	An older term referring to the extent and ease with which a technological innovation can be tested and/or experimented with, i.e. tried, by a prospective customer before committing to a purchase. The term appears to have originated with Rogers and his diffusion of innovation model. Reference [77].
Trial and error strategy	*See* rationalist strategy.
Trialing	*See* Awareness, Trial, Availability, Repeat model.
Trickle-up innovation	*See* reverse innovation.
Triple-helix field theory	*See* triple-helix model.

Triple-helix model	The "*triple-helix model*" is an innovation ecosystem (sociological) model developed by Etzkowitz and Leydesdorff in 1994 to describe the roles and intersections of governments, universities, and industry in advancing knowledge-based economies [259, 260, 262]. By extension, the model has also been applied to the advancement of economies based on innovation (the *innovation-related triple-helix*). Here, governments provide a regulatory framework, science and technology (S&T) policies, programmes, and infrastructure; universities provide new knowledge, understanding, and technology, and industry provides production and deployment of the innovations into the marketplace. In the triple-helix model, universities would be the primary incubators for technology-based start-ups and small- and medium-sized enterprises (SMEs). The helix is a metaphor for an innovation process that is more like a spiral than a straight line or S-curve, and along the spiral there are multiple points of intersection and interaction among the three kinds of organizations. Synonyms include triple-helix field theory (or TH theory), and the ABG Model (referring to academia, business, and government). In a different extension from innovation, there is also a sustainability-related triple-helix in which the public-at-large is substituted for industry, to represent societal concerns about technologies, industry, the environment, and sustainability. *See* Figure 5.4. *See also* innovation ecosystem models, innovation ecosystem, quad-helix model, and N-tuple-helix model.
TRIZ	*See* theory of inventive problem solving.
TRL	*See* technology readiness level.
TRL 1	Technology readiness level 1. Different countries and industrial sectors use somewhat different definitions but on nine-point TRL scales, TRL 1 generally refers to the stage at which basic principles have been observed and reported, and are becoming translated into applied research and development. *See* Table 4.3, Figure 4.4, and technology readiness level.
TRL 2	Technology readiness level 2. Different countries and industrial sectors use somewhat different definitions but on nine-point TRL scales, TRL 2 generally refers to the stage at which practical applications and inventions are being identified. *See* Table 4.3, Figure 4.4, and technology readiness level.
TRL 3	Technology readiness level 3. Different countries and industrial sectors use somewhat different definitions but on nine-point TRL scales, TRL 3 generally refers to the stage at which applied research and development are underway at laboratory scale, including proof of concept. *See* Table 4.3, Figure 4.4, and technology readiness level.
TRL 4	Technology readiness level 4. Different countries and industrial sectors use somewhat different definitions but on nine-point TRL scales, TRL 4 generally refers to the stage at which multiple technological components, if applicable, are integrated and demonstrated to work together, again at laboratory scale. *See* Table 4.3, Figure 4.4, and technology readiness level.
TRL 5	Technology readiness level 5. Different countries and industrial sectors use somewhat different definitions but on nine-point TRL scales, TRL 5 generally refers to the stage at which the technological components are integrated for testing and validation in a simulated and/or realistic

	environment beyond the laboratory. *See* Table 4.3, Figure 4.4, and technology readiness level.
TRL 6	Technology readiness level 6. Different countries and industrial sectors use somewhat different definitions but on nine-point TRL scales, TRL 6 generally refers to the stage at which the integrated technological components in a model or prototype are tested and validated in a simulated and/or realistic environment beyond the laboratory. *See* Table 4.3, Figure 4.4, and technology readiness level.
TRL 7	Technology readiness level 7. Different countries and industrial sectors use somewhat different definitions but on nine-point TRL scales, TRL 7 generally refers to the stage at which a complete prototype, at or near full-scale, is ready for demonstration and/or demonstrated in a realistic operational environment. *See* Table 4.3, Figure 4.4, and technology readiness level.
TRL 8	Technology readiness level 8. Different countries and industrial sectors use somewhat different definitions but on nine-point TRL scales, TRL 8 generally refers to the stage at which a complete technology has been tested and demonstrated to work in its final form and under realistic operational conditions. *See* Table 4.3, Figure 4.4, and technology readiness level.
TRL 9	Technology readiness level 9. Different countries and industrial sectors use somewhat different definitions but on nine-point TRL scales, TRL 9 generally refers to the stage at which a complete technology, in its final form and including any final "fixes", has been proven through deployment in actual operational environments and conditions. *See* Table 4.3, Figure 4.4, and technology readiness level.
TTM	*See* time to market.
Turning point	*See* tipping point.
Type I error	*See* decision-making errors.
Type II error	*See* decision-making errors.

11.22 U

Ubiquitous computing	*See* Internet of Things.
UILO	University-Industry Liaison Office. *See* Industry Liaison Office.
Unarticulated needs	Products, processes, or services that customers might be willing to buy, but which they are unable or unwilling to imagine and/or describe.
Universal success curve	A graph illustrating the 1997 results of research by Stevens and Burley, which led them to conclude that in most industries, it takes about 3,000 initial, undeveloped ideas to produce a single successful commercial product. They also identified that improving success rates as an idea survives various intermediate staged-gates of development, as illustrated in Figure 2.8. Others have found similar trends but somewhat different numbers for different industries, such as an estimate that in the pharmaceutical industry, it could take at least 6,000 ideas to produce a commercially successful product. References [110, 111]. *See also* Rainmaker.

University-Industry Liaison Office: (UILO) *See* Industry Liaison Office.

Upstream innovation: The process of originating, evaluating, and developing concepts for new business products, processes, or services. This can include the process of discovery. *Downstream innovation*, on the other hand, refers to the process of converting such concepts into market-ready products, processes, or services and then introducing them into the marketplace. This can include the process of invention. *See also* innovation and top-down innovation.

Upstream research: *See* research and development.

Upstream studies: *See* research and development.

User innovation: *See* consumer-innovators.

Utility patent: *See* patent.

11.23 V

Valley of death: (Innovation) As technology development proceeds, there is a tendency for financing, usually government financing, to become increasingly difficult to obtain, as the technology is developed from basic and applied research through to speculative concept, engineering development, and proof of concept (i.e. from technology readiness levels TRL 1 through 3). Other kinds of financing such as venture capital, private equity, and debt financing tend to become available as the technology matures from field demonstration and first deployment through to full maturity in competitive markets (i.e. from technology readiness level TRL 9 through commercial readiness index level CRI 6). In between lies the so-called *valley of death*, a series of technology development steps such as prototype development and testing, scale-up, and pilot testing (i.e. from technology readiness levels TRL 4 through 8) for which financing is the most difficult to obtain. See Figure 4.5. Synonyms include the "*technological valley of death*" and "*innovation progression gap.*"

(Technological vs. commercialization) Within the *valley of death*, two subcategories are sometimes distinguished. The *technological valley of death* refers to early-stage technology development phases such as speculative concept development, engineering development, and proof of concept. The *commercialization valley of death* refers to later-stage technology development phases such as full field- or plant-scale commercial demonstrations and first and second customer sales and deployments. The valleys of death are a challenge for innovative organizations of all kinds, but they particularly plague RTOs and start-up companies. See Figure 4.5.

(Entrepreneurship) In entrepreneurship, the valley of death refers to the gap, or period of time, between the initial funding and launch of a start-up company and the point at which it starts generating revenue from sales. During this interval, start-up companies tend to have a high risk of requiring more operating capital than they can afford, which can cause the enterprise to fail. An illustration showing steps in a linearized process leading from the initial business concept through to a successful, sustainable company is sometimes termed a death valley curve. Other synonyms include *"death valley"* and *"funding gap."*

Value chain — A systems view of the series of processes by which a manufacturer begins with starting materials, adds value, and profitably sells a (higher value) final product, process, or service. Each such process comprises a sub-system involving sub-systems, each with its own inputs, use of resources, transformation processes, and outputs. In Porter's model, value chains have five main segments: *"inbound logistics"* (receiving and preparing the raw materials), *"operations"* (converting the raw materials into a finished product, process, or service), *"outbound logistics"* (delivering it to a customer), *"marketing and sales,"* and *"service."* The name has been attributed to Michael Porter, who developed a specific value chain model. Reference [749]. Within a value chain, the processes occurring with the company are sometimes referred-to as *"business logistics management,"* while the process handled by external firms are sometimes referred-to as *"supply chain management."* Such supply chain firms range from raw material suppliers, to assemblers, to shippers, to retailers.

Venture capital — Final-stage, usually substantial, capital for a commercialization process that is typically raised through formal, institutional investment firms. Venture capital is usually used to take a product, process, or service that has already been introduced into the marketplace and to rapidly drive it "up" past the marginal or break-even level and into profitability. *See* 10/5 rule. *See also* capital, sweat equity, seed capital, start-up capital, and pre-venture capital.

Venture capital intensity — The ratio of venture capital investments to gross domestic product in an economy. *See* innovation indicators and Table 8.2.

Venturing — Creating a for-profit business enterprise of some kind. A commercialization strategy that involves venturing and involves creating or expanding a business in order to realize an innovation.

Verhulst curve — *See* S-curve (science).

Verhulst, Pierre François (1804–1849) — A Belgian mathematician known to the innovation world for his development of the sigmoid S-curve, which he originally used to describe population growth. The concept of a growth rate that is initially exponential but then slows and plateaus has since been applied to many other phenomena, including product life cycles. The S-curve is sometimes termed a Verhulst curve. *See* S-curve.

Vertical thinking — A synonym for *Linear sixth generation thinking. See* creative thinking models.

Virtual cluster — *See* business ecosystem.

Visionary — *See* early adopter.

VoC — *See* voice of the customer.

Voice of the customer — (VoC) Feedback of almost any kind received from customers of an existing or proposed product, process, or service. Such feedback can be integrated into any stage of the product development process, from the conception of prospective new products to the marketing and sales of existing ones.

Voucher programme — *See* innovation voucher programme.

VUCA — Volatility, uncertainty, complexity, and ambiguity. The acronym is sometimes used with reference to unpredictability and challenge of the seemingly ever-increasing pace of change in the business sectors, in large and complex organizations, and in global events and affairs in general.

11.24 W

Water-ribbed balloon	A metaphor for a model of the product development process that is very similar to the stage-gate process for managing technological innovation [753].
Waterfall method	A new product development process in which a linear series of developmental steps is followed in sequence, and with each step only being undertaken after the preceding one has been completed. This method seems to have originated in the software industry, but it has been used more broadly in the manufacturing and other industries as well. Modern usage of this method often includes feedback loops that permit making backing up into earlier steps to make changes as issues are discovered downstream. *See also* new product development process and Stage-Gate® product development process [16, 186].
Waves of innovation	*See* Kondratieff waves.
Weak signals	*See* foresight.
WEF	World Economic Forum (Geneva, Switzerland). *See* Global Competitiveness Index and competitiveness drivers.
White space	*See* white space mapping.
White space mapping	(Business processes) In business processes, *white space* usually refers to areas within which policies and/or authorities are undefined or vague. These create inefficiencies and mapping the *white space* can help identify opportunities for improvement and/or creativity. The term *white space* seems to have originated with the concept of the white space between the boxes in an organizational chart. Areas within which policies and/or authorities are defined and/or clear are sometimes referred to as *black space*. (Business development) In business development, *white space* usually refers to either new product/process/service offerings to current customers or to new or existing product/process/service offerings to new customers. Either way, *white space* refers to the aspect that is new. In some usage, the term is used to refer to market niches that are under-served, but this is probably better referred to as *grey space*. In this context, *white space* could also refer to a customer/market opportunity for which an appropriate product/process/service is not available.
Whole product R&D	The practice of keeping research and development (R&D) involved through all stages of a product's launch, evolution, marketing, and life cycle in order to ensure that marketing and customer feedback is used to continuously improve, if possible, the product, process, or service in order to extend its life cycle. The value in keeping the R&D function involved is that it can be difficult for others to discern the difference between a technologically simple change and a substantial technological boundary. Reference [138]. *See also* research and development.
Wild cards	*See* foresight.

Wisdom Hierarchy	An idealized linear model for the process by which information is obtained from data, knowledge from information, and wisdom from knowledge. *Also termed "Data-Information-Knowledge-Wisdom Hierarchy," "DIKW hierarchy," "data hierarchy," "information hierarchy,"* or *"knowledge hierarchy."*
Wizard	*See* Rainmaker.
Working model	A model built to demonstrate that a product or process has been reduced to practice and basically works (proof of concept). Working models are expected to function, but are usually not built to scale and their functioning has usually not been optimized. *See also* prototype. Reference [76].
"Works like" model	A model that illustrates the function of a product, process, or service without necessarily working exactly like the final product will. As such, a "works like" model is more advanced than a mockup, but less representative of the final product than a working model, engineering prototype, or production prototype. The benefits of a "works like" model are the ability to quickly and inexpensively produce a model that can be shown to prospective customers in order to get feedback before finalizing the product design. Similarly, a "looks like" model illustrates the appearance of a product, process, or service without functioning exactly like the final product will. *See also* prototype and fail fast.
World-first	In discussing the introduction to the marketplace of new products, processes, or services, it can sometimes be useful to categorize them as *"firm-first," "country-first,"* or *"world-first."* Some companies' competitive markets are such that country-first is sufficient for their competitive needs, whereas companies that compete globally tend to aim for world-first.

11.25 X, Y, and Z

Year 1 New Product Sales Index	*See* New Product Vitality Index.
Yield	*See* return on investment.
Zizhu chuangxin	Self-determined innovation. *See* indigenous innovation.

References

[1] Godin, B., *Innovation Contested. The Idea of Innovation over the Centuries*, Routledge: New York, 2015.

[2] Schramm, L.L., *Innovation Technology – A Dictionary*, de Gruyter: Berlin, Germany, 2017.

[3] Vaughan, J., "Technological Innovation: Perceptions and Definitions," *Libr. Technol. Rep.*, 2013, *49(7)*, 1–70.

[4] Garcia, R.; Calantone, R., "A Critical Look at Technological Innovation Typology and Innovativeness Terminology: A Literature Review," *J. Prod. Innov. Manag.*, 2002, *19*, 110–132.

[5] Courvisanos, J., "Political Aspects of Innovation," *Res. Policy*, 2009, *38(7)*, 1117–1124.

[6] de Castro, G.M.; Verde, M.D.; Sáez, P.L.; López, J.E.N., *Technological Innovation, An Intellectual Capital-Based View*, Palgrave Macmillan: London, 2010.

[7] Bell, S., "Ideas for Innovation are All around Us," ACRLog, 28 June 2011, http://acrlog.org/2011/06/28/ideas-for-innovation-are-all-around-us/.

[8] Mercier-Laurent, E., *Innovation Ecosystems*, Wiley: New York, 2011.

[9] Godin, B., "The Vocabulary of Innovation: A Lexicon," *Presented at 2nd CASTI Workshop*, Agder, Norway, October 20 2014.

[10] Kotsemir, M.; Abroskin, A., "Innovation Concepts and Typology – An Evolutionary Discussion," *MPRA Paper No. 45400*, Munich Personal RePEc Archive (MPRA), Ludwig Maximilian University: Munich, 2013, p. 51.

[11] Turner, J., *A Defence of the Doctrine and Practice of the Church of England, against Some Modern Innovations . . .*, John Morphew: London, 1712.

[12] Anonymous, *An Abstract of What Is Necessary to Be Known Concerning the Constitution Unigenitus: The Better to Avoid the Errors, and Snares of the Modern Innovators, Amsterdam*, 1724. A translation by P. Plowden of *Abrégé de ce qu'il faut sçavoir touchant la constitution Unigenitus*, Permissa Superiorum: Luxembourg, 1719.

[13] Anonymous, *The Liberties of England Asserted*, in *Opposition to Popery, Slavery, and Modern Innovation*, John Roberts: London, 1714.

[14] Machiavelli, *N., The Prince*, translation of the 1513 edition by W. K. Marriott, Project Gutenberg Edition, EBook #1232, 2006, www.gutenberg.org, accessed 13 November 2025.

[15] Schumpeter, J.A., *Business Cycles. A Theoretical, Historical and Statistical Analysis of the Capitalist Process*, McGraw-Hill: New York, 1939.

[16] Schumpeter, J.A., *Capitalism, Socialism and Democracy*, 3rd Ed, HarperCollins: New York, 1950.

[17] Garcia, R.; Calantone, R., "A Critical Look at Technological Innovation Typology and Innovativeness Terminology," *J. Prod. Innov. Manag.*, 2002, *19*, 110–132.

[18] Schmookler, J., *Invention and Economic Growth*, Harvard University Press: Cambridge, 1966.

[19] Pavitt, K., "Sectoral Patterns of Technical Change: Towards a Taxonomy and a Theory," *Res. Policy*, 1984, *13*, 343–373.

[20] Archibugi, D., "Pavitt's Taxonomy Sixteen Years On: A Review Article," *Econ. Innov. New Technol.*, 2001, *10*, 415–425.

[21] OECD, *ISIC Rev. 3 Technology Intensity Definition. Classification of Manufacturing Industries into Categories Based on R&D Intensities*, OECD Directorate for Science, Technology and Industry, Organisation for Economic Co-operation and Development: Paris, July 7 2011.

[22] Carter, C.F.; Williams, B.R., *Industry and Technical Progress: Factors Governing the Speed of Application of Science*, Oxford University Press: London, 1957.

[23] Carter, C.F.; Williams, B.R., *Investment in Innovation*, Oxford University Press: London, 1958.

[24] Macaulay, L.A.; Miles, I.; Wilby, J.; Tan, Y.L.; Zhao, L.; Theodoulidis, B., Eds., *Case Studies in Service Innovation*, Springer Science+Business: New York, 2012.

[25] Mustafa, E., *Service Innovation*, Routledge: Oxford, 2019.

https://doi.org/10.1515/9783112206720-012

[26] OECD, *Oslo Manual, Guidelines for Collecting and Interpreting Innovation Data*, 3rd Ed, Organisation for Economic Co-Operation and Development (OECD): Paris, 2005.

[27] ISO, "Innovation Management – Fundamentals and Vocabulary," in *ISO Standard 56000:2025*, International Organization for Standardization: Geneva, 2025.

[28] Soon, O., "Dr. Geoff Nicholson, the 'Father of Post-it Notes', on 3M & Innovation," Hardwarezone Internet blog, 22 March 2013, http://www.hardwarezone.com.sg/feature-dr-geoff-nicholson-father-post-it-notes-3m-innovation, accessed 13 November 2025.

[29] Fagerberg, J., "Innovation: A Guide to the Literature," in *The Oxford Handbook of Innovation*, Fagerberg, J.; Mowery, D.C., (Eds.), Online Ed, Oxford University Press: Oxford, UK, 2009, pp. 1–26.

[30] Deming, W.E., *Out of the Crisis*, Massachusetts Institute of Technology: Cambridge, 1982.

[31] Slater, R., *Jack Welch and the GE Way: Management Insights and Leadership Secrets of the Legendary CEO*, McGraw-Hill: New York, 1998.

[32] Chandy, R.; Prabhu, J., "Innovation Typologies," in *Wiley International Encyclopedia of Marketing*, Online Ed., Part 5, Wiley: New York, 2010.

[33] Doner, R.F., (Ed.), *Explaining Institutional Innovation. Case Studies from Latin America and East Asia*, Social Science Research Council: Washington, 2010.

[34] Fagerberg, J.; Mowery, D.C.; Nelson, R.R., (Eds.), *The Oxford Handbook of Innovation*, Oxford University Press: Oxford, UK, 2004.

[35] Nobel Media, *Robert M. Solow – Biographical*, Nobel Media: Stockholm, 2014, Accessed 13 July 2015 at, http://www.nobelprize.org/nobel_prizes/economic-sciences/laureates/1987/solow-bio.html. accessed 13 November 2025.

[36] The Economist, "The Growth of Growth Theory," *The Economist*, 18 May 2006.

[37] Acemoglu, D., *Introduction to Modern Economic Growth*, Chapter 2, Princeton University Press: Princeton, 2009, pp. 26–76.

[38] Romer, P.M., "Increasing Returns and Long-Run Growth," *J. Polit. Econ.*, 1986, *94*(*5*), 1002–1037.

[39] Lucas, R.E., "On the Mechanics of Economic Development," *J. Monet. Econ.*, 1988, *22*(*1*), 3–42.

[40] Romer, P.M., "Endogenous Technological Change," *J. Polit. Econ.*, 1990, *98*(*5*), S71–S102.

[41] Krugman, P., "The New Growth Fizzle," in *New York Times*, 18 August 2013.

[42] Nobel Media, "The Sveriges Riksbank Prize in Economic Sciences in Memory of Alfred Nobel 2025," Nobel Prize Outreach 2025, 13 Oct 2025, https://www.nobelprize.org/prizes/economic-sciences /2025/summary/. accessed 13 November 2025.

[43] Ference Weicker & Co, "Promoting Research and Innovation in British Columbia," in *Report for BC Ministry of Advanced Education*, Victoria, BC, 7 September 2005, p. 78.

[44] Hervas-Olivera, J.-L.; Sempere-Ripolla, F.; Boronat-Molla, C.; Rojasb, R., "Technological Innovation without R&D: Unfolding the Extra Gains of Management Innovations on Technological Performance," *Technol. Anal. Strateg. Mgmt.*, 2015, *27*(*1*), 19–38.

[45] Porter, M.E., "Canada at the Crossroads: The Reality of a New Competitive Environment," in *Report for the Business Council on National Issues and the Government of Canada*, Ottawa, October 1991, p. 107.

[46] Howitt, P., "Innovation, Competition and Growth: A Schumpeterian Perspective on Canada's Economy," *Report, C.D*, Howe Institute: Toronto, ON, 2007.

[47] Norman, A.L., *Information Society*, University of Texas: Austin, 25 Aug 2009, http://www.laits.utexas.edu/~anorman/61N/Text/000IS. accessed 13 November. 2025.

[48] Briley, G.C., "A History of Refrigeration," *Am. Soc. Heat. Refrig. Air-Cond. Eng. J.*, *November* 2004, S31–S34.

[49] Nagengast, B., "It's a Cool Story. Refrigeration and Air Conditioning in the 20th Century," *Mech. Eng.*, May 2000, 56–63.

[50] Kusiak, A., "Put Innovation Science at the Heart of Discovery," *Nature*, 2016, *530*(*7590*), 255.

[51] Brown, T., *Change by Design: How Design Thinking Transforms Organizations and Inspires Innovation*, Harper Business: New York, 2009.
[52] Courvisanos, J., "Technological Innovation: Galbraith, the Post Keynesians, and a Heterodox Future," *J. Post Keynes. Econ.*, 2005, *28(1)*, 85–104.
[53] Standage, T., (Ed.), *The Future of Technology*, Profile Books: London, 2005.
[54] Israel, P., *Edison. A Life of Invention*, Wiley: New York, 1998.
[55] Homburg, E., "The Emergence of Research Laboratories in the Dyestuffs Industry, 1870–1900," *Br. J. Hist. Sci.*, 1992, *25(1)*, 91–111.
[56] Pithan, D.M., *Corporate Research Laboratories and the History of Innovation*, Routledge: New York, 2022.
[57] Kelly, P.; Kranzberg, M.; Rossini, F.A.; Baker, N.R.; Tarpley, F.A.; Mitzner, M., "Technological Innovation: A Critical Review of Current Knowledge, Volume 1. The Ecology of Innovation," *Advanced Technology and Science Studies Group Report*, Georgia Tech: Atlanta, February 1975.
[58] Holland, M., *Industrial Explorers*, Harper & Bros.: New York, 1928.
[59] Godin, B., "The Linear Model of Innovation (II): Maurice Holland and the Research Cycle," *Working Paper No. 3 Project on the Intellectual History of Innovation, Institut National de la Recherche Scientifique*, (INRS): Montreal, 2009, p. 20.
[60] Holland, M., "Research, Science and Invention," in *A Century of Industrial Progress*, Wile, F.W., (Ed.), Doubleday: New York, 1928, pp. 312–334.
[61] Betz, F., *Technological Innovation. Competitive Advantage from Change*, 3rd Ed, Wiley: Hoboken, 2011.
[62] Godin, B., "In the Shadow of Schumpeter: W. Rupert Maclaurin and the Study of Technological Innovation," *Minerva*, 2008, *46(3)*, 343–360.
[63] Kline, S.J.; Rosenberg, N., "An Overview of Innovation," in *The Positive Sum Strategy: Harnessing Technology for Economic Growth*, Laudan, R.; Rosenberg, N., (Eds.), National Academies Press: Washington, 1986, pp. 275–305.
[64] Badulin, N.A., "Innovation Snail, Triple Helix, and Other Circular Processes in Economics," in *Innovation-2012: Collection of Materials of the VIII All-Russian School-Conference of Students, Graduate Students and Young Scientists with International Participation*, Soldatov, A.N.; Minkov, S.L., (Eds.), Vol. 1, TUSUR Publ.: Tomsk, 2012, pp. 7–29.
[65] Ferrara, M.H.; LaMeau, M.P., Eds., *Innovation Masters: History's Best Examples of Business Transformation*, Gale: Detroit, 2012.
[66] McKetta, J.J., (Ed.), *Petroleum Processing Handbook*, CRC Press/Dekker: New York, 1992.
[67] Kline, S.J., "Innovation Is Not a Linear Process," *Res. Manag.*, 1985, *Jul.–Aug*, 36–45.
[68] Vincenti, W.G., "Technological Knowledge without Science: The Innovation of Flush Riveting in American Airplanes, Ca. 1930 to Ca. 1950," *Technol. Cult.*, 1984, *25(July)*, 540–576.
[69] Vincenti, W.G., "The Air Propeller Tests of W.F. Durand and E.P. Leslie: A Case Study in Technological Methodology," *Technol. Cult.*, 1979, *20(Oct)*, 712–751.
[70] Ridley, M., *The Evolution of Everything. How New Ideas Emerge*, Harper: New York, 2015.
[71] Petroski, H., *The Evolution of Useful Things*, Vintage Books: New York, 1994.
[72] Altshuller, G., *And Suddenly the Inventor Appeared*, English translation, Technical Innovation Center: Worcester, MA, 2004.
[73] Merton, R.K., "Resistance to the Systematic Study of Multiple Discoveries in Science," *Eur. J. Sociol.*, 1963, *4(2)*, 237–282.
[74] Epstein, R.C., "Industrial Invention: Heroic, or Systematic?," *Q. J. Econ.*, 1926, *40(2)*, 232–272.
[75] Pontin, J., "The Rules of Innovation," *Technol. Rev.*, 2005, *108(5)*, 12.
[76] Lux, D.; Rorke, M., *From Invention to Innovation, Report DOE/GO-10099-810*, U.S. Dept. of Energy: Washington, DC, 1999.
[77] Rogers, E.M., *Diffusion of Innovations*, 5th Ed, Free Press: New York, 2003.

[78] Erixon, R.; Weigel, B., *The Innovation Illusion. How so Little Is Created by so Many Working so Hard*, Yale University Press: New Haven, 2016.
[79] Christensen, C.M., *The Innovator's Dilemma: When New Technologies Cause Great Firms to Fail*, Harvard Business Review Press: Boston, 1997.
[80] DeBruin, J.P.; Corey, J.D., "Innovation, the Economy, and the Federal Laboratories: Principles and Issues in the Innovation Process," *Report SAND 89-3014*, Sandia National Laboratories, U.S. Department of Energy, December 1991.
[81] Verloop, J., *Success in Innovation. Improving the Odds by Understanding the Factors for Unsuccess*, Elsevier: Burlington, MA, 2013.
[82] Godin, B.; Lane, J.P., "A Century of Talks on Research: What Happened to Development and Production?," *Int. J. Transit. Innov. Syst.*, 2012, *2(1)*, 5–13.
[83] Godin, B.; Lane, J.P., "'Pushes and Pulls': The Hi(story) of the Demand Pull Model of Innovation," Working Paper No Vol. 13 *Project on the Intellectual History of Innovation, Institut National de la Recherche Scientifique*, (INRS): Montreal, 2013, p. 39.
[84] Wilson, R.A.; Marcus, S., Eds., *American Greats*, Public Affairs: New York, 1999, p. 70.
[85] Von Hippel, E.; Ogawa, S.; De Jong, J.P.J., "The Age of the Consumer-Innovator," *MIT Sloan Manage. Rev.*, 2011, *53(1)*, 27–35.
[86] Abernathy, W.J.; Clark, K.B., "Innovation: Mapping the Winds of Creative Destruction," *Res. Policy*, 1985, *14(1)*, 3–22.
[87] Henderson, R.M.; Clark, K.B., "Architectural Innovation: The Reconfiguration of Existing Product Technologies and the Failure of Established Firms," *Adm. Sci. Q.*, 1990, *35(1)*, 9–30.
[88] Popadiuk, S.; Choo, C.W., "Innovation and Knowledge Creation: How are These Concepts Related?," *Int. J. Inf. Mgmt.*, 2006, *26*, 302–312.
[89] Sherwin, C.W.; Isenson, R.S., "Project Hindsight," *Science*, *1967*, 1571–1577.
[90] Defense Research and Engineering, *"Project Hindsight," Final Report*, Defense Research and Engineering: Washington, October 1969.
[91] Hoffman, J.T., Ed., *A History of Innovation. U.S. Army Adaptation in War and Peace*, United States Army: Washington, DC, 2009.
[92] Lyons, J.; Chait, R.; Long, D., "Critical Technology Events in the Development of Selected Army Weapons Systems: A Summary of 'Project Hindsight Revisited'," *Defense & Technology Paper DTP-035*, National Defense University Center for Technology and National Security Policy: Washington, September 2006.
[93] Chait, R.; Lyons, J.; Long, D., "Critical Technology Events in the Development of the Abrams Tank: Project Hindsight Revisited," *Defense & Technology Paper DTP-022*, National Defense University Center for Technology and National Security Policy: Washington, December 2005.
[94] Mansfield, E., "Academic Research and Industrial Innovation," *Res. Policy*, 1991, *20*, 1–12.
[95] McFetridge, D.G., "Innovation and the Productivity Problem," *IRPP Choices*, 2008, *14(3)*, 3–37.
[96] Usher, A.P., *A History of Mechanical Inventions*, 2nd Ed, Harvard Univ. Press: Cambridge, MA, 1954.
[97] Usher, A.P., "Technical Change and Capital Formation," in *Capital Formation and Economic Growth. A Conference of the Universities-National Bureau Committee for Economic Research*, National Bureau of Economic Research, (Ed.), Princeton University Press: Princeton, 1955, pp. 521–548.
[98] Rothwell, R., "Successful Industrial Innovation: Critical Factors for the 1990s," *R&D Manag.*, 1992, *22 (3)*, 221–239.
[99] Myers, S.; Marquis, D.G., "Successful Industrial Innovations: A Study of Factors Underlying Innovation in Selected Firms," *Technical Report*, National Science Foundation: Washington, DC, 1969, p. 124.
[100] Scherer, F.M., "Demand-Pull and Technological Innovation: Schmookler Revisited," *J. Ind. Econ.*, 1982, *30(3)*, 225–237.

[101] Mowery, D.; Rosenberg, N., "Influence of Market Demand upon Innovation – Critical-Review of Some Recent Empirical Studies," *Res. Pol*, 1979, *8(2)*, 102–153.
[102] Marquis, D.G., "The Anatomy of Successful Innovations," in *Readings in the Management of Innovation*, Tushman, M.L.; Moore, W.L., (Eds.), Pitman Publishing Inc.: Marshfield, Mass, 1982, pp. 42–50.
[103] Salerno, M.S.; De Vasconcelos Gomes, L.A.; Da Silva, D.O.; Bagno, R.B.; Freitas, S.L.T.U., "Innovation Processes: Which Process for Which Project?," *Technovation*, 2015, *35*, 59–70.
[104] Verganti, R., *Design-Driven Innovation. Changing the Rules of Competition by Radically Innovating What Things Mean*, Harvard Business Press: Boston, 2009.
[105] Maclaurin, W.R., "Innovation and Capital Formation in Some American Industries," in *Capital Formation and Economic Growth*, Universities-National Bureau, Princeton University Press: Princeton, 1955, p. 549–578.
[106] Katz, J.S., "What Is a Complex Innovation System?," *PLoS ONE*, 2016, *11(6)*, e0156150, doi: 10.1371/journal.pone.0156150.
[107] Keeley, L.; Walters, H.; Pikkel, R.; Quinn, B., *Ten Types of Innovation: The Discipline of Building Breakthroughs*, Wiley: New York, 2013.
[108] Rosanoff, M.A., "Edison in His Laboratory," *Harpers Mag.*, 1932, *165(987)*, 402–417.
[109] Dyer, F.L.; Martin, T.C., *Edison: His Life and Inventions*, Vol. 2, Harper & Brothers: New York, 1910, pp. 615–616.
[110] Stevens, G.A.; Burley, J., "3,000 Raw Ideas Equals 1 Commercial Success!," *Res. Technol. Mgmt.*, 1997, *40(3)*, 16–27.
[111] Conference Board, *Trading in the Global Ideas Market. 5th Annual Innovation Report*, Report 510-03, Conference Board of Canada: Ottawa, 2002.
[112] Curley, M., "Twelve Principles for Innovation 2.0," *Nature*, 2016, *533(7603)*, 314–316.
[113] Tuff, G.; Wunker, S., "Beacons for Business Model Innovation," *Doblin Research Report*, Deloitte Development: New York, 2014, p. 11.
[114] McDougall, J.R., *Personal Communication, at I-CAN Innovation School*, Toronto, ON, 31 October 2012.
[115] Innovation Leaders Club, "Brewing Success: The Nespresso Story & 3 Innovation Lessons," *YouTube Video Posted April 30*, 2024, https://www.youtube.com/watch?v=zC0I1aXI_Fs&t=201s. accessed 13 November 2025.
[116] Conley, J.G.; Bican, P.M.; Ernst, H., "Value Articulation – A Framework for the Strategic Management of Intellectual Property – The Nespresso Case," *Case Study, Kellogg School of Management*, Northwestern University: Evanston, IL, June 22, 2013, https://www.kellogg.northwestern.edu/Faculty/Conley/htm/VA_Nespresso.pdf. accessed 13 November 2025.
[117] Lachapelle, M., *"Nespresso – Evolution of a $4 Billion Business Model for Coffee," Business Model Design Case Study*, Chartered Professional Accountants of Canada (CPA): Toronto, 2018, p. 12.
[118] Stevens, G.A.; Burley, J., "Piloting the Rocket of Radical Innovation," *Res. Technol. Mgmt.*, 2003, *46(2)*, 16–26.
[119] Thomson, L., *Personality Type. An Owners Manual*, Shambhala Publications: Boston, 1998.
[120] Lundvall, B.-A., "Innovation System Research and Policy. Where It Came from and Where It Might Go," *Presented at CAS Seminar*, Oslo, 4 December 2007.
[121] Fitzgerald, E.A.; Wanker, A.; Schramm, C., *Inside Real Innovation. How the Right Approach Can Move Ideas from R&D to Market – And Get the Economy Moving*, World Scientific Publishing Co.: Singapore, 2010.
[122] Kerzner, H., *Innovation Project Management*, Wiley: Hoboken, NY, 2019.
[123] Rothwell, R.; Zegveld, W., *Reindustrialization and Technology*, Longman: Harlow, 1985.
[124] Usher, A.P., *A History of Mechanical Inventions*, 1st Ed, McGraw-Hill: New York, 1929.
[125] Québec, *Innovation Ouverte, Rapport de Conjoncture 2009*, Conseil de la science et de la technologie: Québec, 2010.

[126] Rothwell, R., "Towards the Fifth-Generation Innovation Process," *Int. Mark. Rev.*, 1994, *11(1)*, 7–31.

[127] Barbieri, J.C.; Álvares, A.C.T., "Sixth Generation Innovation Model: Description of a Success Model," *Rev. Admin. Inovação*, 2016, *13*, 116–127.

[128] Marinova, D.; Phillimore, J., "Innovation Models," in Shavinina, L.V., (Ed.), *The International Handbook on Innovation*, Elsevier: Oxford, 2003, pp. 44–53.

[129] Kucharavy, D.; De Guio, R., "Application of S-Shaped Curves," *Proc. Eng.*, 2011, *9*, 559–572.

[130] Lehfeldt, R.A., "The Normal Law of Progress," *J. R. Stat. Soc.*, 1916, *79(3)*, 329–332.

[131] Foster, R.N., *Innovation: The Attacker's Advantage*, Summit Books: New York, 1986.

[132] Foster, R.N., "Timing Technological Transitions," in *Readings in the Management of Innovation*, Tushman, M.L.; Moore, W.L., (Eds.), 2nd Ed., Harper Business: New York, 1988, pp. 215–228.

[133] Asthana, P., "Jumping the Technology S-Curve," *IEEE Spectr.*, June 1995, 49–54.

[134] Ryu, J.; Byeon, S.C., "Technology Level Evaluation Methodology Based on the Technology Growth Curve," *Technol. Forecast. Soc. Change.*, 2011, *78*, 1049–1059.

[135] Phillips, F., "On S-Curves and Tipping Points," *Technol. Forecast. Soc. Change.*, 2007, *74*, 715–730.

[136] Modis, T., "Strengths and Weaknesses of S-Curves," *Technol. Forecast. Soc. Change.*, 2007, *74*, 866–872.

[137] Scharl, S., "Human Mobility and the Spread of Innovations – Case Studies from Neolithic Central and Southeast Europe," *Open Archaeol.*, 2023, *9(20220325)*, 20.

[138] Huenteler, J.; Schmidt, T.S.; Ossenbrink, J.; Hoffmann, V.H., "Technology Life-Cycles in the Energy Sector – Technological Characteristics and the Role of Deployment for Innovation," *Technol. Forecast. Soc. Change.*, 2016, *104*, 102–121.

[139] Asthana, P., "Jumping the Technology S-Curve," *IEEE Spectr.*, 1995, *June*, 49–54.

[140] Gartner, *Gartner Hype Cycle*, Gartner, Inc: Stamford, CT, 2016, https://www.gartner.com/en/research/methodologies/gartner-hype-cycle. accessed 13 November 2025.

[141] Shi, Y.; Herniman, J., "The Role of Expectation in Innovation Evolution: Exploring Hype Cycles," *Technovation*, 2023, *119*, 102459.

[142] Dean, R.C., "Technical Innovation in the U.S.A," *Mech. Eng.*, *Nov* 1978, 23–32.

[143] Moore, G.A., *Crossing the Chasm*, Harper Collins: New York, 1991.

[144] Atkinson, R.D.; Bennett, R., "The Future of the Internet and Broadband . . . and How to Enable It," *Presented at "Big Ideas Workshop," National Broadband Plan, Federal Communications Commission*, Washington, DC, 3 September 2009, https://d1bcsfjk95uj19.cloudfront.net/files/20090903_The%20Future_of_the_Internet_FCC.pdf. accessed 13 November 2025.

[145] Reich, L.S., *The Making of American Industrial Research. Science and Business at GE and Bell, 1876–1926*, Cambridge University Press: Cambridge, UK, 1985.

[146] I-CAN, "Accessing the System," *Annual Report 2011*, Innoventures Canada Inc. (I-CAN): Ottawa, 2011, p. 28.

[147] Bashir, N., "Knowledge Identification for Open Innovation: A Case Study of Biotechnology Firms," *Bus. Process Manag. J.*, 2024, *30(4)*, 1131–1153.

[148] Kozioł-Nadolna, K.; Świadek, A., "Innovation Process Models with Emphasis on Open Innovation Model," *Folia Oeconomica Stetinensia*, January 2011, 13, doi: 10.2478/v10031-010-0007-5.

[149] Gabison, G.; Pesole, A., *"An Overview of Models of Distributed Innovation," Report EUR 27035 EN, Joint Research Centre*, European Commission: Luxembourg, 2014, p. 29.

[150] Håkansson, H.; Waluszewski, A., Eds., *Knowledge and Innovation in Business and Industry. The Importance of Using Others*, Routledge: New York, 2007.

[151] Jaekel, M.; Wallin, A.; Isomursu, M., "Guiding Networked Innovation Projects Towards Commercial Success – A Case Study of an EU Innovation Programme with Implications for Targeted Open Innovation," *J. Knowl. Econ.*, 2015, *6*, 625–639.

[152] Mäkimattila, M.; Melkas, H.; Uotila, T., "Dynamics of Openness in Innovation Processes – A Case Study in the Finnish Food Industry," *Knowl. Process Manag.*, 2013, *20(4)*, 243–255.

[153] Immelt, J.R.; Govindarajan, V.; Trimble, C., "How GE Is Disrupting Itself," *Harv. Bus. Rev.*, 2009, *87*(*10*), 56–65.
[154] Govindarajan, V.; Trimble, C., *Reverse Innovation: Create Far from Home, Win Everywhere*, Harvard Business Review Press: Boston, MA, 2012.
[155] Von Zedtwitz, M.; Corsi, S.; Søberg, P.V.; Frega, R., "A Typology of Reverse Innovation," *J. Prod. Innov. Manag.*, 2015, *32*(*1*), 12–28.
[156] Chesbrough, H., "Open Innovation and Open Business Models: A New Approach to Industrial Innovation," *Presentation to Joint OECD/Dutch Ministry of Economic Affairs Conference on "Globalization and Open Innovation*, 6 December 2006, https://www.academia.edu/6324588/Open_Innovation_and_Open_Business_Models_A_new_approach_to_industrial_innovation_Presentation_to_Joint_OECD_Dutch_Ministry_of_Economic_Affairs_Conference_on_Globalization_and_Open_Innovation_. accessed 24 November 2025.
[157] Wolcott, R.C.; Lippitz, M.J., *BP's Office of the Chief Technology Officer (B): Driving Open Innovation through an Advocate Team*, Sage Business Cases, SAGE Publications: Thousand Oaks, CA, 2016, doi.org/10.4135/9781473968912.
[158] Chesbrough, H.; Van Alstyne, M., "Permissionless Innovation," *Commun. ACM*, 2015, *58*(*8*), 24–26.
[159] "Grand Challenges," a family of initiatives including Grand Challenges Canada, USAID, and the Bill & Melinda Gates Foundation, 2016, https://grandchallenges.org/grant-opportunities.
[160] Chesbrough, H.W., *Open Innovation: The New Imperative for Creating and Profiting from Technology*, Harvard Business School Press: Boston, 2003.
[161] Huizingh, E.K.R.E., "Open Innovation: State of the Art and Future Perspectives," *Technovation*, 2011, *31*(*1*), 2–9.
[162] Chesbrough, H.; Vanhaverbeke, W.; West, J., Eds., *Open Innovation: Researching a New Paradigm*, Oxford University Press: UK, 2006.
[163] Roelandt, T.J.A.; Hertog, P.D., "Cluster Analysis and Cluster-Based Policy Making: The State of the Art," in *Boosting Innovation: The Cluster Approach*, Organisation for Economic Co-Operation and Development (OECD): Paris, 1999, p. 413–427.
[164] Chesbrough, H., "GE's Ecomagination Challenge: An Experiment in Open Innovation," *Calif. Manage. Rev.*, 2012, *54*(*3*), 140–154.
[165] Buyse, K., "Five Generations of Innovation Models," *Blog Post, Stratagem*, Sept. 29 2012, https://stratagemgr.wordpress.com/2012/09/29/rothwells-generations-innovation-models/. accessed 24 November 2025.
[166] Meissner, D.; Kotsemir, M., "Conceptualizing the Innovation Process Towards the 'Active Innovation Paradigm'- Trends and Outlook," *J. Innov. Entrep.*, 2016, *5*(*14*), 18, doi: 10.1186/s13731-016-0042-z.
[167] Stefanovska Ceravolo, L.J.; Polenakovikj, R.; Dzidrov, M., "Summary of Innovation Models on a Company Level – Creating a Framework for an Innovation Model that Will Increase a Company's Innovation Activity," *Sci. Bus. Soc.*, 2016, *1*(*6*), 23–26.
[168] Mintzberg, H., *The Rise and Fall of Strategic Planning*, Simon and Schuster: New York, 1994.
[169] Florida, R., *The Rise of the Creative Class*, Basic Books: New York, 2002.
[170] Osborn, A.F., *Your Creative Power, How to Use Imagination*, Charles Scribner's Sons: New York, 1949.
[171] Osborn, A.F., *Applied Imagination: Principles and Procedures of Creative Problem-Solving*, Charles Scribner's Sons: New York, 1953.
[172] Isaksen, S.; Treffinger, D., "Celebrating 50 Years of Reflective Practice: Versions of Creative Problem Solving," *J. Creat. Behav.*, 2004, *38*(*2*), 1–27.
[173] Parnes, S.J., *Creative Behavior Guidebook*, Scribner: New York, 1967.
[174] de Bono, E., *The Use of Lateral Thinking*, Jonathan Cape: London, 1967. also published as *New Think*, Basic Books: New York.
[175] de Bono, E., *Lateral Thinking for Management*, American Management Association: New York, 1972.

[176] de Bono, E., *Six Thinking Hats: An Essential Approach to Business Management*, Little Brown and Co.: New York, 1985.

[177] de Bono Group, *"Six Thinking Hats," Corporate Web Page*, The de Bono Group: Lancaster, PA, 2017, https://www.debonogroup.com/services/core-programs/six-thinking-hats/. accessed 13 November 2025.

[178] Gordon, W.J.J., *Synectics, the Development of Creative Capacity*, Harper and Row: New York, 1961.

[179] Nolan, V., "Whatever Happened to Synectics?," *Creat. Innov. Manag.*, 2003, *12(1)*, 24–27.

[180] Johansson, F., *The Medici Effect. Breakthrough Insights at the Intersection of Ideas, Concepts and Cultures*, Harvard Business School Press: Boston, 2004.

[181] Roberts, E.B., Ed., *Innovation. Driving Product, Process, and Market Change*, Jossey-Bass: San Francisco, 2002.

[182] Johnston, R.E.; Bate, J.D., *The Power of Strategy Innovation*, American Management Association: New York, 2003.

[183] O'Sullivan, D.; Dooley, L., *Applying Innovation*, SAGE Publications: Thousand Oaks, CA, 2009.

[184] Franklin, S.W., *The Cult of Creativity*, University of Chicago Press: Chicago, 2023.

[185] Ford, G.S.; Koutsky, T.M.; Spiwak, L.J., *A Valley of Death in the Innovation Sequence*, Phoenix Center for Advanced Legal and Economic Public Policy Studies: Washington, September 2007, p. 43.

[186] Touhill, C.J.; Touhill, G.J.; O'Riordan, T.A., *Commercialization of Innovative Technologies*, Wiley-Interscience: New York, 2008.

[187] Huang, G.Q.; Mak, K.L., "Developing a Generic Design for X Shell," *J. Eng. Des.*, 1997, *8(3)*, 251–260.

[188] Huang, G.Q.; Mak, K.L., "The DFX Shell: A Generic Framework for Applying `design for X' (DFX) Tools," *Int. J. Comput. Integr. Manuf.*, 1998, *11(6)*, 475–484.

[189] Donaldson, K.M.; Ishii, K.; Sheppard, S.D., "Customer Value Chain Analysis," *Res. Eng. Des.*, 2006, *16*, 174–183.

[190] Freiler, L., "Alpha Vs. Beta Testing," in *Blog Post*, Centercode: Laguna Hills, CA, 17 January 2011, https://www.centercode.com/blog/alpha-vs-beta-testing. accessed 13 November 2025.

[191] QATestLab, "Difference between Alpha, Beta, and Gamma Testing" Knowledge Center Blog, 21 August 2018, https://qatestlab.com/resources/knowledge-center/alpha-beta-gamma/, accessed 13 November 2025.

[192] Katz, G., "Rethinking the Product Development Funnel," *Visions*, 2011, *July*, 24–31.

[193] Schaufeld, J., *Commercializing Innovation. Turning Technology Breakthroughs into Products*, Springer: New York, 2015.

[194] Mankins, J.C., "Technology Readiness Assessments: A Retrospective," *Acta Astronaut.*, 2009, *65*, 1216–1223.

[195] Canada, P.W.G.S., "Technology Readiness Levels," *Public Works and Government Services*, Canada: Ottawa, 2013, https://buyandsell.gc.ca/initiatives-and-programs/build-in-canada-innovation-program-bcip/program-specifics/technology-readiness-levels.

[196] EARTO, *The TRL Scale as a Research & Innovation Policy Tool, EARTO Recommendations*, European Association of Research and Technology Organizations: Brussels, 2014.

[197] ARENA, "Technology Readiness Levels for Renewable Energy Sectors," *Australian Renewable Energy Agency*, Canberra, February 2014.

[198] ISO, "Space Systems – Definition of the Technology Readiness Levels (Trls) and Their Criteria of Assessment," *ISO Standard 16290*, 2013, International Organization for Standardization: Geneva, 2013.

[199] Lyons, J.; Chait, R.; Long, D., "Critical Technology Events in the Development of Selected Army Weapons Systems: A Summary of 'Project Hindsight Revisited,'," *Defense & Technology Paper DTP-035*, National Defense University Center for Technology and National Security Policy: Washington, September 2006.

[200] ARENA, "Commercial Readiness Index for Renewable Energy Sectors," *Australian Renewable Energy Agency*, Canberra, February 2014.

[201] Yeo, W.; Kimb, S.; Park, H.; Kang, J., "A Bibliometric Method for Measuring the Degree of Technological Innovation," *Technol. Forecast. Soc. Change*, 2015, 95, 152–162.

[202] Thota, H.; Munir, Z., *Palgrave Key Concepts: Key Concepts in Innovation*, Macmillan: New York, 2011.

[203] Bank of England, *Financing of Technology-Based Small Firms*, Bank of England: London, February 2001.

[204] Libecap, G.D.; Thursby, M.C., Eds., *Technological Innovation: Generating Economic Results*, Emerald: Bingley, UK, 2008.

[205] Bound, K.; Mulgan, G., (Eds.), "A Compendium of Innovation Methods," *Report*, Nesta: London, UK, March 12 2019, https://www.nesta.org.uk/report/compendium-innovation-methods/. accessed 13 November 2025.

[206] Yadav, V.; Goyal, P., "User Innovation and Entrepreneurship: Case Studies from Rural India," *J. Innov. Entrep.*, 2015, *4*(*5*), 20.

[207] Mitra, J., *Entrepreneurship, Innovation and Regional Development. An Introduction*, 2nd Ed, Routledge: New York, 2020.

[208] Rogers, E.M.; Shoemaker, F.F., *Communication of Innovations: A Cross-Cultural Approach*, 2nd Ed, Free Press: New York, 1971.

[209] Rogers, E.M., "Rise of the Classical Diffusion Model," *Curr. Contents*, 1991, *28*, 16.

[210] Mahajan, V.; Peterson, R.A., *Models for Innovation Diffusion*, SAGE Publications: Newbury Park, 2011.

[211] MacVaugh, J.; Schiavone, F., "Limits to the Diffusion of Innovation. A Literature Review and Integrative Model," *Eur. J. Innov. Manag.*, 2010, *13*(*2*), 197–221.

[212] Schiavone, F.; MacVaugh, J., "Nonadoption of New Technology," in *The Dark Side of Technological Innovation*, Ran, B., (Ed.), Information Age Publishing: Charlotte, NC, 2013, pp. 123–148.

[213] Robertson, T.S., "The Process of Innovation and the Diffusion of Innovation," *J. Mark.*, 1967, *31*, 14–19.

[214] Mansfield, E., "Technical Change and the Rate of Imitation," *Econometrica*, 1961, *29*(*4*), 741–766.

[215] Rogers, E.M.; Singhal, A.; Quinlan, M.M., "Diffusion of Innovations," in *An Integrated Approach to Communication Theory and Research*, Stacks, D.; Salwen, M., (Eds.), 2nd Ed, Routledge: New York, 2009, pp. 418–434.

[216] Howard, J.A.; Moore, W.L., "Changes in Consumer Behavior over the Product Life Cycle," in *Readings in the Management of Innovation*, Tushman, M.L.; Moore, W.L., (Eds.), 2nd Ed, Harper Business: New York, 1988, pp. 343–351.

[217] Gladwell, M., *The Tipping Point*, Little, Brown and Co.: New York, 2002.

[218] Maloney, C., "The Secret to Accelerating Diffusion of Innovation: The 16% Rule Explained," in *Innovate or Die*, Internet Blog, 10 May 2010, https://innovateordie.com.au/2010/05/10/the-secret-to-accelerating-diffusion-of-innovation-the-16-rule-explained/. accessed 23 November 2025.

[219] Cooper, J.; Greenberg, D.; Zuk, J., *"Reshaping the Funnel: Making Innovation More Profitable for High-Tech Manufacturers," IBM Institute for Business Value Report*, IBM Corporation: Somers, NY, 2002, p. 19.

[220] Parasuraman, A., "Technology Readiness Index (TRI). A Multiple-Item Scale to Measure Readiness to Embrace New Technologies," *J. Serv. Res.*, 2000, *2*(*4*), 307–320.

[221] Parasuraman, A.; Colby, C.L., "An Updated and Streamlined Technology Readiness Index. TRI 2.0," *J. Serv. Res.*, published online before print 22 June 2014, doi: 10.1177/1094670514539730.

[222] Lee, Y.-H.; Hsieh, Y.-C.; Hsu, C.-N., "Adding Innovation Diffusion Theory to the Technology Acceptance Model: Supporting Employees' Intentions to Use E-Learning Systems," *Educ. Technol. Soc.*, 2011, *14*(*4*), 124–137.

[223] Cooper, R.G., *Winning at New Products: Creating Value through Innovation*, 4th Ed, Basic Books: New York, 2011.

[224] Cooper, R.G., "New Products – What Separates the Winners from the Losers and What Drives Success," in *The PDMA Handbook of New Product Development*, Kahn, K.B., (Ed.), 3rd Ed, Wiley: New York, 2013, pp. 3–34.

[225] Cooper, R.G., "Third Generation New Product Process," *J. Prod. Innov. Manag.*, 1994, *11(1)*, 3–14.

[226] Bowers, J.; Khorakian, A., "Managing Risk in Innovation Projects," in *The Dark Side of Technological Innovation*, Ran, B., (Ed.), Information Age Publishing: Charlotte, NC, 2013, pp. 278–298.

[227] Cooper, R.G., "Where are All the Breakthrough New Products?," *Res. Technol. Mgmt.*, 2013, *56(5)*, 25–32.

[228] Cooper, R.G.; Edgett, S.J.; Kleinschmidt, E.J., *Portfolio Management for New Products*, 2nd Ed, Perseus Publishing: Cambridge, MA, 2001.

[229] Ajamian, G.M.; Koen, P.A., "Technology Stage-Gate™: A Structured Process for Managing High-Risk New Technology Projects," in *The PDMA ToolBook for New Product Development*, Belliveau, P.; Griffin, A.; Sommermeyer, S., (Eds.), Wiley: New York, 2002, pp. 29.

[230] Johnston, C.; Grant, M., *"Innovation Management for Established Businesses. Management Matters," Report 6368*, Conference Board of Canada: Ottawa, August 2014.

[231] Royce, W., "Managing the Development of Large Software Systems," *Proc. IEEE WESCON*, 1970, *26 (Aug.)*, 1–9.

[232] Bell, T.E.; Thayer, T.A., "Software Requirements: Are They Really a Problem?," in *Proc. 2nd Internat. Conf. Software Engineering*, IEEE Computer Society Press, 1976, pp. 61–68.

[233] Marquis, A.; Searle, M.; Jovin, D., "Bosch: Scaling Large Company Innovation for Strategic Advantage," in *The Berkeley Haas Case Series*, University of California: Berkeley Haas School of Business, 2022, pp. 15.

[234] Lewis, T., "Corporate R&D in the Age of E-Commerce," *Res. Technol. Mgmt.*, 2000, *43(6)*, 16–20.

[235] Maxwell, J.C., *Failing Forward*, Thomas Nelson Publishers: New York, 2000.

[236] WD-40 Co., *History: Fascinating Facts You Never Learned in School*, WD-40 Company: San Diego, 2015, http://wd40.com/cool-stuff/history.

[237] Comedy, Y.L.; Grama, S., "The Critical Role of Failure in the Innovation Process: How Failures Help Inventors Succeed," *Technol. Innov*, 2016, *17*, 169–176.

[238] Ford, H.; Crowther, S., *My Life and Work*, Garden City Publishing Co: Garden City, NY, 1922, pp. 19–20.

[239] Ates, A.; Suppayah, K., "Disciplined Innovation: A Case Study of the Amazon Working Backwards Approach to Internal Corporate Venturing," *Res. Technol. Mgmt.*, 2024, *67(3)*, 23–33.

[240] Farson, R.; Keyes, R., *The Innovation Paradox: The Success of Failure, the Failure of Success*, Free Press: New York, 2003.

[241] Davila, T.; Epstein, M.J., *The Innovation Paradox: Why Good Businesses Kill Breakthroughs and How They Can Change*, Berrett-Koehler Publishers: San Francisco, 2014.

[242] Khan, Z.; Joseph, K., "Embracing the Paradoxes of Innovation," *Stanford Soc. Innov. Rev.*, 2013, *Summer*, 21–23.

[243] Conference Board of Canada, *"Collaborating for Innovation. 2nd Annual Innovation Report," Report 303-00*, The Conference Board of Canada: Ottawa, 2000.

[244] Smith, K.; Dietrichs, E.; Nås, S.O., *The Norwegian National Innovation System: A Preliminary Overview and Assessment*, Organisation for Economic Co-Operation and Development (OECD): Paris, 1995.

[245] Roelandt, T.J.A., "Cluster Analysis and Cluster-Based Policy Making in OECD Countries: An Introduction to the Theme," in *Boosting Innovation: The Cluster Approach*, Organisation for Economic Co-Operation and Development (OECD): Paris, 1999, p. 9–23.

[246] Edquist, E., Ed., *Systems of Innovation. Technologies, Institutions and Organizations*, Routledge: Oxford, 1997.

[247] Adner, R.; Kapoor, R., "Value Creation in Innovation Ecosystems: How the Structure of Technological Interdependence Affects Firm Performance in New Technology Generations," *Strateg. Manag. J.*, 2010, *31*, 306–333.
[248] Hsu, C.-W., "Formation of Industrial Innovation Mechanisms through the Research Institute," *Technovation*, 2005, *25*, 1317–1329.
[249] OECD, *National Innovation Systems*, Organisation for Economic Co-Operation and Development (OECD): Paris, 1997.
[250] Freeman, C., "The 'National System of Innovation' in Historical Perspective," *Cambridge J. Econ.*, 1995, *19*, 5–24.
[251] Lundvall, B.-A.; Johnson, B.; Andersen, E.S.; Dalum, B., "National Systems of Production, Innovation and Competence Building," *Res. Policy*, 2002, *31*, 213–231.
[252] Lundvall, B.-A., "National Innovation Systems – Analytical Concept and Development Tool," *Ind. Innov.*, 2007, *14(1)*, 95–119.
[253] Cooke, P., *"Introduction," in Regional Innovation Systems. The Role of Governances in a Globalized World*, Braczyk, H.-J.; Cooke, P.; Heidenreich, M., Eds., UCL Press: London, UK, 1998, pp. 2–25.
[254] Asheim, B.T.; Smith, H.L.; Oughton, C., "Regional Innovation Systems: Theory, Empirics and Policy," *Reg. Stud.*, 2011, *45(7)*, 875–891.
[255] OECD, *Regions and Innovation Policy*, Organisation for Economic Co-Operation and Development (OECD): Paris, 2011.
[256] Asheim, B.; Cooke, P.; Martin, R., Eds., *Clusters and Regional Development. Critical Reflections and Explorations*, Routledge: Oxford, 2006.
[257] CICC Global Institute, *The Rise of China's Innovation Economy*, Springer Nature Singapore: Singapore, 2022.
[258] Storper, M., *The Regional World – Territorial Development in a Global Economy*, Guilford Press: New York, 1997.
[259] Etzkowitz, H.; Leydesdorff, L., "The Triple Helix-University-Industry-Government Relations: A Laboratory for Knowledge Based Economic Development," *EASST Rev.*, 1995, *14(1)*, 14–19.
[260] Leydesdorff, L., "The Knowledge-Based Economy and the Triple Helix Model," *Annu. Rev. Inf. Sci. Technol.*, 2010, *44*, 367–417.
[261] Etzkowitz, H.; Leydesdorff, L., "The Dynamics of Innovation: From National Systems and 'Mode 2' to a Triple Helix of University-Industry-Government Relations," *Res. Policy*, 2000, *29*, 109–123.
[262] Etzkowitz, H., "Innovation in Innovation: The Triple Helix of University-Industry-Government Relations," *Soc. Sci. Inf.*, 2003, *42(3)*, 293–337.
[263] Potter, J.; Miranda, G., Eds., *Clusters, Innovation, and Entrepreneurship*, Organisation for Economic Co-Operation and Development (OECD): Paris, 2009.
[264] Asplund, C., "Beyond 'Triple Helix' – Towards 'Quad Helix'," internet blog post, 22 March 2012, Bearing Consulting, http://blog.bearing-consulting.com/2012/03/22/beyond-triple-helix-towards-quad-helix/
[265] Wilson, E.J., "Forms and Dynamics of Leadership for a Knowledge Society: The Quad," in *Center for International Development and Conflict Management*, University of Maryland: College Park, 2003, https://www.semanticscholar.org/paper/Forms-and-Dynamics-of-Leadership-for-a-Knowledge-Wilson/946823f6d987de53e47cc8534e3dfa38d7c90e7a. accessed 13 November 2025.
[266] Leydesdorff, L., "The Triple Helix, Quadruple Helix, . . ., and an N-tuple of Helices: Explanatory Models for Analyzing the Knowledge-based Economy," *J. Knowl. Econ.*, 2012, *3(1)*, 25–35.
[267] UNESCO, "Toward Knowledge Societies," *UNESCO World Report*, United Nations Educational, Scientific and Cultural Organization: Paris, 2005.
[268] Leydesdorff, L., "N-Tuple of Helices," in *Encyclopedia of Creativity, Innovation, and Entrepreneurship*, Carayannis, E.G., (Ed.), Springer: New York, 2013, pp. 1400–1402.

[269] Ahonen, L.; Hämäläinen, T., "CLIQ: A Practical Approach to the Quadruple Helix and More Open Innovation," in *Sustaining Innovation: Collaboration Models for a Complex World*, MacGregor, S.P.; Carleton, T., (Eds.), Ch. 2, Springer: New York, 2012, pp. 15–29.
[270] Dedehayir, O.; Seppänen, M., "Birth and Expansion of Innovation Ecosystems: A Case Study of Copper Production," *J. Technol. Manag. Innov.*, 2015, *10(2)*, 145–153.
[271] Owen, R.; Vedanthacharia, L.N.; Hussainc, J., "The Role of the University Entrepreneurial Ecosystem in Entrepreneurial Finance: Case Studies of UK Innovation Knowledge Centres," *Venture Cap.*, 2024, *26(3)*, 351–375.
[272] Yasin, N.; Gilani, S.A.M., "Assessing the Current State of University-Based Business Incubators in Canada," *Ind. High. Educ.*, 2023, *37(3)*, 359–369.
[273] OECD, *"A Guiding Framework for Entrepreneurial Universities*, Organisation for Economic Co-Operation and Development (OECD): Paris, 2012.
[274] Dalziel, M., "4th Pillar Organizations in Canada," *Report for Prime Minister's Advisory Council on Science and Technology*, Government of Canada, 15 September 2005, http://papers.ssrn.com/sol3/papers.cfm?abstract_id=1518863. accessed 23 November 2025.
[275] Dalziel, M., "Why Do Innovation Intermediaries Exist?," in *Proc. Summer Conference 2010 on Opening up Innovation: Strategy, Organization and Technology, Imperial College London Business School*, 16–18 June 2010, p. 15.
[276] Howells, J., "Intermediation and the Role of Intermediaries in Innovation," *Res. Policy*, 2006, *35*, 715–728.
[277] Howard Partners Pty, "Study of the Role of Intermediaries in Support of Innovation," *Report for Department of Industry, Tourism and Resources*, April 2007.
[278] OECD, "Policy Brief on Incubators and Accelerators that Support Inclusive Entrepreneurship," *Organisation for Economic Co-operation and Development*, Paris, France, 2019.
[279] OECD, *"Business Incubation. International Case Studies"*, Organisation for Economic Co-operation and Development: Paris, France, 1999.
[280] OECD, "Promoting Start-Ups and Scale-Ups in Denmark's Sector Strongholds and Emerging Industries," Organisation for Economic Co-operation and Development, Paris, France, 2022.
[281] Roberts, P.S.; Schmid, J., "Government-Led Innovation Acceleration: Case Studies of US Federal Government Innovation and Technology Acceleration Organizations," *Rev. Policy Res.*, 2022, *39*, 353–378.
[282] Suvinen, N.; Konttinen, J.; Nieminen, M., "How Necessary are Intermediary Organizations in the Commercialization of Research?," *Eur. Plan. Stud.*, 2010, *18(9)*, 1365–1389.
[283] Yildiz, H.B.; Henchion, M.; Hewitt-Dundas, N., *Exploring Innovation Intermediaries*, unpublished manuscript, Queen's University Management School: Belfast, Northern Ireland, 2012, pp. 26.
[284] NRC, "Concierge," *Program Brochure*, National Research Council of Canada: Ottawa, 2016.
[285] Wright, J.D., "The Missing Link in Canada's Innovation Chain," *Res. Money, April 16* 2003, 8.
[286] EARTO, *RTOs – Research and Technology Organisations*, European Association of Research and Technology Organizations: Brussels, 2014, http://www.earto.eu/about-rtos/. accessed 13 November 2025.
[287] ACIL Tasman, "Assessment of CSIRO Impact & Value," *Report Prepared as Input to CSIRO's Lapsing Program Review*, ACIL Tasman Pty Ltd: Melbourne, Australia, July 2010, p. 325.
[288] Conference Board of Canada, *"Innovation Challenge Paper #4," Report 349-02*, The Conference Board of Canada: Ottawa, 2002.
[289] Conference Board of Canada, "Networking Canada's Innovation Networks," *Report*, The Conference Board of Canada: Ottawa, September 2001.
[290] Schramm, L.L., *Research and Development on the Prairies: A History of the Saskatchewan Research Council*, Saskatchewan Research Council, Saskatoon, and Amazon.com Inc.: Seattle, WA, 2016.

[291] Grier, D.; Mengu, M., "Best Practices for Management of Research & Technology Organizations," in *Proceedings, IAMOT Conference, IAMOT Conference: Orlando, FL, 1998.*

[292] Hauser, H., "The Current and Future Role of Technology and Innovation Centres in the UK," Report for the Secretary of State, Department for Business, Innovation and Skills: London, UK, March 2010.

[293] Harding, R., "Competition and Collaboration in German Technology Transfer," *Eur. Manag. J.*, 2002, *20(5)*, 470–485.

[294] TNO, "TNO Innovation for Life," The Hague, the Netherlands, 2017, https://www.tno.nl/en/about-tno/, accessed 13 November 2025.

[295] NRC, "National Research Council Canada," National Research Council: Ottawa, 2025, https://nrc.canada.ca/en/corporate/corporate-overview, accessed 13 November 2025.

[296] EARTO, European Association of Research and Technology Organizations, http://www.earto.eu/, accessed 13 November 2025.

[297] WAITRO, World Association of Industrial and Technological Research Organizations, https://waitro.org, accessed 13 November 2025.

[298] U.S. DOE, *National Laboratories*, U.S. Dept. of Energy: Washington, 2017, https://energy.gov/national-laboratories.

[299] Wikipedia, *United States Department of Energy National Laboratories*, Wikimedia Foundation: San Francisco, 2017, https://en.wikipedia.org/wiki/United_States_Department_of_Energy_national_laboratories. accessed 23 November 2025.

[300] Klika, C.; Longinidou, E., "The Role of Innovation Labs in the Public Sector: Concepts and Conditions for Successful Transformation," in *Briefing Report*, European Institute of Public Administration (EIPA): Maastricht, The Netherlands, 2024, p. 14.

[301] OECD, *National Innovation Systems: Report on Pilot Case Studies*, Organisation for Economic Co-operation and Development: Paris, France, 1996.

[302] Lundvall, B.-A., "National Systems of Innovation: Towards a Theory of Innovation and Interactive Learning," in *The Learning Economy and the Economics of Hope*, Lundvall, B.-A., Anthem Press: London, 2015, pp. 85–106.

[303] Schmoch, U.; Rammer, C.; Legler, H., Eds., *National Systems of Innovation in Comparison. Structure and Performance Indicators for Knowledge Societies*, Springer: Dordrecht, 2006.

[304] Parliament (Australia), "Pathways to Technological Innovation," *Report of the House of Representatives Standing Committee on Science and Innovation*, Parliament of the Commonwealth of Australia: Canberra, 2006, https://aphref.aph.gov.au/house/committee/scin/pathways/report/fullreport.pdf. accessed 13 November 2025.

[305] NEC, "A Strategy for American Innovation," National Economic Council and Office of Science and Technology Policy, The White House: Washington, October 2015, https://obamawhitehouse.archives.gov/sites/default/files/strategy_for_american_innovation_october_2015.pdf, accessed 13 November 2025.

[306] Vonortas, N.S.; Rouge, P.C.; Aridi, A., Eds., *Innovation Policy. A Practical Introduction*, Springer: New York, 2015.

[307] OECD, *"Innovation Vouchers," OECD Innovation Policy Platform*, Organisation for Economic Co-Operation and Development (OECD): Paris, February, 2010.

[308] Conference Board of Canada, "Performance and Potential 1997," *Report*, The Conference Board of Canada: Ottawa, 1997.

[309] Howells, J., "Research and Technology Outsourcing and Innovation Systems: An Exploratory Analysis," *Ind. Innov.*, 1999, *6*, 111–129.

[310] Nelson, R.R.; Langlois, R.N., "Industrial Innovation Policy: Lessons from American History," in *Readings in the Management of Innovation*, Tushman, M.L.; Moore, W.L., (Eds.), 2nd Ed, Harper Business: New York, 1988, pp. 661–665.

[311] Boekholt, P.; Thuriaux, B., "Public Policies to Facilitate Clusters: Background, Rationale and Policy Practices in International Perspective," in *Boosting Innovation: The Cluster Approach*, Organisation for Economic Co-Operation and Development (OECD): Paris, 1999, p. 381–412.

[312] Institute for Competitiveness & Prosperity "Clusters in Ontario, "Creating an Ecosystem for Prosperity," *Working Paper 26*, Institute for Competitiveness & Prosperity: Toronto, 2016, https://www.researchgate.net/publication/332869196_Clusters_in_Ontario_Creating_an_ecosystem_for_prosperity. accessed 13 November 2025.

[313] Huggins, R., "The Evolution of Knowledge Clusters. Progress and Policy," *Econ. Dev. Q.*, 2008, *22*(*4*), 277–289.

[314] Furr, N.; O'Keeffe, K.; Dyer, J.H., "Managing Multiparty Innovation," *Harv. Bus. Rev.*, 2016, 76–83.

[315] Klimova, N.; Kozyrev, O.; Babkin, E., *Innovation in Clusters Understanding Universities, Special Economic Zones, and Modeling*, Springer: Heidelberg, 2016.

[316] Porter, M.E. "Clusters and the New Economics of Competition," *Harv. Bus. Rev.*, 1998, *76*(*6*), 77–90.

[317] Harrison, R.T.; Cooper, S.Y.; Mason, C.M., "Entrepreneurial Activity and the Dynamics of Technology-Based Cluster Development: The Case of Ottawa," *Urban Stud.*, 2004, *41*(*5/6*), 1045–1070.

[318] Wolfe, D.A.; Lucas, M., Eds., *Global Networks and Local Linkages*, McGill-Queens University Press: Montreal & Kingston, 2005.

[319] Holbrook, J.A.; Arthurs, D.; Cassidy, E., "Understanding the Vancouver Hydrogen and Fuel Cells Cluster: A Case Study of Public Laboratories and Private Research," *Eur. Plan. Stud.*, 2010, *18*(*2*), 317–328.

[320] Juglar, C., *Des Crises Commerciales Et Leur Retour Periodique En France, En Angleterre, Et Aux Etats-Unis*, Guillaumin: Paris, 1862.

[321] Kitchin, J., "Cycles and Trends in Economic Factors," *Rev. Econ. Stat.*, 1923, *5*(*1*), 10–16.

[322] Gordon, R.J., *The Rise and Fall of American Growth*, Princeton University Press: Princeton, 2016.

[323] Nagengast, B.; Groff, G., "Keeping Cool. A Timeline of Refrigeration and Air Conditioning," *Am. Soc. Heat. Refrig. Air-Cond. Eng. J., March* 2007, 64–68.

[324] Launius, R.D., Ed., *Innovation and the Development of Flight*, Texas A & M University Press: College Station, 1999.

[325] Williams, J.R.; Weiss, M., "Competitive Innovation in the U.S. Airline Industry," in *The Dark Side of Technological Innovation*, Ran, B., (Ed.), Information Age Publishing: Charlotte, NC, 2013, pp. 31–44.

[326] Malerba, F.; Nelson, R.R.; Orsenigo, L.; Winter, S.G., *Innovation and the Evolution of Industries. History-Friendly Models*, Cambridge University Press: Cambridge, 2016.

[327] Nagengast, B., "A History of Refrigeration," *Am. Soc. Heat. Refrig. Air-Cond. Eng. J., November* 2004, S11–S19.

[328] Hoffman, J.T., Ed., *A History of Innovation. U.S. Army Adaptation in War and Peace*, Center of Military History, United States Army: Washington, 2009.

[329] Sabel, C.F.; Herrigel, G.; Kazis, R.; Deeg, R., "How to Keep Mature Industries Innovative," in *Readings in the Management of Innovation*, Tushman, M.L.; Moore, W.L., (Eds.), 2nd Ed, Harper Business: New York, 1988, pp. 651–660.

[330] Rosenbloom, R.S.; Cusumano, M.A., "Technological Pioneering and Competitive Advantage: The Birth of the VCR Industry," in *Readings in the Management of Innovation*, Tushman, M.L.; Moore, W.L., (Eds.), 2nd Ed, Harper Business: New York, 1988, pp. 3–22.

[331] Howells, J., *The Management of Innovation and Technology: The Shaping of Technology and Institutions of the Market Economy*, SAGE Publications: London, 2005.

[332] Graham, M.W., *The Business of Research. RCA and the VideoDisc*, Cambridge University Press: Cambridge, 1986.

[333] Zolla, G.A., "Information Technology Diffusion: A Comparative Case Study of Intranet Adoption," *Proc. Portland Internat. Conf. Mgmt. Engineering and Technology (PICMET 99)*, Portland, OR, 25–29 July 1999, p. 6.

[334] Silva, H.P.; Viana, A.L., "Health Technology Diffusion in Developing Countries: A Case Study of CT Scanners in Brazil," *Health Policy Plan.*, 2011, *26(5)*, 385–394.

[335] Neij, L.; Andersen, P.D., "A Comparative Assessment of Wind Turbine Innovation and Diffusion Policies. Historical Case Studies of Energy Technology Innovation," in *The Global Energy Assessment*, Grubler, A.; Aguayo, F.; Gallagher, K.S.; Hekkert, M.; Jiang, K.; Mytelka, L.; Neij, L.; Nemet, G.; Wilson, C., (Eds.), Chapter 24, Cambridge University Press: Cambridge, 2012, pp. 1665–1744.

[336] Montalvo, C., "Cleaner Technology Diffusion: Case Studies, Modeling and Policy," *J. Clean. Prod.*, 2008, *16(1)*, S1–S6.

[337] Haselip, J.; Nygaard, I.; Hansen, U.; Ackom, E., Eds., *Diffusion of Renewable Energy Technologies: Case Studies of Enabling Frameworks in Developing Countries*, Report, Technology Transfer Perspectives Series, UNEP Risø Centre: Denmark, 2011.

[338] Kondratieff, N., *Long Wave Cycle*, Translation of the original Russian book of 1925, E.P. Dutton: New York, 1984.

[339] Hecht, J., "Are Long Waves 50 Years? Reexamining Economic and Financial Long Wave Periodicities in Kondratieff and Schumpeter," *Rev. Polit. Econ.*, *2023*, 35, doi.org/10.1080/09538259.2023.2280803.

[340] Grinin, L.; Korotayev, A.; Tausch, A., *Economic Cycles, Crises, and the Global Periphery*, Springer: Cham, Switzerland, 2016.

[341] Kuznets, S., *Secular Movements in Production and Prices. Their Nature and Their Bearing upon Cyclical Fluctuations*, Houghton Mifflin: Boston, 1930.

[342] Marchetti, C., "Society as a Learning System: Discovery, Invention, and Innovation Cycles Revisited," *Technol. Forecast. Soc. Change.*, 1980, *18*, 267–282.

[343] Wilenius, M.; Kurki, S., "Surfing the Sixth Wave. Exploring the Next 40 Years of Global Change," Finland Futures Research Centre, University of Turku: Helsinki, Dec 2012, p. 127.

[344] Kondriatieffwave, "Kondratieff Winter," Web page, KondratieffWave.com, https://www.kondratieffwave.com/kondratieff-winter/, accessed 13 November 2025.

[345] Ferasso, M.; Bergamaschi, E.A., "Kondratieff's Economic Waves and Future Scenarios Planning: An Approach for Organizations," *Technol. Innov. Mgmt. Rev.*, 2020, *10(2)*, 51–61.

[346] UNESCO, "Towards Knowledge Societies," *UNESCO World Report*, United Nations Educational, Scientific and Cultural Organization: Paris, 2005.

[347] Florida, R., *The Rise of the Creative Class*, Basic Books: New York, 2002. See also *The Rise of the Creative Class – Revisited*, Basic Books: New York, 2012.

[348] Parihar, S.; Kumari, M., "A Review Paper on "IOT", Its Smart Applications, Challenges, & Related Future Technologies," in *AIP Conf. Proc. 2782, June 15* 2023, p. 020001, doi.org/10.1063/5.0154758.

[349] Duda, J.; Gasior, A., Eds., *Industry 4.0. A Global Perspective*, Routledge: New York, 2022.

[350] Elngar, A.A.; Thillaiarasu, N.; Saravanan, T.; Balas, V.E., Eds., *Technology Innovation Pillars for Industry 4.0 Challenges, Improvements, and Case Studies*, CRC Press: Boca Raton, 2025.

[351] Wooldridge, M., *A Brief History of Artificial Intelligence: What It Is, Where We Are, and Where We are Going*, Flatiron Books: New York, 2021.

[352] Turing, A.M., "Computing Machinery and Intelligence," *Mind (New Ser.)*, 1950, *59(236)*, 433–460.

[353] Mitchell, M., *Artificial Intelligence. A Guide for Thinking Humans*, Picador: New York, 2019.

[354] Karin, R.; Galar, D.; Kumar, U., *AI Factory. Theories, Applications and Case Studies*, CRC Press: Boca Raton, 2023.

[355] Lutkevich, B., *"AI Winter," Search Enterprise AI Blog Post*, Informa TechTarget: London, UK, Aug. 26 2024, https://www.techtarget.com/searchenterpriseai/definition/AI-winter. accessed 13 November 2025.

[356] Transforma Insights, "AI & Machine Learning," *Blog Post*, Echo Nomad Ltd.: Reading, U.K, undated, accessed Sept. 16 2025, https://transformainsights.com/ai-machine-learning. accessed 13 November 2025.

[357] Rotman, D., "Tech Slowdown Threatens the American Dream," *MIT Technol. Rev.*, 2016, *119(3)*, 82–86.

[358] Smil, V., *Made in the USA. The Rise and Retreat of American Manufacturing*, MIT Press: Cambridge, MA, 2013.
[359] Maddison, A., *The World Economy. Vol. 1 A Millennial Perspective. Vol. 2 Historical Statistics*, Organisation for Economic Co-Operation and Development (OECD): Paris, 2006.
[360] Maddison, A., *Chinese Economic Performance in the Long Run. 2nd Ed., 960-2030 AD*, Organisation for Economic Co-Operation and Development (OECD): Paris, 2007.
[361] Smil, V., *Invention and Innovation. A Brief History of Hype and Failure*, MIT Press: Cambridge, MA, 2023.
[362] Breque, M.; De Nul, L.; Petridis, A., "Industry 5.0: Towards a Sustainable, Human-Centric and Resilient European Industry," in *R&I Paper Series Policy Brief*, European Commission: Luxembourg, 2021, p. 48.
[363] Kraaijenbrink, J., "The Strategic Leadership Playbook," *Effectual Strategy Press*, 2024, *1*.
[364] Govindarajan, V.; Trimble, C., *The Other Side of Innovation. Solving the Execution Challenge*, Harvard Business School Publishing: Boston, 2010.
[365] Baker, K., *Innovation: Management Benchmark Study*, Office of Science, U.S. Dept. of Energy: Washington, 2002.
[366] Grant, M., *"Approaches to Innovation Management: Innovation and Canadian Business Survey 2014," Report 6652*, Conference Board of Canada: Ottawa, December 2014.
[367] PwC, *Breakthrough Innovation and Growth*, Report 3641-10, PricewaterhouseCoopers Ltd. (PwC): New York, 2013.
[368] Christensen, C.M.; Kaufman, S.P.; Shih, W.C., "Innovation Killers. How Financial Tools Destroy Your Capacity to Do New Things," *Harv. Bus. Rev.*, 2008, *86(1)*, 98–105.
[369] Abernathy, W.J.; Utterback, J.M., "Innovation over Time and in Historical Context," in *Readings in the Management of Innovation*, Tushman, M.L.; Moore, W.L., (Eds.), 2nd Ed, Harper Business: New York, 1988, pp. 25–36.
[370] Troy, K.L., *"Making Innovation Work. From Strategy to Practice," Research Report R-1348-04-RR*, Conference Board of Canada: Ottawa, 2004.
[371] Hamel, G., *Leading the Revolution*, Harvard Business School Press: Brighton, MA, 2000.
[372] Hamel., G., "The Why, What, and How of Management Innovation," *Harv. Bus. Rev.*, 2006, *Feb*, 72–84.
[373] Davis, P., "Innovation White Paper," *Report*, Scanlon Leadership Development Network: East Lansing, MI, January 2008.
[374] Abbey, A., "The Strategic Management of Technological Innovation," *Ind. Manage.*, 1989, *31*(*5*), 16–19.
[375] Volberda, H.W.; Van Den Bosch, F.A.J.; Heij, C.V., "Management Innovation: Management as Fertile Ground for Innovation," *Eur. Mgmt. Rev*, 2013, *10*, 1–15.
[376] Grant, G., *Purpose-Driven Innovation Leadership for Sustainable Development*, Routledge: New York, 2024.
[377] Maital, S.; Seshadri, D.V.R., *Innovation Management: Strategies, Concepts and Tools for Growth and Profit*, SAGE Publications: New Delhi, 2007.
[378] Dodgson, M.; Gann, D.; Phillips, N., "Organizational Learning and the Technology of Foolishness: The Case of Virtual Worlds at IBM," *Organ. Sci.*, 2013, *24*(*5*), 1358–1376.
[379] Peters, T., "The Mythology of Innovation, or a Skunkworks Tale, Part II," in *Readings in the Management of Innovation*, Tushman, M.L.; Moore, W.L., (Eds.), 2nd Ed, Harper Business: New York, 1988, pp. 138–147.
[380] AMA, *The Quest for Innovation. A Global Study of Innovation Management 2006–2016*, American Management Association: New York, 2006.
[381] Jaruzelski, B.; Dehoff, K., "How the Top Innovators Keep Winning," *Strategy+business*, 2010, *61* (*Winter*), 16.

[382] Maidique, M.A.; Hayes, R.H., "Executive Leadership and the Management of Innovation and Change," in *Readings in the Management of Innovation*, Tushman, M.L.; Moore, W.L., (Eds.), 2nd Ed, Harper Business: New York, 1988, pp. 689–704.

[383] Quinn, J.B., "Managing Strategic Change," in *Readings in the Management of Innovation*, Tushman, M.L.; Moore, W.L., (Eds.), 2nd Ed., Harper Business: New York, 1988, pp. 733–751.

[384] Jonsson, I., *The Political Economy of Innovation and Entrepreneurship: From Theories to Practice*, Ashgate Publishing: Farnham, UK, 2015.

[385] Koops, B.-J.; Oosterlaken, I.; Romijn, H.; Swierstra, T.; van den Hoven, J., Eds., *Responsible Innovation 2. Concepts, Approaches, and Applications*, Springer International: Cham, 2015.

[386] Chubakova, V.D.; Drobkova, O.S., "Regulatory Requirements Integration for the Sustainable Innovation Classification," in *7th Internat. Youth Conf. On Radio Electronics, Electrical and Power Engineering (REEPE)*, Institute of Electrical and Electronics Engineers (IEEE): New York, 2025, p. 6, doi: 10.1109/REEPE63962.2025.10971063.

[387] Gohr, C.F.; Rodrigues, B.C.B., "Dynamic Capabilities as Enablers for Sustainability-Oriented Innovation: A Case Study in A Footwear Industry," *Bus. Strateg. Environ.*, 2025, *34*, 3114–3140.

[388] Filho, W.L.; Sigahi, T.F.A.C.; Anholon, R.; Rampasso, I.S.; Platje, J.J.; Velazquez, L.; Salvia, A.L., "Sustainable Innovation: Case Studies from Product Development," *Sustain. Dev.*, 2025, *33*, 3086–3098.

[389] Mousavi, S.; Bossink, B.; Van Vliet, M., "Microfoundations of Companies' Dynamic Capabilities for Environmentally Sustainable Innovation: Case Study Insights from High-Tech Innovation in Science-Based Companies," *Bus. Strateg. Environ.*, 2019, *28*, 366–387.

[390] Maidique, M.A.; Patch, P., "Corporate Strategy and Technological Policy," in *Readings in the Management of Innovation*, Tushman, M.L.; Moore, W.L., (Eds.), 2nd Ed, Harper Business: New York, 1988, pp. 236–248.

[391] Blank, S., "You're Better off Being a Fast Follower than an Originator," Blog post, Business Insider, 5 October 2010, http://www.businessinsider.com/youre-better-off-being-a-fast-follower-than-an-originator-2010-10, accessed 13 November 2025.

[392] Nagji, B.; Tuff, G., "Managing Your Innovation Portfolio," *Harv. Bus. Rev., May* 2012, 66–74.

[393] Wai, C., "There are Three Types of Innovation. Here's How to Manage Them," Blog post, Fast Company, Mansueto Ventures, 10 Nov. 2011, https://www.fastcodesign.com/1665186/there-are-three-types-of-innovation-heres-how-to-manage-them, accessed 14 November 2025.

[394] Chen, J.; Yin, X.; Mei, L., "Holistic Innovation: An Emerging Innovation Paradigm," *Int. J. Innov. Stud.*, 2018, *2*, 1–13.

[395] Chen, J., *Holistic Innovation. Innovation Paradigm Explorations in the New Era*, Springer: Singapore, 2023.

[396] Damberg, S.; Lucius, Z.K.; Blenski, B., Eds., *Holistic Innovation. The Next Generation of Innovation Management?*, Springer: Cham, Switzerland, 2025.

[397] Chen., J., *Enterprise Innovation Ecosystem. Theory and Practice*, Springer: Singapore, 2023.

[398] Ansoff, I., "Strategies for Diversification," *Harv. Bus. Rev.*, 1957, *35*(5), 113–124.

[399] Deschamps, J.-P.; Nelson, B., *Innovation Governance: How Top Management Organizes and Mobilizes for Innovation*, Jossey-Bass: New York, 2014.

[400] Conference Board, " *'Don't Overlook the "Mes' Medium-sized Firms Show the Way to Global Success. 6th Annual Innovation Report," Report 670-04*, Conference Board of Canada: Ottawa, 2005.

[401] Tushman, M.L.; Newman, W.H.; Romanelli, E., "Convergence and Upheaval: Managing the Unsteady Pace of Organizational Evolution," in *Readings in the Management of Innovation*, Tushman, M.L.; Moore, W.L., (Eds.), 2nd Ed, Harper Business: New York, 1988, pp. 705–717.

[402] Smith, G.D.; Wright, J.E., "Alcoa Goes Back to the Future," in *Readings in the Management of Innovation*, Tushman, M.L.; Moore, W.L., (Eds.), 2nd Ed, Harper Business: New York, 1988, pp. 752–759.

[403] Schon, D.A., "Champions for Radical New Inventions," *Harv. Bus. Rev.*, 1963, *41(2)*, 77–86.

[404] Chakrabarti, A.K., "The Role of Champion in Product Innovation," *Calif. Manage. Rev.*, 1974, *17(2)*, 58–62.

[405] Markham, S.K., "A Longitudinal Examination of How Champions Influence Others to Support Their Projects," *J. Prod. Innov. Manag.*, 1998, *15(5)*, 490–504.

[406] Maidique, M.A., "Venturing and Organization Learning," in *Readings in the Management of Innovation*, Tushman, M.L.; Moore, W.L., (Eds.), 2nd Ed, Harper Business: New York, 1988, pp. 565–584.

[407] Goduscheit, R.C., "Innovation Promoters – A Multiple Case Study," *Ind. Mark. Manag.*, 2014, *43*, 525–534.

[408] Curnow, R.C.; Moring, G.G., "'Project SAPPHO': A Study in Industrial Innovation," *Futures*, 1968, *1(2)*, 82–90.

[409] Rothwell, R.; Freeman, C.; Horlsey, A.; Jervis, V.T.P.; Robertson, A.B.; Townsend, J., "SAPPHO Updated – Project SAPPHO Phase II," *Res. Policy*, 1974, *3(3)*, 258–291.

[410] Radosevic, S.; Yoruk, E., *SAPPHO Revisited: Factors of Innovation Success in Knowledge-Intensive Enterprises in Central and Eastern Europe*, DRUID Working Paper 12-11, Danish Research Unit for Industrial Dynamics (DRUID), 2012, 30, https://wp.druid.dk/wp/20120011.pdf. accessed 14 November 2025.

[411] Trias de Bes, F.; Kotler, P., *Winning at Innovation: The A- to -F Model*, Palgrave MacMillan: New York, 2011.

[412] Conference Board, *"It's People Who Innovate," Report 6069*, Conference Board of Canada: Ottawa, 2014.

[413] Hill, C.W.L.; Jones, G.R.; Schilling, M.A., *Strategic Management: Theory & Cases: An Integrated Approach*, 11th Ed, Cengage Learning: Boston, 2014.

[414] Cohn, S.; Good, B., *"Innovation Measurement Metrics and Practices," Report 7035*, Conference Board of Canada: Ottawa, April 2015.

[415] Conference Board of Canada, *"Innovation Challenge Paper #1," Report 346-02*, The Conference Board of Canada: Ottawa, 2002.

[416] Roth, L.M., "A Critical Examination of the Dual Ladder Approach to Career Advancement," in *Readings in the Management of Innovation*, Tushman, M.L.; Moore, W.L., (Eds.), 2nd Ed, Harper Business: New York, 1988, pp. 275–292.

[417] Sawka, K., "Department of Surprise Avoidance," *Compet. Intell. Mag.*, 2002, *5(5)*, 32–34.

[418] Fleisher, C.S.; Bensoussan, B.E., *Business and Competitive Analysis: Effective Application of New and Classic Methods*, 2nd Ed, Pearson Education: Upper Saddle River, NJ, 2015.

[419] Matelski, J., "Driving Innovation Through Business Intelligence," *Blog Post, Database Trends and Applications*, 6 April New Providence: New Jersey, Unisphere Media 2015, http://www.dbta.com/Columns/IOUG-Insight/Driving-Innovation-Through-Business-Intelligence-102959.aspx. accessed 14 November 2025.

[420] Bose, R., "Competitive Intelligence Process and Tools for Intelligence Analysis," *Ind. Manag. Data Syst.*, 2008, *108(4)*, 510–528.

[421] Irvine, J.; Martin, B.R., *Foresight in Science: Picking the Winners*, Pinter: London, 1984.

[422] Andersen, A.D.; Andersen, P.D., "Innovation System Foresight," *Technol. Forecast. Soc. Change*, 2014, *88*, 276–286.

[423] Johnston, R., "Historical Review of the Development of Future-Oriented Technology Analysis," in *Future-Oriented Technology Analysis*, Cagnin, C.; Keenan, M.; Johnston, R.; Scapolo, F.; Barré, R., (Eds.), Springer-Verlag: Berlin, 2008, pp. 17–23.

[424] Chan, C.; Daim, T., "Exploring the Impact of Technology Foresight Studies on Innovation: Case of BRIC Countries," *Futures*, 2012, *44*, 618–630.

[425] Popper, R., "Critical Factors Influencing the Selection of Foresight Methods," *myForesight, April* 2011, 13–26.

[426] Tegart, G.; Johnston, R., "Some Advances in the Practice of Foresight," in *Proc. EU-US Seminar: New Technology Foresight, Forecasting & Assessment Methods*, Seville, 13–14 May 2004, pp. 30–46.

[427] Georghiou, L.; Keenan, M.; Miles, I., "Assessing the Impact of the UK's Evolving National Foresight Programme," *Int. J. Foresight Innov. Policy*, 2010, *6(1/2/3)*, 131–150.

[428] Georghiou, L.; Keenan, M., "Evaluation of National Foresight Activities: Assessing Rationale, Process and Impact," *Technol. Forecast. Soc. Change*, 2006, *73(7)*, 761–777.

[429] Popper, R.; Georghiou, L.; Miles, I.; Keenan, M., *Evaluating Foresight: Fully-Fledged Evaluation of the Colombian Technology Foresight Programme (CTFP)*, Research Monograph, Universidad del Valle: Santiago de Cali, Colombia, October 2010.

[430] Popper, R.; Medina, J., "Foresight in Latin America," in *Handbook of Technology Foresight: Concepts and Practice*, Georghiou, L.; Cassingena, H.J.; Keenan, M.; Miles, I.; Popper, R., (Eds.), Edward Elgar: Cheltenham, 2008, pp. 256–286.

[431] Rohrbeck, R.; Gemünden, H.G., "Corporate Foresight: Its Three Roles in Enhancing the Innovation Capacity of a Firm," *Technol. Forecast. Soc. Change*, 2011, *78*, 231–243.

[432] Pataki, B.; Bíró-Szigeti, S.; Szalkai, Z., "Company-Level Technology Roadmapping," in *The Dark Side of Technological Innovation*, Ran, B., (Ed.), Information Age Publishing: Charlotte, NC, 2013, pp. 65–83.

[433] Bernal, L.; Dornberger, U.; Torres, O.; Byrnes, T., "Technology Roadmapping Handbook," in *Report of the International SEPT Program*, Leipzig University: Leipzig, Germany, June 2009.

[434] Garcia, M.L.; Bray, O.H., *"Fundamentals of Technology Roadmapping," Report SAND97-0665, Sandia National Laboratories*, available from US National Technical Information Service: Springfield, VA, April 1997.

[435] US DOE, "Technology Roadmaps," *Office of Energy Efficiency & Renewable Energy*, U.S. Department of Energy: Washington, 2016, https://energy.gov/eere/ssl/technology-roadmaps.

[436] ACR, *Oil Sands Technology Roadmap Unlocking the Potential*, Alberta Chamber of Resources: Edmonton, January 2004, https://canadacommons.ca/artifacts/35408314/oil-sands-technology-roadmap/36307889/. accessed 14 November 2025.

[437] Arden, W.M., "The International Technology Roadmap for Semiconductors. Perspectives and Challenges for the Next 15 Years," *Curr. Opin. Solid State Mater. Sci.*, 2002, *6(5)*, 371–377.

[438] Phaal, R.; Farrukh, C.J.P.; Probert, D.R., "Technology Roadmapping – A Planning Framework for Evolution and Revolution," *Technol. Forecast. Soc. Change*, 2004, *71(1–2)*, 5–26.

[439] Moehrle, M.G.; Isenmann, R.; Phaal, R., Eds., *Technology Roadmapping for Strategy and Innovation*, Springer: Heidelberg, 2013.

[440] Willyard, C.H.; McClees, C.W., "Motorola's Technology Roadmap Process," *Res. Manag.*, 1987, *30(5)*, 13–19.

[441] Conference Board, *"Culture and Innovation. The Secret Sauce," Report 14-170*, Conference Board of Canada: Ottawa, 2013.

[442] Katz, R.; Allen, T.J., "Investigating the Not Invented Here (NIH) Syndrome: A Look at the Performance, Tenure, and Communication Patterns of 50 R&D Project Groups," in *Readings in the Management of Innovation*, Tushman, M.L.; Moore, W.L., (Eds.), 2nd Ed, Harper Business: New York, 1988, pp. 293–309.

[443] Cohn, S.; Good, B., *"Metrics for Firm-Level Business Innovation in Canada," Report 14-169*, Conference Board of Canada: Ottawa, December 2013.

[444] Cohn, S.; Good, B., *"The State of Firm-Level Innovation in Canada," Report 14-046*, Conference Board of Canada: Ottawa, July 2013.

[445] Ebert, C., "Industry Trends 2024: Innovation Needs Competence," *Computer*, 2024, *April*, 93–98.

[446] Van der Duin, P.; Trott, P.; Ortt, R., "Failed Technology Management: Introducing 'Future Technology Myopia' and How to Address It," *Technol. Forecast. Soc. Change*, 2024, *198*, 122927, 8.

[447] Legace, M., *Gerstner: Changing Culture at IBM – Lou Gerstner Discusses Changing the Culture at IBM*, Harvard Business School: Boston, Blog post, 9 December 2002, Accessed 7 May 2013 at http://hbswk.hbs.edu/archive/3209.html.

[448] Bart, C., *20 Questions Directors Should Ask About Strategy*, 2nd Ed, Canadian Institute of Chartered Accountants: Toronto, 2006, p. 23.

[449] BCG, "Reaching New Heights in Uncertain Times," *Most Innovative Companies 2023 Report*, Boston Consulting Group: Boston, May 2023, p. 28.

[450] Adams, S., "Is Apple the World's Most Innovative Company?" *Forbes*, 8 December 2015, https://www.forbes.com/sites/susanadams/2015/12/08/is-apple-the-worlds-most-innovati, accessed 14 November 2025.

[451] Strategy&, "The Global Innovation 1000 Study. Investigating Trends at the World's 1000 Largest Corporate R&D Spenders," Strategy&: New York, 2018, https://www.strategyand.pwc.com/gx/en/insights/innovation1000.html, accessed 14 November 2025.

[452] Bowen, H.K.; Purrington, C., "Corning: 156 Years of Innovation," *Case Study, Harv. Bus. Rev*, Cambridge, 5 March 2008, https://hbr.org/product/corning-156-years-of-innovation/608108-PDF-ENG.

[453] Miller, R.; Cote, M., *Innovation Reinvented. Six Games that Drive Growth*, University of Toronto Press: Toronto, 2012.

[454] Garud, R.; Nayyar, P.R.; Shapira, Z.B., "Beating the Odds: Towards a Theory of Technological Innovation," in *Technological Innovation. Oversights and Foresights*, Garud, R.; Nayyar, P.R.; Shapira, Z.B., (Eds.), Cambridge University Press: Cambridge, UK, 1997, pp. 345–354.

[455] 3M, *3M Innovation Story: Uncommon Connections, Innovative Solutions*, 3M Inc.: Minneapolis, 2011.

[456] Apple, *Designed by Apple in California*, Apple Inc.: Cupertino, CA, 2016.

[457] Robischon, N., "Why Amazon Is the World's Most Innovative Company of 2017," Blog post, Fast Company, Mansueto Ventures, 13 February 2017, https://www.fastcompany.com/3067455/why-amazon-is-the-worlds-most-innovative-company-of-2017, accessed 14 November 2025.

[458] Gewirtz, D., "Why Amazon Is the King of Innovation: Online Retail Mastery," Blog post, ZD Net, CBS Interactive, 27 May 2016, http://www.zdnet.com/article/why-amazon-is-the-king-of-innovation-the-store-stars/, accessed 14 November 2025.

[459] Schory, G., "An inside Look at eBay's Global Innovation Strategy," Online article, *The Wall Street Journal*, 25 August 2014, http://deloitte.wsj.com/cio/2014/08/25/an-inside-look-at-ebays-global-innovation-strategy/, accessed 14 November 2025.

[460] Aaker, J.; Yong, S.; Dodson, J.; Sharkey, T.; Howard, D., "eBay: Designing Culture Change. The Innovation Playbook," in *Case Study M364*, Stanford Graduate School of Business: Stanford, 2016, https://www.gsb.stanford.edu/faculty-research/case-studies/ebay-designing-culture-change-innovation-playbook. accessed 14 November 2025.

[461] Slater, S.F.; Mohr, J.J.; Sengupta, S., "Radical Product Innovation Capability: Literature Review, Synthesis, and Illustrative Research Propositions," *J. Prod. Innov. Manag.*, 2014, *31*(*3*), 552–566.

[462] Welch, J., *Jack. Straight from the Gut*, Warner Business Books: New York, 2001.

[463] Magee, D., *Jeff Immelt and the New GE Way: Innovation, Transformation and Winning in the 21st Century*, McGraw-Hill: New York, 2009.

[464] Steiber, A., *The Google Model: Managing Continuous Innovation in a Rapidly Changing World*, Springer International: Basel, 2014.

[465] Leong, K.C., "Google Reveals Its 9 Principles of Innovation," Blog post, Fast Company, Mansueto Ventures, 20 November 2013, https://www.fastcompany.com/3021956/googles-nine-principles-of-innovation, accessed 14 November 2025.

[466] He, L., "Google's Secrets of Innovation: Empowering Its Employees," Blog Post Forbes, 29 March 2013, https://www.forbes.com/sites/laurahe/2013/03/29/googles-secrets-of-innovation-empowering-its-employees/#750bf44357e7, accessed 14 November 2025.

[467] Hawn, C., “Lessons on Innovation from Microsoft,” Blog post, Fast Company, Mansueto Ventures, 1 December 2004, https://www.fastcompany.com/51161/lessons-innovation-microsoft, accessed 14 November 2025.

[468] Egan, B., “Microsoft Is Sporting a New Attitude – It’s Called Innovation,” *Forbes*, 30 April 2015, https://www.forbes.com/sites/bobegan/2015/04/30/microsoft-is-sporting-a-new-attitude/#144cdce4572f, accessed 14 November 2025.

[469] Chao, R.; Kavadias, S., “Innovation Strategy at Microsoft: Clouds on the Horizon,” in *Case Study, Harv. Bus. Rev*, Cambridge, 16 June 2011, https://hbr.org/product/innovation-strategy-at-microsoft-clouds-on-the-horizon/UV5965-PDF-ENG. accessed 14 November 2025.

[470] Procter & Gamble, *“P&G. A Company History. 1837 – Today,” Report*, Procter & Gamble: Cincinnati, OH, 2006, p. 11.

[471] Ozkan, N.N., “An Example of Open Innovation: P&G,” *Proc. Soc. Behav. Sci.*, 2015, *195*, 1496–1502.

[472] Dyer, J.; Gregersen, H., “How Procter & Gamble Keeps Its Innovation Edge,” *Forbes*, 12 April 2012, https://www.forbes.com/sites/innovatorsdna/2012/04/12/how-procter-gamble-keeps-its-innovation-edge/#7101124015cb, accessed 14 November 2025.

[473] Surowiecki, J., “The Open Secret of Success,” in *The New Yorker*, 12 May 2008, The Financial Page, http://www.newyorker.com/magazine/2008/05/12/the-open-secret-of-success. accessed 14 November 2025.

[474] Abdulmalek, F.A.; Rajgopal, J., “Analyzing the Benefits of Lean Manufacturing and Value Stream Mapping via Simulation: A Process Sector Case Study,” *Int. J. Prod. Econ.*, 2007, *107*(*1*), 223–236.

[475] Rapp, J.D., “Inside Whirlpool’s Innovation Machine,” Blog post, Management Innovation eXchange, 23 January 2013, http://www.managementexchange.com/story/inside-whirlpools-innovation-machine, accessed 14 November 2025.

[476] Tennant Snyder, N.; Duarte, D.L., *Unleashing Innovation: How Whirlpool Transformed an Industry*, Jossey-Bass: San Francisco, 2008.

[477] Godin, B., “The Emergence of Science and Technology Indicators: Why Did Governments Supplement Statistics with Indicators?,” *Working Paper No. 8, Project on the Intellectual History of Innovation*, Institut National de la Recherche Scientifique (INRS): Montreal, 2001.

[478] OTA, *“Research Funding as an Investment: Can We Measure the Returns?” Technical Memorandum*, OTA-TMSET-36, Office of Technology Assessment, U.S. Congress: Washington, DC, April 1986.

[479] Conference Board of Canada, *“Building the Future. 1st Annual Innovation Report,” Report 274-99*, The Conference Board of Canada: Ottawa, 1999.

[480] Conference Board, *“Who Dimmed the Lights? Canada’s Declining Global Competitiveness Ranking,” Report 13-120*, Conference Board of Canada: Ottawa, September 2012.

[481] Conference Board, *“Performance and Potential 2003–04: Defining the Canadian Advantage,” Special Report*, The Conference Board of Canada: Ottawa, 2003.

[482] Conference Board, *“Including Innovation in Regulatory Frameworks. 4th Annual Innovation Report,” Report 361-02*, Conference Board of Canada: Ottawa, 2002.

[483] OECD, *Main Science and Technology Indicators*, Organisation for Economic Co-Operation and Development (OECD): Paris, May 2010.

[484] Crawford, C.M.; Tellis, G.J., “The Technological Innovation Controversy,” *Bus. Horiz.*, 1981, *24*(*4*), 76–88.

[485] OECD, *Frascati Manual, Proposed Standard Practice for Surveys on Research and Experimental Development*, Organisation for Economic Co-Operation and Development (OECD): Paris, 2002.

[486] OECD, *Main Science and Technology Indicators*, Organisation for Economic Co-Operation and Development: Paris, France, 2014.

[487] Ginarte, J.C.; Park, W.G., “Determinants of Patent Rights: A Cross National Study,” *Res. Policy*, 1997, *26*, 283–301.

[488] Park, W.G., “International Patent Protection: 1960–2005,” *Res. Policy*, 2008, *37*, 761–766.

[489] Council of Canadian Academies, "Innovation and Business Strategy: Why Canada Falls Short," *Expert Panel on Business Innovation Report*, The Council of Canadian Academies: Ottawa, ON, 2009.

[490] Niosi, J., Ed., *Technology and National Competitiveness*, McGill-Queen's University Press: Montreal & Kingston, 1991.

[491] *The Global Competitiveness Report*, World Economic Forum: Geneva, Switzerland, 2014.

[492] Conference Board of Canada, *Ease of Entrepreneurship Index*, Conference Board of Canada: Ottawa, 2016, https://www.conferenceboard.ca/hcp/entrepreneurship.aspx. accessed 14 November 2025.

[493] OECD, *Measuring Innovation. A New Perspective*, Organisation for Economic Co-Operation and Development (OECD): Paris, 2010.

[494] Conference Board of Canada, *"Index of Corporate Innovation. From the Starting Gate to Global Best," Briefing Report 366-02*, Conference Board of Canada: Ottawa, 2002.

[495] Conference Board of Canada, "The State of Firm-Level Innovation in Canada," in *Report 14-046*, Conference Board of Canada: Ottawa, 2013.

[496] Kim, -S.-S., "Measuring Firm-Level Innovation Capability of Small and Medium Sized Enterprises with Composite Indicators," *Acad. Entrep. J.*, 2013, *19(3)*, 1–24.

[497] Gupta, P., *The Six Sigma Business Scorecard*, 2nd Ed., McGraw-Hill: New York, 2007.

[498] Krstevski, D.; Mancheski, G., "SMEs Open Innovation Management: Strategy Map for Innovation Driven Company," *Econ. Dev.*, 2016, *1–2*, 195–210.

[499] Sattler, M., *Excellence in Innovation Management. A Meta-Analytic Review on the Predictors of Innovation Performance*, Gabler Verlag: Wiesbaden, 2011.

[500] Tadeu, H.F.B.; Silva, J.T.M., "Management Indicators and Measurement of Innovation: Review of the Literature," *Bus. Manag. Dyn.*, 2014, *3(10)*, 52–58.

[501] Thistle, M., *"The Inner Ring. The Early History of the National Research Council of Canada*, University of Toronto Press: Toronto, 1966.

[502] INO, *Annual Report 1994*, National Optics Institute: Sainte-Foy, QC, 1994.

[503] INO, *A Sustainable Economic Impact*, National Optics Institute: Québec City, QC, accessed July 28 2025, https://www.ino.ca/en/impact/. accessed 14 November 2025.

[504] INO, "35 Years of Shedding Light," *Annual Report 2023/24*, National Optics Institute: Québec City, QC, 2024.

[505] INO, *24/5 Annual Report*, National Optics Institute: Québec City, QC, 2025.

[506] INO, *Annual Report 2020/21. Shedding Light*, National Optics Institute: Québec City, QC, 2021.

[507] INO, *Annual Report 2021/22*, National Optics Institute: Québec City, QC, 2022.

[508] Quantino, *Make Your Great Ideas Shine*, National Optics Institute: Québec City, QC, accessed July 28 2025, https://www.quantino.ca/en/about/.

[509] Quantino, "Incubatees. Our Purpose and Our Pride!" Québec City, QC, 2025, https://www.quantino.ca/en/incubatees/, accessed 14 November 2025.

[510] Council of Canadian Academies, *"The State of Industrial R&D in Canada," Report*, Council of Canadian Academies: Ottawa, 2013.

[511] Kramers, J., "The Alberta Research Council's Performance Measures," *Paper Presented at Conference Board of Canada R&D Impact Network Workshop*, Ottawa, ON, 11 September 1998.

[512] Kramers, J., "The Alberta Research Council's Performance Measures," in *Presented at Conference Board of Canada R&D Impact Network Workshop*, London, ON, 5 November 1999.

[513] Nyirfa, W.; Schramm, L.L., "Mandate Effectiveness at the Saskatchewan Research Council," in *WAITRO News*, World Association of Industrial Technology and Research Organizations, *June* 2005, pp. 4–5.

[514] Schramm, L.L.; Nyirfa, W.; Grismer, K.; Kramers, J., "Research and Development Impact Assessment for Innovation-enabling Organizations," *Can. Public Adm.*, 2011, *54(4)*, 567–581.

[515] Arnold, E.; Clark, J.; Jávorka, Z., "Impacts of European RTOs. A Study of Social and Economic Impacts of Research and Technology Organisations," *Report to EARTO*, Technopolis Ltd.: Brighton, UK, 20 October 2010.

[516] Schubert, T.; Darold, D.; Will, M., "Measuring the Causal Economic Effects of Scientific Research. Evidence from the Staggered Foundation of the SENAI Innovation Institutes in Brazil," in *Discussion Paper Innovation Systems and Policy Analysis, No. 87*, Fraunhofer ISI: Karlsruhe, 2024, p. 27.

[517] Koelma, G., "$140 Million Government Funding Cuts to the CSIRO Will Have Huge Impacts on Scientific Research in Australia," news.com.au, 30 May 2014, http://www.news.com.au/finance/economy/federal-budget/140-million-government-funding-cuts-to-the-csiro-will-have-huge-impacts-on-scientific-research-in-australia/news-story/26d9a3f1e457930b205575e9c09a38d8, accessed 14 November 2025.

[518] Nogrady, B., "Australia: Balancing Act," *Nature*, 2016, *537(7618)*, S14–S15.

[519] Jeffery, S., "Election 2016: Former Chief Scientist Penny Sackett Condemns CSIRO Cuts," The Canberra Times, 18 June 2016, http://www.canberratimes.com.au/act-news/election-2016-former-chief-scientist-penny-sackett-condemns-csiro-cuts-20160617-gplr5v.html, accessed 14 November 2025.

[520] Goodhart, C., "Problems of Monetary Management: The U.K. Experience," in *Inflation, Depression, and Economic Policy in the West*, Courakis, A.S., (Ed.), Rowman & Littlefield: Lanham, MD, 1981, pp. 111–146.

[521] Schwab, K., Ed., *"The Global Competitiveness Report 2016–2017," Insight Report*, World Economic Forum: Geneva, 2016.

[522] Schwab, K., Ed., *"The Global Competitiveness Report 2019," Insight Report*, World Economic Forum: Geneva, 2019.

[523] NSF, "Research and Development: U.S. Trends and International Comparisons," in *Science and Engineering Indicators U.S*, National Science Foundation, National Center for Science and Engineering Statistics: Alexandria, VA, 2025, https://ncses.nsf.gov/pubs/nsb20246/cross-national-comparisons-of-r-d-performance. accessed 24 November 2025.

[524] OECD, "Main Science and Technology Indicators," Organization for Economic Co-Operation and Development: Paris, 2016, http://www.oecd.org/science/msti.htm, accessed 14 November 2025.

[525] Ledford, H., "Riding the CRISPR Wave. Biologists are Embracing the Power of Gene-Editing Tools to Explore Genomes," *Nature*, 2016, *531*, 156–159.

[526] Talbot, D., "Precise Gene Editing in Plants," *MIT Technol. Rev.*, 2016, *119(2)*, 41.

[527] European Commission, *"Kets: Time to Act," Final Report*, High-Level Expert Group on Key Enabling Technologies, European Commission: Brussels, June 2015.

[528] NWO, "Key Enabling Technologies," Dutch Research Council: The Hague 2025, 3 pp., https://www.nwo.nl/en/key-enabling-technologies, accessed 14 November 2025.

[529] Schramm, L.L., "The 'Invisible Domain' from Colloids and Interfaces to Nanosystems," in *Reference Collection in Materials Science and Materials Engineering*, Elsevier: London, 2024, p. 19, https://doi.org/10.1016/B978-0-323-95486-0.00024-7,. accessed 20 November 2025.

[530] Ratnesh, R.K.; Kumar, R.; Singh, J.; Chandra, R., Eds., *Nanotechnology Applications for Industry 4.0*, CRC Press: Boca Raton, 2026.

[531] Brown-Martin, G., "The Future of Work – 3 Mega-Trends," Innovation Culture Blog Post, 17 March 2017, https://medium.com/innovation-culture/the-future-of-work-3-mega-trends-a60cef7aae09, accessed 20 November 2025.

[532] Sindhwani, N.; Anand, R.; George, A.S.; Pandey, D., Eds., *Robotics and Automation in Industry 4.0. Smart Industries and Intelligent Technologies*, CRC Press: Boca Raton, 2024.

[533] Sharma, A.; Jain, A.; Sharma, P.; Roy, M., Eds., *Recent Trends and Best Practices in Industry 4.0*, River Publishers: Gistrup, Denmark, 2023.

[534] Bali, V.; Bhatnagar, V.; Aggarwal, D.; Bali, S.; Diván, M.J., Eds., *Cyber-Physical, IoT, and Autonomous Systems in Industry 4.0*, CRC Press: Boca Raton, 2022.

[535] Vinge, V., "The Coming Technological Singularity: How to Survive in the Post-Human Era," in *Vision 21. Interdisciplinary Science and Engineering in the Era of Cyberspace, 1993 Symposium Proceedings, NASA Conf. Publ. No. 10129*, NASA Lewis Research Center, March 1993.

[536] Butler, D., "Tomorrow's World," *Nature*, 2016, *530(7591)*, 398–399.

[537] Simonite, T., "AI Software Learns to Make AI Software," *MIT Technol. Rev.*, 2017, *120(3)*, 16.

[538] Knight, W., "The Dark Secret at the Heart of AI," *MIT Technol. Rev.*, 2017, *120(3)*, 54–63.

[539] Swope, C.; Bingen, K.A.; Young, M.; Lafave, K., *Space Threat Assessment 2025*, Centre for Strategic and International Studies: Washington, DC, 2025.

[540] NATO, "NATO Commercial Space Strategy," *Official Text*, North Atlantic Treaty Organization: Brussels, updated 24 June 2025, https://www.nato.int/en/about-us/official-texts-and-resources/official-texts/2025/02/13/nato-commercial-space-strategy?selectedLocale=. accessed 14 November 2025.

[541] Weber, R.M., "Where No One Has Gone Before: When Science Fiction Inspires Technology," *J. Financ. Serv. Prof.*, 2016, *70(4)*, 43–46.

[542] Tsekleves, E., "Science Fiction as Fact: How Desires Drive Discoveries," in *The Guardian*, U.S. Edition, 13 August 2015, www.theguardian.com/media-network/2015/aug/13/science-fiction-reality-predicts-future-technology. accessed 14 November 2025.

[543] Perkowitz, S., "Boldly Going for 50 Years," *Nature*, 2016, *537(7619)*, 165–166.

[544] Toffler, A., *Future Shock*, Random House: New York, 1970.

[545] Walter, K., "Star Trek-Inspired Tricorder Device Can Test, Diagnose Diseases," *R&D World*, July 28 2017, 1.

[546] Frith, K.H., "The Star Trek Tricorder Is Here: Handheld Imaging Technology," *Nurs. Educ. Perspect.*, 2021, *42(2)*, 130.

[547] IBM, *IBM Watson to Watsonx*, IBM Corporation: Armonk, NY, 2023, https://www.ibm.com/watson/. accessed 14 November 2025.

[548] Apple, *Siri*, Apple: Cupertino, CA, 2025, https://www.apple.com/ca/siri/. accessed 14 November 2025.

[549] Patterson, B., "Amazon Shut down Alexa.com. What Does that Mean for Alexa?," in *TechHive Blog Post*, IDG Communications, Inc.: Needham, MA, November 23 2022, https://www.techhive.com/article/579886/amazon-is-shutting-down-alexa-com.html. accessed 14 November 2025.

[550] Grzybowski, A.; Pawlikowska–Łagód, K.; Lambert, W.C., "A History of Artificial Intelligence," *Clin. Dermatol.*, 2024, *42*, 221–229.

[551] Garber, M., "What 3D-Printed Cake Tastes Like," *The Atlantic*, 8 January 2014, https://www.theatlantic.com/technology/archive/2014/01/what-3d-printed-cake-tastes-like/282904/, accessed 14 November 2025.

[552] Mitchell, T.; Brynjolfsson, E., "Track How Technology Is Transforming Work," *Nature*, 2017, *544(7650)*, 290–292.

[553] Newswire feed, "This Is How the First World Humanoid Robot Games Start in China," in *CE Noticias Financieras*, English ed, ContentEngine LLC: Miami, August 14 2025.

[554] Milojicic, D., "Technology Predictions 2025," *Computer, Jan* 2025, 88–90.

[555] Rotman, D., "The 3-D Printer That Could Finally Change Manufacturing," *MIT Technol. Rev.*, 2017, *120(3)*, 42–53.

[556] Schurig, D.; Mock, J.J.; Justice, B.J.; Cummer, S.A.; Pendry, J.B.; Starr, A.F.; Smith, D.R., "Metamaterial Electromagnetic Cloak at Microwave Frequencies," *Science*, 2006, *314(5801)*, 977–980.

[557] Ni, X.; Wong, Z.J.; Mrejen, M.; Wang, Y.; Zhang, X., "An Ultrathin Invisibility Skin Cloak for Visible Light," *Science*, 2015, *349(6254)*, 1310–1314.

[558] Shvedov, V.; Davoyan, A.R.; Hnatovsky, C.; Engheta, N.; Krolikowski, W., "A Long-Range Polarization-Controlled Optical Tractor Beam," *Nat. Photonics*, 2014, *8*, 846–850.

[559] Takesue, H.; Dyer, S.D.; Stevens, M.J.; Verma, V.; Mirin, R.P.; Nam, S.W., "Quantum Teleportation over 100 Km of Fiber Using Highly Efficient Superconducting Nanowire Single-Photon Detectors," *Optica*, 2015, *2*, 832–835.

[560] Edwards, B.C., "The Space Elevator," Phase II Final Report for NASA Institute for Advanced Concepts (NIAC) program, 1 March 2003, http://www.niac.usra.edu/files/studies/final_report/472Edwards.pdf, accessed 14 November 2025.

[561] Li, C.; Wang, C.; Wei, Y.; Lin, Y., "China's Present and Future Lunar Exploration Program," *Science*, 2019, *365(6450)*, 238–239.

[562] Harvey, B., *China in Space. The Great Leap Forward*, 2nd Ed, Springer Nature Switzerland: Cham, 2019.

[563] Jones, A., "China, Russia Open Moon Base Project to International Partners, Early Details Emerge," *SpaceNews*, April 26, 2021, https://spacenews.com/china-russia-open-moon-base-project-to-international-partners-early-details-emerge/

[564] National Museum of Australia, "Forerunner of CSIRO Founded," Sept.18, 2024, https://www.nma.gov.au/defining-moments/resources/csiro-founded, accessed 14 November 2025.

[565] Currie, G.; Graham, J., *The Origins of CSIRO. Science and the Commonwealth Government 1901–1926*, Commonwealth Scientific and Industrial Research Organization: Melbourne, 1966.

[566] Schedvin, C.B., *Shaping Science and Industry. A History of Australia's Council for Scientific and Industrial Research, 1926–49*, Allen & Unwin: North Sydney, Australia, 1987.

[567] Thompson, L.-J.; Gilding, M.; Spurling, T.H.; Simpson, G.; Elsum, I.R., "The Paradox of Public Science and Global Business: CSIRO, Commercialisation and the National System of Innovation in Australia," *Innov.: Manag. Policy Pract.*, 2011, *13(3)*, 327–340.

[568] Mina, A.; Connell, D.; Hughes, A., "Models of Technology Development in Intermediate Research Organisations," *Centre for Business Research Working Paper No. 396*, University of Cambridge: Cambridge, December 2009.

[569] Zylberberg, E., "Beyond RTO Benchmarking: Towards a Typology of Innovation Intermediaries," *MIT-IPC Working Paper 17-002*, MIT: Cambridge, MA, June 2017.

[570] Piore, M.; Cardoso, C.F., "Beyond the Silicon Valley Consensus," in *Innovation in Brazil. Advancing Development in the 21st Century*, Reynolds, E.B.; Schneider, B.R.; Zylberberg, E., (Eds.), Routledge: London, 2019, pp. 211–230.

[571] ARC, *The Alberta Research Council – A Historical Perspective*, Alberta Research Council: Edmonton, 1906, p. 25.

[572] Middleton, G.R.; Zhang, S.Y., "Characterizing the Wood Attributes of Canadian Tree Species: A Thirty-Year Chronicle," *For. Chron.*, 2009, *85(3)*, 392–400.

[573] Eggleston, W., *National Research in Canada, the NRC 1916 – 1966*, Clarke, Irwin & Co.: Toronto, 1978.

[574] Martínez-Vela, C., "Benchmarking Research and Technology Organizations (Rtos): A Comparative Analysis," in *MIT-IPC Working Paper 16-005*, Massachusetts Institute of Technology: Cambridge, MA, June 2016, p. 23.

[575] Schramm, L.L.; Cook, E., *Research and Development in New Brunswick. A History of the Research and Productivity Council*, New Brunswick Research and Productivity Council: Fredericton, 2020.

[576] Palmer, E.F., *Horticultural Experiment Station and Products Laboratory: The First Fifty Years, 1906 to 1956*, Ontario Department of Agriculture: Toronto, 1956, https://atrium.lib.uoguelph.ca/xmlui/handle/10214/3076. accessed 20 November 2025.

[577] Loughton, A.; Chudyk, R.V.; Wanner, J.A., Eds., *Celebrating A Century of Success 1906 – 2006, Agricultural Experimental Station, Vineland*, University of Guelph: Guelph, ON, 2006.

[578] Solberg, E.; Larsen, K.; Wiig, O.; Aagaard, K.; Sivertsen, G., "Markets for Applied Research. A Comparative Analysis of R&D-Systems in Five Countries," *Report 46/2012*, Nordisk institutt for studier av innovasjon, forskning og utdanning: Oslo, 2012.

[579] Åström, T.; Eriksson, M.-L.; Niklasson, L.; Arnold, E., *International Comparison of Five Institute Systems*, Danish Ministry of Science, Technology and Innovation: Copenhagen, 2008, as cited in Zylberberg (2017) above.

[580] Suominen, A.; Kauppinen, H.; Hyytinen, K., "'Gold', 'Ribbon' or 'Puzzle': What Motivates Researchers to Work in Research and Technology Organizations," *Technol. Forecast. Soc. Change*, 2021, *170*, 120882.

[581] Loikkanen, T.; Hyytinen, K.; Konttinen, J., "Public Research and Technology Organisations in Transition-The Case of Finland," *Sci. Technol. Soc.*, 2011, *16(1)*, 75–98.

[582] Conreur, G., "Une Brève Histoire du CNRS," (A Brief History of the CNRS), Radio France Blog Post, 2009 (Updated January 22, 2016), https://www.radiofrance.fr/franceculture/une-breve-histoire-du-cnrs-3023702, accessed 20 November 2025.

[583] The Journal "La Revue Pour L'histoire du CNRS" Published Articles on the History of the CNRS (And of French Science in General) from Its Inception in 1999 through 2010. To some extent, the journal's mission is carried on in "Histoire de la recherche contemporaine."

[584] Burgos, V., "Chronologie du CNRS," (Chronology of the CNRS), Committee for the History of the CNRS: France, July 25, 2025, https://eman-archives.org/ComiteHistoireCNRS/chronologie-du-cnrs, accessed 14 November 2025.

[585] Belot, R., "The Political and Cultural Revolution of the CNRS: An Attempt at the Systematic Organisation of Research in Opposition to 'The Academic Spirit," in *Sciences in the Universities of Europe, Nineteenth and Twentieth Centuries*, Simões, A.; Diogo, M.P.; Gavroglu, K., (Eds.), Springer: New York, 2015, pp. 245–263.

[586] Kang, B., "Innovation Processes in Public Research Institutes: AIST, Fraunhofer and ITRI Case Studies," *Sci. Technol. Soc.*, 2021, *26(3)*, 433–458.

[587] Rush, H.; Hobday, M.; Bessant, J.; Arnold, E., "Strategies for Best Practice in Research and Technology Institutes: An Overview of a Benchmarking Exercise," *R&D Manag.*, 1995, *25(1)*, 17–31.

[588] Arnold, E.; Rush, H.; Bessant, J.; Hobday, M., "Strategic Planning in Research and Technology Institutes," *R&D Manag.*, 1998, *28(2)*, 89–100.

[589] Tiwari, V.M., "CSIR-National Geophysical Research Institute – 60 Years of Enduring Scientific Contributions," *J. Geol. Soc. India, Oct.* 2020, *96*, 319–324.

[590] Abrol, D., "Publicly Funded Research and Policy Reforms in India: Lessons from the Council of Scientific and Industrial Research (CSIR)," *Contemp. Perspect.*, 2007, *1(2)*, 58–88.

[591] Clara, M., "CITER. Info Centre – Knitwear and Textile," *Case Study Report for UNIDO/Italy Programme for SME Development*, United Nations Industrial Development Organization (UNIDO): Vienna, February 1999. p. 17.

[592] van Rooij, A., "Gaps and Plugs: TNO, and the Problems of Getting Knowledge Out of Laboratories," *Minerva*, 2013, *51(1)*, 25–48.

[593] Nygaard, P.; Grove, K.; Aven, H.B., *Teknologibyggerne (The Technology Builders). SINTEF 1950 – 2025*, Fagbokforlaget: Bergen, 2025, In Norwegian.

[594] Thurner, T.W.; Zaichenko, S., "Research and Technology Organizations (Rtos) in the Primary Sector," *Eur. J. Innov. Manag.*, 2014, *17(3)*, 292–310.

[595] Hallonsten, O., "Development and Transformation of the Third Sector of R&D in Sweden, 1942–2017," *Sci. Public Policy*, 2018, *45(5)*, 634–644.

[596] Arnold, E.; Brown, N.; Eriksson, A.; Jansson, T.; Muscio, A.; Nählinder, J.; Zaman, R., "The Role of Industrial Research Institutes in the National Innovation System," in *VINNOVA Analysis VA 2007:12, Verket För Innovatonssystem*, VINNOVA; Swedish Governmental Agency for Innovation Systems): Stockholm, 2007, p. 194.

[597] ITRI, *See the History of the Deep Ploughing 50*, Industrial Technology Research Institute: Zhudong, Taiwan, https://50th.itri.org.tw/history/.

[598] Breznitz, D., *Innovation in Real Places*, Oxford University Press: New York, 2021, Ch.6.

[599] Intarakumnerd, P., "Two Models of Research Technology Organisations in Asia," *Sci. Technol. Soc.*, 2011, *16(1)*, 11–28.

[600] Sheikheldin, G.H., "Research and Technology Organizations as Super Intermediaries: A Conceptual Framework for Policy and A Case Study from Tanzania," *Front. Res. Metr. Anal.*, 2021, *6(691247)*, 14.

[601] Yood, C.N., *Argonne National Laboratory and the Emergence of Computer and Computational Science, 1946–1992*, Thesis, Department of History, Pennsylvania State University: University Park, PA, 2005, p. 367.

[602] Boehm, G.A.W.; Groner, A., *Science in the Service of Mankind: The Battelle Story*, Heath and Co: Lexington, 1972.

[603] Pillai, R.S.; Gregory, C.J.; Phippen, W.; Stark, J.T., *"How We Do Our Work Is as Important as What We Do. A History of Los Alamos National Laboratory Operations," Report LA-UR-22-30602*, Los Alamos National Laboratory: Los Alamos, NM, 2022, p. 105.

[604] Tchakerian, C.L., *"Manhattan, Mars, MANIAC and More. A History of Los Alamos National Laboratory," Report LA-UR-24-25091*, Los Alamos National Laboratory: Los Alamos, NM, 2025, p. 65.

[605] Quist, A.S., "A History of Classified Activities at Oak Ridge National Laboratory," in *Report ORCA-7*, Oak Ridge National Laboratory: Oak Ridge, TN, September 29 2000. p. 35.

[606] Company Histories, "SRI International Inc.," Company-Histories.com, 2022, https://www.company-histories.com/SRI-International-Inc-Company-History.html, accessed 14 November 2025.

[607] Vagtborg, H., *The Story of Southwest Research Center. A Private, Nonprofit*, Scientific Research Adventure, University of Texas Press: Austin, 1973.

[608] Goffin, K.; Åhlström, P.; Bianchi, M.; Richtnér, A., "Perspective: State-of-the-Art: The Quality of Case Study Research in Innovation Management," *J. Prod. Innov. Manag.*, 2019, *36*(*5*), 586–615.

[609] Cherroun, R.; Debla, F., "Hospital at Home: A New Model of Healthcare Services Innovation – Case Study of Hakim Saadane Hospital, Biskra City, Algeria," *Univ. J. Public Health*, 2024, *12*(*1*), 17–27.

[610] Amatullo, M.; Boyer, B.; May, J.; Shea, A., Eds., *Design for Social Innovation Case Studies from around the World*, Routledge: New York, 2022.

[611] Kim, H.-M.; Sabri, S.; Kent, A., Eds., *Smart Cities for Technological and Social Innovation: Case Studies, Current Trends, and Future Steps*, Academic Press: London, 2021.

[612] DESA, *Innovation for Sustainable Development. Local Case Studies from Africa*, Dept. Economic and Social Affairs (DESA), United Nations: New York, 2008.

[613] Sandrey, R.; Vink, N., "Trade and Innovation Project Case Study 4: Deregulation, Trade Reform and Innovation in the South African Agriculture Sector," *OECD Trade Policy Working Paper No. 76*, Organisation for Economic Co-Operation and Development: Paris, 2008.

[614] Valeri, M., Ed, *Innovation Strategies and Organizational Culture in Tourism. Concepts and Case Studies on Knowledge Sharing*, Routledge: London, 2024.

[615] Tarnovskaya, V.; Hånell, S.M.; Tolstoy, D., "Proactive Corporate Sustainability via Social Innovation – A Case Study of the Hennes & Mauritz Grand Challenge in Bangladesh," *Sustainability*, 2022, *14*, 599–624.

[616] Correia, A.; Lopes, J.D.; Portugal, M., Eds., *International Case Studies in Innovation and Entrepreneurship in Tourism, Routledge: London*, 2024.

[617] CICC Global Institute, *The Rise of China's Innovation Economy*, Springer Nature Singapore: Singapore, 2022.

[618] Li, W.; Wu, W.; Yu, B.; Foo, C., "Is China Transmuting to Fast Overtake the USA in Innovation? R&D Case-Studies in Advanced Technology Manufacturing," *Chin. Manag. Stud.*, 2015, *9*(*1*), 8–26.

[619] Wan, Y.; Zeng, G.; Wang, J.; Zou, L., "'Buzz-and-pipeline' Dynamics in AI Innovation Network: A Case Study of Zhangjiang National Innovation Demonstration Zone, Shanghai," *Land*, 2025, *14*, 114–132.

[620] Zhang, X.; Chen, S.; Wang, X., "How Can Technology Leverage University Teaching & Learning Innovation? A Longitudinal Case Study of Diffusion of Technology Innovation from the Knowledge Creation Perspective," *Educ. Inf. Technol.*, 2023, *28*, 15543–15569.

[621] Luo, S.; Chan, -C.K.-C., "Reclaiming "Territories": Two Case Studies of Trade Union Innovation Projects in South China," *China Rev.*, 2020, *20(1)*, 135–163.

[622] Lu, C.; Liu, S., "Cultural Tourism O2O Business Model Innovation-A Case Study of Ctrip," *J. Electron. Commer. Organ.*, 2016, *14(2)*, 16–31.

[623] Shi, Z.; Wu, Y.; Chiu, Y.; Shi, C.; Na, X., "Comparing the Efficiency of Regional Knowledge Innovation and Technological Innovation: A Case Study of China," *Technol. Econ. Dev. Econ.*, 2022, *28*(5), 1392–1418.

[624] Shi, J.; Xiao, Z., "Research on the Impact of Inter-Industry Innovation Networks on Collaborative Innovation Performance: A Case Study of Strategic Emerging Industries," *Systems*, 2024, *12*, 211–235.

[625] Jiang, H.; Cao, Y., "Practice and Reference of Enterprise Quality Management Innovation – Case Study of Chinese Enterprises," in *Proc. 2nd Internat. Conf. E-Commerce and Internet Technology*, IEEE: New York, 2021, p. 138–141.

[626] Matheson Connell, C., "Pursuing Three Horizons of Growth – Three Cases: Bombardier (Canada), Disney (US) and Hutchison Whampoa (China)," *Bus. Strateg. Ser.*, 2007, *8*(*1*), 14–25.

[627] Sonne, L., "The Usefulness of Networks: A Study of Social Innovation in India," in *New Frontiers in Social Innovation Research*, Nicholls, A.; Simon, J.; Gabriel, M., (Eds.), Palgrave Macmillan: Houndmills, 2015, pp. 212–232.

[628] Nicholls, A.; Simon, J.; Gabriel, M., Eds., *New Frontiers in Social Innovation Research*, Palgrave Macmillan: Hampshire, 2015.

[629] Cheah, S.L.-Y., Ed., *Social Innovation. Asian Case Studies of Innovating for the Common Good*, Routledge: Oxon, 2020.

[630] Phani, B.V.; Khandekar, S., Eds., *Innovation, Incubation and Entrepreneurship. Case Studies from IIT Kanpur*, Springer: Singapore, 2017.

[631] Ghoshal, S.; Roy, R.K., "Anatomy of an Innovation in IT Service," *J. Inf. Technol. Teach. Cases, 2020*, doi: 10.1057/S41266-017-0018-2.

[632] Pandiarajan, V., *Business Innovation. A Case Study Approach*, Routledge: Oxon, 2022.

[633] OECD, *Governance of Innovation Systems. Volume 2: Case Studies in Innovation Policy*, Organisation for Economic Co-Operation and Development: Paris, 2005.

[634] OECD, *Governance of Innovation Systems. Volume 3: Case Studies in Cross-Sectoral Policy*, Organisation for Economic Co-Operation and Development: Paris, 2005.

[635] Walaszczyk, L., "Strategic Research Programs in the Area of Technical Innovations – Case Study Analyses," *Found. Manag.*, 2016, *8*, 139–150.

[636] Honda, A.; Oliveira, S.R.de A.; Ridde, V.; Zinszer, K.; Gautier, L., "Attributes and Organizational Factors that Enabled Innovation in Health Care Service Delivery during the COVID-19 Pandemic – Case Studies from Brazil, Canada and Japan," *Health Syst. Reform*, 2023, 9:2, #2176022, 12 pp.

[637] Kodama, M.; Shibata, T., "Strategy Transformation through Strategic Innovation Capability – A Case Study of Fanuc," *R&D Manag.*, 2014, *44*(*1*), 75–103.

[638] Yew, L.K., "From Family Firm to Japanese OEM in Tacit Knowledge Innovation: A Case Study of Sin Kwang Plastic Resources in Malaysia," *J. Organ. Change Manag.*, 2023, *36*(*6*), 966–984.

[639] Lee, B.; Kang, J., "The Power of Peripheral Innovation. A Case Study of Amore Pacific's Cushion Foundation," *Res. Technol. Manag.*, 2016, *July/Aug*, 21–28.

[640] Kim, J.; Rim, S.; Han, S.; Park, A., "Seoul City's Social Innovation Strategy: New Models of Communication to Strengthen Citizen Engagement," in *New Frontiers in Social Innovation Research*, Nicholls, A.; Simon, J.; Gabriel, M., (Eds.), Palgrave Macmillan: Houndmills, 2015, pp. 170–191.

[641] Onodera, O.; Kim, H.E., "Trade and Innovation Project Case Study 5: Trade and Innovation in the Korean Information and Communication Technology Sector," *OECD Trade Policy Working Paper No. 77*, Organisation for Economic Co-Operation and Development: Paris, 2008.

[642] Hsu, S.-M.; Hsieh, P.-H.; Yuan, S.-T., "Roles of 'Small- and Medium-Sized Enterprises' in Service Industry Innovation: A Case Study on Leisure Agriculture Service in Tourism Regional Innovation," *Serv. Ind. J.*, 2013, *33*(*11*), 1068–1088.
[643] Ng, W.-K.; Hsu, F.-T.; Chen, C.-L., "The Impacts of Digital Technology on Service Design and Experience Innovation: Case Study of Taiwan's Cultural Heritage under the COVID-19 Pandemic," *Systems*, 2022, *10*, 184–200.
[644] Shang, S.S.C.; Chiu, L.S.L., "A RACE Pathway for Inventing and Sustaining Mobile Payment Innovation – A Case Study of A Leading Bank in Taiwan," *Asia Pac. Manag. Rev.*, 2023, *28*, 401–409.
[645] Shih, T.; Hu, M.-C.; Chou, J.C.-P., "Navigating an Emerging Innovation Ecosystem: A Case Study of Fuel Cell Innovation in Taiwan," *Asian J. Technol. Innov., May* 2024, 14, doi: 10.1080/19761597.2024.2352734.
[646] Hsiao, Y.-C.; Wu, M.-H.; Li, S.C., "Elevated Performance of the Smart City – A Case Study of the IoT by Innovation Mode," *IEEE Trans. Eng. Manag.*, 2021, *68*(*5*), 1461–1475.
[647] Thuyen, T.T.N.; Bich, L.N., "Green Innovation Practices: A Case Study of Vietnamese Manufacturing Companies," *Cogent Bus. Manag.*, 2024, *11*(*1*), 2333603, 14.
[648] Tuni, A.; Gutteridge, F.; Ijomah, W.L.; Mirpourian, M., "Risks in Circular Business Models Innovation: A Cross-Industrial Case Study for Composite Materials," *Bus. Strateg. Environ.*, 2024, *33*, 2771–2787.
[649] Krause, J., "Factors Hampering Innovation Activities – Case Study from the Czech Republic," *J. Technol. Manag. Innov.*, 2016, *11*(*4*), 75–79.
[650] Zizlavsky, O.; Vaverkova, P., "Innovation Scorecard Design: A Process Innovation Case Study," *Int. Adv. Econ. Res.*, 2018, *24*, 299–300.
[651] Hyrkäs, P.; Haukipuro, L.; Väinämö, S.; Iivari, M.; Sachinopoulou, A.; Majava, J., "Collaborative Innovation in Healthcare: A Case Study of Hospitals as Innovation Platforms," *Int. J. Value Chain Manag.*, 2020, *11*(*1*), 24–41.
[652] Pihlajamaa, M.; Merisalo, M., "Organizing Innovation Contests for Public Procurement of Innovation – A Case Study of Smart City Hackathons in Tampere, Finland," *Eur. Plan. Stud.*, 2021, *29* (*10*), 1906–1924.
[653] Lesser, C., "Market Openness, Trade Liberalisation and Innovation Capacity in the Finnish Telecom Equipment Industry: Trade and Innovation Project – Case Study No. 1," in *OECD Trade Policy Papers, No. 73*, Organisation for Economic Co-Operation and Development: Paris, 2008.
[654] Buchmann, T., *The Evolution of Innovation Networks. An Automotive Case Study*, Springer Gabler: Wiesbaden, 2015.
[655] Chesbrough, H., "Innovating Innovation @tk," in *The BerkeleyHaas Case Series*, University of California: Berkeley.Haas School of Business, 2018, p. 18.
[656] Hurmelinna-Laukkanen, P.; Yang, J., "Long-Lasting Appropriation Success of an Innovation? A Comparative Case Study of Bayer's Aspirin and Roundup," in *Proc. XXXIII ISPIM Innovation Conference*, LUT Scientific and Expertise Publications: Lahti, Finland, 2022, p. 17.
[657] Russo, M.; Caloffi, A.; Rossi, F.; Righi, R., "Innovation Intermediaries and Performance-Based Incentives: A Case Study of Regional Innovation Poles," *Sci. Public Policy*, 2018, *46*(*1*), 1–12.
[658] Kofler, I.; Marcher, A., "Inter-Organizational Networks of Small and Medium-Sized Enterprises (SME) in the Field of Innovation: A Case Study of South Tyrol," *J. Small Bus. Entrep.*, 2018, *30*(*1*), 9–25.
[659] Cicerelli, F.; Ravetti, C., "Sustainability, Resilience and Innovation in Industrial Electronics: A Case Study of Internal, Supply Chain and External Complexity," *J. Econ. Interact. Coord.*, 2024, *19*, 343–372.
[660] Iandoli, L., "Olivetti: The Innovation Factory," in *Sage Business Cases*, SAGE Publications: Thousand Oaks, CA, 2021, p. 24, //doi.org/10.4135/9781529761887.
[661] Kovács, J.K.; Zoltán, E.S., "Rural Enterprise Hub Supporting Rural Entrepreneurship and Innovation – Case Studies from Hungary," *Eur. Countrys.*, 2017, *9*(*3*), 473–485.

[662] Lit, F.C.; Huijben, J.C.C.M.; Cloodt, M.M.A.H.; Paredis, E., "Business Model Innovation in Circular Start-Ups: Overcoming Barriers in the Circular Plastics Economy," *Int. Small Bus. J. Res. Entrep.*, 2024, *42*(*4*), 506–550.

[663] Westnes, P.; Hatakenakab, S.; Gjelsvika, M.; Lester, R.K., "The Role of Universities in Strengthening Local Capabilities for Innovation – A Comparative Case Study," *High. Educ. Policy*, 2009, *22*, 483–503.

[664] Steinmo, M., "Collaboration for Innovation: A Case Study on How Social Capital Mitigates Collaborative Challenges in University–Industry Research Alliances," *Ind. Innov.*, 2015, *22*(*7*), 597–624.

[665] Aslesen, H.W.; Isaksen, A.; Karlsen, J., "Modes of Innovation and Differentiated Responses to Globalisation – A Case Study of Innovation Modes in the Agder Region, Norway," *J. Knowl. Econ.*, 2012, *3*, 389–05.

[666] de Moura, I.S., "Export Performance of the Textile Industry: Influence of Financial Resources and Productivity," *In Industry and Innovation: Textile Industry*, Martins, J.M. (Ed.), Springer: Cham, 2024, pp. 317–337.

[667] Franco, M.; Esteves, L.; Rodrigues, M., "Clusters as a Mechanism of Sharing Knowledge and Innovation: Case Study from a Network Approach," *Glob. Bus. Rev.*, 2024, *25*(*2*), 377–400.

[668] Martins, J.M., Ed., *Industry and Innovation: Textile Industry*, Springer: Cham, 2024.

[669] Santos, G.; Gomes, S.; Braga, V.; Braga, A.; Lima, V.; Teixeira, P.; Sá, J.C., "Value Creation through Quality and Innovation – A Case Study on Portugal," *TQM J.*, 2019, *31*(*6*), 928–947.

[670] Bartlett, D.; Trifilova, A., "Green Technology and Eco-Innovation. Seven Case-Studies from a Russian Manufacturing Context," *J. Manuf. Technol. Manag.*, 2010, *21*(*8*), 910–929.

[671] Scarpellini, S.; Ortega-Lapiedra, R.; Marco-Fondevila, M.; Aranda-Usón, A., "Human Capital in the Eco-Innovative Firms: A Case Study of Eco-Innovation Projects," *Int. J. Entrep. Behav. Res.*, 2017, *23*(*6*), 919–933.

[672] Hellström, D.; Nilsson, F., "Logistics-Driven Packaging Innovation: A Case Study at IKEA," *Int. J. Retail Distrib. Manag.*, 2011, *39*(*9*), 638–657.

[673] Martin, H., "The Scope of Regional Innovation Policy to Realize Transformative Change – A Case Study of the Chemicals Industry in Western Sweden," *Eur. Plan. Stud.*, 2020, *28*(*12*), 2409–2427.

[674] Mulvale, G.; Green, J.; Robert, G.; Larkin, M.; Vackerberg, N.; Kjellström, S.; Hossain, P.; Moll, S.; Lim, E.; Craythorne, S.-L., "Adopting, Implementing and Assimilating Coproduced Health and Social Care Innovations Involving Structurally Vulnerable Populations: Findings from a Longitudinal, Multiple Case Study Design in Canada, Scotland and Sweden," *Health Res. Policy Syst.*, 2024, *22*, 42–61.

[675] Erikshammar, J.; Gull, A.-L.; Stehn, L., "Innovation Ecosystems beyond Construction Projects – A Case Study of Swedish Cross Laminated Timber Building Networks," *Wood Mater. Sci. Eng.*, *2024*, 14, doi: 10.1080/17480272.2024.2376180.

[676] Barber, S.; French, C.; Matthews, R.; Lovett, D.; Rollinson, T.; Husson, F.; Turley, M.; Reed, J., "The Role of Patients and Carers in Diffusing A Health-Care Innovation: A Case Study of 'My Medication Passport,'," *Health Expect.*, 2019, *22*, 676–687.

[677] Pattinson, S.; Preece, D., "Communities of Practice, Knowledge Acquisition and Innovation: A Case Study of Science-Based SMEs," *J. Knowl. Manag.*, 2014, *18*(*1*), 107–120.

[678] Wilson-Medhurst, S., "Knowledge Production within the Innovation System: A Case Study from the United Kingdom," *High. Educ. Manag. Policy*, 2010, *22*(*2*), 121–135.

[679] Kaye, R.; Hawkridge, D., Eds., *Learning & Teaching for Business. Case Studies of Successful Innovation*, Kogan Page: London, 2003.

[680] Hoeber, L.; Hoeber, O., "Determinants of an Innovation Process: A Case Study of Technological Innovation in A Community Sport Organization," *J. Sport Manag.*, 2012, *26*, 213–223.

[681] Litvak, I.A.; Maule, C.J., " Canadian Entrepreneurship and Innovation. Six Case Studies," *Technological Innovation Studies Program, Research Report No. 86*, Dept. Regional Industrial Expansion, Government of Canada: Ottawa, Feb 1982.

[682] Martin, M.J.C.; Rosson, P.J., “Four Cases on the Management of Technological Innovation and Entrepreneurship,” *Technological Innovation Studies Program, Research Report No. 85a*, Dept. Regional Industrial Expansion, Government of Canada: Ottawa, Nov 1982.

[683] Martin, M.J.C.; Rosson, P.J., “Further Cases on the Management of Technological Innovation and Entrepreneurship,” *Technological Innovation Studies Program, Research Report No. 108*, Dept. Regional Industrial Expansion, Government of Canada: Ottawa, Feb 1986.

[684] Charles, M.E.; Mackay, D., “Case Studies of Industrial Innovation in Canada,” in *Technological Innovation Studies Program*, Research Report, Dept. Industry, Trade, and Commerce, Government of Canada: Ottawa, May, 1975.

[685] Knight, R.M.; Lemon, J.C., “A Study of Small and Medium Sized Canadian Technology Based Companies,” *Technological Innovation Studies Program*, Research Report, Dept. Industry, Trade, and Commerce, Government of Canada: Ottawa, Sept, 1978.

[686] Ondrack, D.A., “Innovation and Performance of Small and Medium Firms: A Reanalysis of Data on A Sample of Nineteen Small and Medium Firms in the Machinery Industry,” *Technological Innovation Studies Program*, Research Report No. 74 Dept. Regional Industrial Expansion, Government of Canada: Ottawa, May, 1980.

[687] Public Policy Forum, *“At the Vanguard of Canadian Innovation. A Compilation of Co-operative Case Studies,” Report*, Canada’s Public Policy Forum: Ottawa, October 2014.

[688] Rumball, D., “Case Studies of Collaborative Innovation in Canadian Small Firms,” *Report, Small Business Policy Branch*, Industry Canada: Ottawa, May 2007.

[689] Tapon, F.; Osborne, T.J., *Nuclear Power at Ontario Hydro: A Case Study in Technological Innovation*, Ontario Economic Council Report: Toronto, 1980, p. 52.

[690] Industry Canada, *The Practice of Innovation. Seven Canadian Firms in Profile*, Report, Industry Canada: Ottawa, 2003, p. 78.

[691] Industry Canada, *The Innovation Journey of Six Canadian Firms. The Practice of Innovation II*, Report, Industry Canada: Ottawa, 2004, p. 81.

[692] Globerman, S., “Technological Diffusion in Canadian Manufacturing Industries,” *Technological Innovation Studies Program*, Research Report Dept. Industry, Trade, and Commerce, Government of Canada: Ottawa, April 1974.

[693] Leemans, D., “Technological Innovation in the Nondestructive Testing Industry,” *Technological Innovation Studies Program, Research Report No. 101*, Dept. Regional Industrial Expansion, Government of Canada: Ottawa, Feb 1985.

[694] Carayannopoulos, S., “Research in Motion: A Small Firm Commercializing A New Technology,” *Entrep. Theory Pract.*, 2005, *Mar*, 1042–2587.

[695] Westley, F.; Antadze, N.; Riddell, D.J.; Robinson, K.; Geobey, S., “Five Configurations for Scaling up Social Innovation: Case Examples of Nonprofit Organizations from Canada,” *J. Appl. Behav. Sci.*, 2014, *50(3)*, 234–260.

[696] Babin, R.; Smith, D., “Open Banking and Regulation: Please Advise the Government,” *J. Inf. Technol. Teach. Cases*, 2023, *12(2)*, 16, doi.org/10.1177/20438869221082316.

[697] Mason, C., “The Role of Community Entrepreneurs in Depleted Communities: A Case Study of Father Greg Macleod’s ‘Action Research’ in Cape Breton,” *Local Econ.*, 2022, *37(4)*, 317–327.

[698] MacLeod, G.; Ling, C., “Banking Community Assets: A New Model of Community-Based Economic Development,” in *CRC Case Studies, CRC (Community Research Connections) Research*, Royal Roads University: Victoria, Published online August 31 2009, https://www.crcresearch.org/community-research-connections/crc-case-studies/banking-community-assets-new-model-community-based. accessed 14 November 2025.

[699] Vinodrai, T.; Nader, B.; Zavarella, C., “Manufacturing Space for Inclusive Innovation? A Study of Makerspaces in Southern Ontario,” *Local Econ.*, 2021, *36(3)*, 205–223.

[700] Marabelli, M.; Newell, S.; Gogan, J., "Pilot-Testing a Pediatric Complex Care Coordination Service," *J. Inf. Technol. Teach. Cases*, 2016, *6*, 45–55.
[701] Ensign, P.C.; Lanney, A., "After Innovation: Protecting Intellectual Property During Production – Canica Design Inc," *Entrep. Innov.*, 2008, *9(4)*, 269–274.
[702] INO, "Case Studies," *A Web-based Compendium of 43 Cases Studies (As of the Time of Writing) Involving Market-pull Innovations Involving Opto-electronics and Spanning Aerospace, Agriculture, Agrifood, Environment, Infrastructure, Life Sciences, Manufacturing, and Medical Devices*, L'Institut national d'optique (INO, National Optics Institute: Québec City, accessed July 26 2025, https://www.ino.ca/en/resources/case-studies/#results.
[703] Therrien, P.; Anderson, F., "Community Innovation: Innovation Performance of Manufacturing Firms in Canadian Communities," in *Working Paper*, Statistics Canada: Ottawa, 2004, p. 10.
[704] Diaz-Perez, C.; Arechavala-Vargas, R., "Regional Systems of Innovation in Canada: Two Case Studies," in *Proc. Portland Internat. Conf. Mgmt. Engineering and Technology (PICMET 2006), Portland, OR*, 2006, pp. 127–133.
[705] Doloreux, D.; Dionne, S., "Is Regional Innovation System Development Possible in Peripheral Regions? Some Evidence from the Case of La Pocatière, Canada," *Entrep. Reg. Dev.*, 2008, *20(3)*, 259–283.
[706] Breton, M.; Marshall, E.G.; Deslauriers, V.; Smithman, M.A.; Moritz, L.R.; Buote, R.; Morrison, B.; Christian, E.K.; McKay, M.; Stringer, K.; Godard-Sebillotte, C.; Sourial, N.; Laberge, M.; MacKenzie, A.; Isenor, J.E.; Duhoux, A.; Ashcroft, R.; Mathews, M.; Cossette, B.; Hudon, C.; McDougall, B.; Guénette, L.; Kirkwood, R.; Green, M.E., "COVID-19 – An Opportunity to Improve Access to Primary Care through Organizational Innovations? A Qualitative Multiple Case Study in Quebec and Nova Scotia (Canada)," *BMC Health Serv. Res.*, 2022, *22(759)*, 18.
[707] Rowe, A.M., "Gender and Innovation Policy in Canada and Sweden," *Int. J. Gend. Entrep.*, 2018, *10(4)*, 344–360.
[708] DFO, *"Case Study: Reducing the Impacts of Ocean Noise from Vessels – Government of Canada Support for Innovation in the Marine Transport Sector," Report*, Fisheries and Oceans Canada: Ottawa, 2024.
[709] Anastakis, D., "Hubris, Nepotism, and Failure: The Bricklin Car Company and the Question of Inevitability," *BEH Online*, 2010, *8*, 26, https://thebhc.org/hubris-nepotism-and-failure-bricklin-car-company-and-question-inevitability. accessed 14 November 2025.
[710] Campbell, T.; Fuhr, H., Eds., *Leadership and Innovation in Subnational Government. Case Studies from Latin America*, World Bank: Washington, 2004.
[711] McConnell, J.J.; Schwartz, E.S., "The Origin of LYONs: A Case Study in Financial Innovation," *J. Appl. Corp. Finance*, 2023, *35*, 20–26.
[712] Bradley, E.H.; Webster, T.R.; Baker, D.; Schlesinger, M.; Inouye, S.K., "After Adoption: Sustaining the Innovation. A Case Study of Disseminating the Hospital Elder Life Program," *J. Am. Geriatr. Soc.*, 2005, *53*, 1455–1461.
[713] Copp, N.; Zanella, A., *Discovery, Innovation, and Risk. Case Studies in Science and Technology*, MIT Press: Cambridge, 1993.
[714] Madrigal-Sánchez, J.; Quesada-Pineda, H., "Innovation: Case Study among Wood, Energy and Medical Firms," *Bus. Process Manag. J.*, 2012, *18(6)*, 898–918.
[715] Babin, R.; Schuster, C., "Building Innovation into the Outsourcing Relationship: A Case Study," *J. Inf. Technol. Teach. Cases*, 2012, *2*, 98–103.
[716] Agnihotri, A.; Bhattacharya, S., *indieDwell: Inclusive Business Model Innovation*, Sage Business Cases, SAGE Publications: Thousand Oaks, CA, 2023, doi.org/10.4135/9781529618655.
[717] Cantone, L.; Testa, P.; Cantone, G.F., *Strategic Outsourcing, Innovation and Global Supply Chains A Case Study from the Aviation Industry*, Routledge: New York, 2023.

[718] Wolfe, J.W., *Innovation and Entrepreneurship in Established Organizations: Micro Cases on Lntrapreneurship*, Sage Business Cases, SAGE Publications: Thousand Oaks, CA, 2019, p. 10, doi: 10.4135/9781526466594.

[719] Håkanson, L.; Caessens, P.; Macaulay, S., "InnovationXchange: A Case Study in Innovation Intermediation," *Innov. Manag. Policy Pract.*, 2011, *13*, 261–274.

[720] Vitalis, V., "Domestic Reform, Trade, Innovation and Growth in New Zealand's Agriculture Sector," in *OECD Trade Policy Working Paper No. 74*, Organisation for Economic Co-Operation and Development: Paris, 2008.

[721] Nunes, M.P.; Russo, A.P., "Analysis of Business Models Innovation – A Multiple Case Study," *Innov. Manag. Rev.*, 2019, *16(1)*, 17–35.

[722] Pires, A.M.B.; Teixeira, F.L.C.; Filho, H.N.H.; Oliveira, S.R.G., "Innovation in Innovation Management: The Experience of Petrobras Centers and Networks of Excellence Program," *J. Technol. Manag. Innov., Special Issue* 2013, *8*, 49–60.

[723] Chu, D.; Andreassi, T., "Management of Technological Innovation. Case Studies in Biotechnology Companies in Brazil," *J. Iberoam. Acad. Manag.*, 2011, *9(1)*, 7–31.

[724] Rehder, A.; Souza, J.V.; Marx, R.; Salerno, M.S., "Emerging Field or Passing Fashion? A Case Study of Agile-Stage-Gate Model in Innovation Processes," *Rev. Gest.*, 2023, *30*(*4*), 362–386.

[725] Rincón, V.G.; Bayon, M.C.; Zambrano, J.J.A.; Vasquez, J.M., "Exploring the Configuration of Institutional Practices – A Case Study of Innovation Implementation in Healthcare," *Sustainability*, 2022, *14*, 3282–3300.

[726] Benavides, A.M.; Yu, A.S.O.; Nascimento, P.T.S.; Camargo, A.S., "Innovation in Innovation System in a Large Service Corporation in Colombia," in *Proc. Portland Internat. Conf. Mgmt. Technol. Age (PICMET 2015)*, Portland, OR, 2015, pp. 905–911.

[727] Tjornbo, O., "Can Collective Intelligence Produce Social Innovation?," in *New Frontiers in Social Innovation Research*, Nicholls, A.; Simon, J.; Gabriel, M., (Eds.), Palgrave Macmillan: Houndmills, 2015, pp. 192–211.

[728] Planes-Satorra, S.; Paunov, C., "Inclusive Innovation Policies: Lessons from International Case Studies," *OECD Science*, Technology, and Industry Working Paper 2017/02, Organisation for Economic Co-Operation and Development: Paris, 2017.

[729] Nicholls, A.; Simon, J.; Gabriel, M., Eds., *New Frontiers in Social Innovation Research*, Palgrave Macmillan: Houndmills, 2015.

[730] Lopes Cardozo, N.J.; Ward, S.H., "The Interplay of the Innovation Cycle, Build Time, Lifetime, and Deployment Rate of New Energy Technologies: A Case Study of Nuclear Fusion Energy," *Oxf. Open Energy*, oiae005 2024, *3*, 12.

[731] Kohl, R.D., "Key Factors for Advancing Innovations to Scale: Evidence from Multiple Country Case Studies of Agricultural Innovations," *Front. Sustain. Food Syst.*, 2023, *7*, 1053152.

[732] MindDen, "The 10 Most Important Technological Innovations in History," *Blog Post*, MindDen: Alicante, June 29 2023, https://www.mindden.com/en/the-10-most-important-technological-innovations-in-history/. accessed 14 November 2025.

[733] Telefónica, "9 Technological Inventions that Have Changed the World," Blog post, Telefónica: Madrid, June 30, 2023, https://www.telefonica.com/en/communication-room/blog/9-technological-inventions-changed-world/, accessed 14 November 2025.

[734] Leggett, J.; Wolchover, N., *22 Inventions that Changed the World*, LiveScience blog post, Future US Inc.: New York, December 17 2024, https://www.livescience.com/33749-top-10-inventions-changed-world.html. accessed 14 November 2025.

[735] Radford, L., "The Greatest Inventions in the past 1000 Years," *eHistory Blog Post*, Ohio State University: Columbus, 2025, https://ehistory.osu.edu/articles/greatest-inventions-past-1000-years. accessed 14 November 2025.

[736] Tibbals, G., *The Guinness Book of Innovations. The 20th Century from Aerosol to Zip*, Guiness Publishing: Enfield, UK, 1994.

[737] Glasco, J., *The Evolving Language of Innovation, A Glossary of Terms*, GCA Thought Leader Publications: USA, 2013.

[738] Leydesdorff, L.; Meyer, M., "The Decline of University Patenting and the End of the Bayh-Dole Effect," *Scientometrics*, 2010, *83*(*2*), 355–362.

[739] Bush, V., *Science, the Endless Frontier. A Report to the President on A Program for Postwar Scientific Research*, July the National Science Foundation: Washington, reprinted July 1960 by 1945, https://ia600207.us.archive.org/12/items/scienceendlessfr00unit/scienceendlessfr00unit.pdf.

[740] Reingold, N., *American National Biography*, Oxford University Press: Oxford, 2010.

[741] Milford, R., "Design for Future Reuse," *Low Carbon Materials Processing, Working Paper W2*, Cambridge University: Cambridge, September 2010, p. 3.

[742] Vanderbilt, B.M., *Thomas Edison, Chemist*, American Chemical Society: Washington, DC, 1971.

[743] OECD, *"Embracing Innovation in Government. Global Trends," Report*, Organisation for Economic Co-Operation and Development: Paris, February 2017, pp. 31–78.

[744] Rabe, C.B., *The Innovation Killer*, American Management Association: New York, 2006.

[745] Sehested, C.; Sonnenberg, H., *Lean Innovation. A Fast Path from Knowledge to Value*, Springer-Verlag: Berlin, 2011.

[746] Browne, C.A., "Liebig and the Law of the Minimum," in *Liebig and after Liebig: A Century of Progress in Agricultural Chemistry*, Moulton, F.R., (Ed.), Am. Assoc. Adv. Sci.: Washington, DC, 1942, pp. 71–82.

[747] Halbinger, M.A., "The Role of Makerspaces in Supporting Consumer Innovation and Diffusion: An Empirical Analysis," *Res. Policy*, 2018, *47*, 2028–2036.

[748] Zakoth, D.; Mauroner, O.; Emes, J., "The Role of Makerspaces in Innovation Processes: An Exploratory Study," *R&D Manag.*, 2024, *54*(*2*), 398–428.

[749] Porter, M., *Competitive Advantage: Creating and Sustaining Superior Performance*, The Free Press: NY, 1985.

[750] McGrath, M.E., *Setting the PACE in Product Development*, 2nd Ed, Taylor & Francis: Milton Park, UK, 2011.

[751] Prahalad, C.K., "The Innovation Sandbox," *Strategy+business*, 2006, *44*, 62–73.

[752] Schramm, L.L., *Nano- and Microtechnology from A – Z: From Nanosystems to Colloids and Interfaces*, Wiley-VCH: Weinheim, Germany, 2014.

[753] Attar, H.; Shahabi, S.M.R., "Beyond 'Funnel' and 'Fireworks': 'Water Ribbed Balloon' as a New Metaphorical Approach to Innovation-in-Practice," *Iran. J. Manag. Stud.*, 2014, *7*(*1*), 67–93.

Index

https://doi.org/10.1515/9783112206720-013

www.ingramcontent.com/pod-product-compliance
Lightning Source LLC
Chambersburg PA
CBHW061155240926
51846CB00006B/1018
9783119147255